AF412881

HADRONS AT FINITE TEMPERATURE

High energy laboratories are performing experiments in heavy ion collisions to explore the structure of matter at high temperature and density. This elementary book explains the basic ideas involved in the theoretical analysis of these experimental data. It first develops two topics needed for this purpose, namely hadron interactions and thermal field theory. Chiral perturbation theory is developed to describe hadron interactions and thermal field theory is formulated in the real time method. In particular, spectral form of thermal propagators is derived for fields of arbitrary spin and used to calculate loop integrals. These developments are then applied to find quark condensate and hadron parameters in medium, including dilepton production. Finally, the non-equilibrium method of statistical field theory to calculate transport coefficients is reviewed. With technical details explained in the text and appendices, this book should be accessible to researchers as well as graduate students interested in thermal field theory.

SAMIRNATH MALLIK is Emeritus Scientist at Saha Institute of Nuclear Physics. He worked at the University of Berne, the University of Karlsruhe and Tata Institute of Fundamental Research. His research interests are thermal field theory and heavy ion collisions.

SOURAV SARKAR is Scientist at Variable Energy Cyclotron Centre, Kolkata and Associate Professor, Homi Bhabha National Institute. His research interests concern applications of thermal field theory, mainly to study transport phenomena and hadronic properties in hot and dense medium.

CAMBRIDGE MONOGRAPHS ON MATHEMATICAL PHYSICS

General Editors: P. V. Landshoff, D. R. Nelson, S. Weinberg

S. J. Aarseth *Gravitational N-Body Simulations: Tools and Algorithms*[†]

J. Ambjørn, B. Durhuus and T. Jonsson *Quantum Geometry: A Statistical Field Theory Approach*[†]

A. M. Anile *Relativistic Fluids and Magneto-fluids: With Applications in Astrophysics and Plasma Physics*

J. A. de Azcárraga and J. M. Izquierdo *Lie Groups, Lie Algebras, Cohomology and Some Applications in Physics*[†]

O. Babelon, D. Bernard and M. Talon *Introduction to Classical Integrable Systems*[†]

F. Bastianelli and P. van Nieuwenhuizen *Path Integrals and Anomalies in Curved Space*[†]

D. Baumann and L. McAllister *Inflation and String Theory*

V. Belinski and E. Verdaguer *Gravitational Solitons*[†]

J. Bernstein *Kinetic Theory in the Expanding Universe*[†]

G. F. Bertsch and R. A. Broglia *Oscillations in Finite Quantum Systems*[†]

N. D. Birrell and P. C. W. Davies *Quantum Fields in Curved Space*[†]

K. Bolejko, A. Krasiński, C. Hellaby and M-N. Célérier *Structures in the Universe by Exact Methods: Formation, Evolution, Interactions*

D. M. Brink *Semi-Classical Methods for Nucleus-Nucleus Scattering*[†]

M. Burgess *Classical Covariant Fields*[†]

E. A. Calzetta and B.-L. B. Hu *Nonequilibrium Quantum Field Theory*

S. Carlip *Quantum Gravity in 2+1 Dimensions*[†]

P. Cartier and C. DeWitt-Morette *Functional Integration: Action and Symmetries*[†]

J. C. Collins *Renormalization: An Introduction to Renormalization, the Renormalization Group and the Operator-Product Expansion*[†]

P. D. B. Collins *An Introduction to Regge Theory and High Energy Physics*[†]

M. Creutz *Quarks, Gluons and Lattices*[†]

P. D. D'Eath *Supersymmetric Quantum Cosmology*[†]

J. Dereziński and C. Gérard *Mathematics of Quantization and Quantum Fields*

F. de Felice and D. Bini *Classical Measurements in Curved Space-Times*

F. de Felice and C. J. S Clarke *Relativity on Curved Manifolds*[†]

B. DeWitt *Supermanifolds, 2ⁿᵈ edition*[†]

P. G. O. Freund *Introduction to Supersymmetry*[†]

F. G. Friedlander *The Wave Equation on a Curved Space-Time*[†]

J. L. Friedman and N. Stergioulas *Rotating Relativistic Stars*

Y. Frishman and J. Sonnenschein *Non-Perturbative Field Theory: From Two Dimensional Conformal Field Theory to QCD in Four Dimensions*

J. A. Fuchs *Affine Lie Algebras and Quantum Groups: An Introduction, with Applications in Conformal Field Theory*[†]

J. Fuchs and C. Schweigert *Symmetries, Lie Algebras and Representations: A Graduate Course for Physicists*[†]

Y. Fujii and K. Maeda *The Scalar-Tensor Theory of Gravitation*[†]

J. A. H. Futterman, F. A. Handler and R. A. Matzner *Scattering from Black Holes*[†]

A. S. Galperin, E. A. Ivanov, V. I. Ogievetsky and E. S. Sokatchev *Harmonic Superspace*[†]

R. Gambini and J. Pullin *Loops, Knots, Gauge Theories and Quantum Gravity*[†]

T. Gannon *Moonshine beyond the Monster: The Bridge Connecting Algebra, Modular Forms and Physics*[†]

M. Göckeler and T. Schücker *Differential Geometry, Gauge Theories, and Gravity*[†]

C. Gómez, M. Ruiz-Altaba and G. Sierra *Quantum Groups in Two-Dimensional Physics*[†]

M. B. Green, J. H. Schwarz and E. Witten *Superstring Theory Volume 1: Introduction*

M. B. Green, J. H. Schwarz and E. Witten *Superstring Theory Volume 2: Loop Amplitudes, Anomalies and Phenomenology*

V. N. Gribov *The Theory of Complex Angular Momenta: Gribov Lectures on Theoretical Physics*[†]

J. B. Griffiths and J. Podolský *Exact Space-Times in Einstein's General Relativity*[†]

S. W. Hawking and G. F. R. Ellis *The Large Scale Structure of Space-Time*[†]

F. Iachello and A. Arima *The Interacting Boson Model*[†]

F. Iachello and P. van Isacker *The Interacting Boson-Fermion Model*[†]

C. Itzykson and J. M. Drouffe *Statistical Field Theory Volume 1: From Brownian Motion to Renormalization and Lattice Gauge Theory*[†]

C. Itzykson and J. M. Drouffe *Statistical Field Theory Volume 2: Strong Coupling, Monte Carlo Methods, Conformal Field Theory and Random Systems*[†]

G. Jaroszkiewicz *Principles of Discrete Time Mechanics*

C. V. Johnson *D-Branes*[†]

P. S. Joshi *Gravitational Collapse and Spacetime Singularities*[†]

J. I. Kapusta and C. Gale *Finite-Temperature Field Theory: Principles and Applications, 2nd edition*[†]

V. E. Korepin, N. M. Bogoliubov and A. G. Izergin *Quantum Inverse Scattering Method and Correlation Functions*[†]

M. Le Bellac *Thermal Field Theory*[†]

Y. Makeenko *Methods of Contemporary Gauge Theory*[†]

N. Manton and P. Sutcliffe *Topological Solitons*[†]

N. H. March *Liquid Metals: Concepts and Theory*[†]

I. Montvay and G. Münster *Quantum Fields on a Lattice*[†]

L. O'Raifeartaigh *Group Structure of Gauge Theories*[†]

T. Ortín *Gravity and Strings, 2nd edition*

A. M. Ozorio de Almeida *Hamiltonian Systems: Chaos and Quantization*[†]

L. Parker and D. Toms *Quantum Field Theory in Curved Spacetime: Quantized Fields and Gravity*

R. Penrose and W. Rindler *Spinors and Space-Time Volume 1: Two-Spinor Calculus and Relativistic Fields*[†]

R. Penrose and W. Rindler *Spinors and Space-Time Volume 2: Spinor and Twistor Methods in Space-Time Geometry*[†]

S. Pokorski *Gauge Field Theories, 2nd edition*[†]

J. Polchinski *String Theory Volume 1: An Introduction to the Bosonic String*[†]

J. Polchinski *String Theory Volume 2: Superstring Theory and Beyond*[†]

J. C. Polkinghorne *Models of High Energy Processes*[†]

V. N. Popov *Functional Integrals and Collective Excitations*[†]

L. V. Prokhorov and S. V. Shabanov *Hamiltonian Mechanics of Gauge Systems*

A. Recknagel and V. Schiomerus *Boundary Conformal Field Theory and the Worldsheet Approach to D-Branes*

R. J. Rivers *Path Integral Methods in Quantum Field Theory*[†]

R. G. Roberts *The Structure of the Proton: Deep Inelastic Scattering*[†]

C. Rovelli *Quantum Gravity*[†]

W. C. Saslaw *Gravitational Physics of Stellar and Galactic Systems*[†]

R. N. Sen *Causality, Measurement Theory and the Differentiable Structure of Space-Time*

M. Shifman and A. Yung *Supersymmetric Solitons*

H. Stephani, D. Kramer, M. MacCallum, C. Hoenselaers and E. Herlt *Exact Solutions of Einstein's Field Equations, 2nd edition*[†]

J. Stewart *Advanced General Relativity*[†]

J. C. Taylor *Gauge Theories of Weak Interactions*[†]

T. Thiemann *Modern Canonical Quantum General Relativity*[†]

D. J. Toms *The Schwinger Action Principle and Effective Action*[†]

A. Vilenkin and E. P. S. Shellard *Cosmic Strings and Other Topological Defects*[†]

R. S. Ward and R. O. Wells, Jr *Twistor Geometry and Field Theory*[†]

E. J. Weinberg *Classical Solutions in Quantum Field Theory: Solitons and Instantons in High Energy Physics*

J. R. Wilson and G. J. Mathews *Relativistic Numerical Hydrodynamics*[†]

T. Ortín *Gravity and Strings*

D. Baumann and L. McAllister *Inflation and String Theory*

S. Raychaudhuri and K. Sridhar *Particle Physics of Brane Worlds and Extra Dimensions*

J. Kroon *Conformal Methods in General Relativity*

[†] Available in paperback

Hadrons at Finite Temperature

SAMIRNATH MALLIK
Saha Institute of Nuclear Physics

SOURAV SARKAR
Variable Energy Cyclotron Center, Kolkata

CAMBRIDGE
UNIVERSITY PRESS

University Printing House, Cambridge CB2 8BS, United Kingdom

Cambridge University Press is part of the University of Cambridge.

It furthers the University's mission by disseminating knowledge in the pursuit of
education, learning and research at the highest international levels of excellence.

www.cambridge.org
Information on this title: www.cambridge.org/9781107145313

First published 2016

Printed in the United States of America by Sheridan Books, Inc.

A catalogue record for this publication is available from the British Library

Library of Congress Cataloging-in-Publication Data
Names: Mallik, Samirnath, author. | Sarkar, S. (Sourav), author.
Title: Hadrons at finite temperature / Samirnath Mallik (Saha
Institute of Nuclear Physics), Sourav Sarkar (Variable Energy
Cyclotron Center, Kolkata).
Other titles: Cambridge monographs on mathematical physics.
Description: Cambridge, United Kingdom ; New York,
NY : Cambridge University Press, 2016. |
Series: Cambridge monographs on mathematical physics |
Includes bibliographical references and index.
Identifiers: LCCN 2016032396 | ISBN 9781107145313 (hardback) |
ISBN 1107145317 (hardback)
Subjects: LCSH: Hadrons. | Thermodynamics. | Field theory (Physics) | Broken
symmetry (Physics) | Heavy ion collisions.
Classification: LCC QC793.5.H32 M35 2016 | DDC 539.7/216–dc23
LC record available at https://lccn.loc.gov/2016032396

ISBN 978-1-107-14531-3 Hardback

To the memory of my parents —S.M.

Contents

Preface

This book is an elementary introduction to hadronic properties at finite temperature (and chemical potential). Its contents may be divided into three parts.

The first part, comprising the first three chapters, develops the (vacuum) theory of hadronic interactions at low energy. As propagators play a major role in applications, we review in Chapter 1 in detail the vacuum propagators for fields of different spins. This provides the background in which to discuss later (in Chapter 4) the form of thermal propagators. The next two chapters are devoted to strong interaction dynamics. Chapter 2 describes the phenomenon of spontaneous symmetry breaking leading to Goldstone bosons. It prepares the ground for chiral perturbation theory in Chapter 3. We review this theory in some detail, including Goldstone as well as (heavy) non-Goldstone fields interacting with the Goldstone fields. Chapters 2 and 3 constitute an elementary yet general introduction to the effective theory of strong interactions at low energy.

The second part consists of the next two chapters and presents the equilibrium thermal field theory. Broadly speaking, there are two formulations of this theory, concerning so-called real and imaginary time. Though the imaginary time formulation is used more often than that of real time, we choose the latter in this book and try to show some of its advantages. In Chapter 4 we derive in elementary ways the thermal matrix propagators for fields of low spin and go on to obtain the spectral representation of complete propagators for fields of arbitrary spin. In Chapter 5 the thermal perturbation theory is developed in relation to the matrix structure of the propagators.

The third part consists of the last four chapters. Here we apply the methods developed to study different thermal one- and two-point functions in the hadronic phase using chiral perturbation theory. In Chapter 6 we obtain the temperature dependence of a number of hadronic parameters to one loop. Some two-loop results are presented in Chapter 7. In Chapter 8 we derive the dilepton cross-section measured in heavy ion collisions. Finally, Chapter 9 describes the (non-equilibrium) transport phenomenon, whose discussion can be reduced to fluctuations in thermal equilibrium. Clearly the range of applications of thermal field theory in particle physics is much wider. Many such applications are covered in books (mentioned in Chapter 4), though mostly in imaginary time formulation.

The first two parts are independent of each other, but it is the thread of spectral representations which binds all the three parts together. In the first part it appears in the rigorous proof of the Goldstone theorem. In the second part the thermal propagators for general fields are derived as spectral representations, which are then applied to problems in the third part. Our treatment of general fields demonstrates the belief that spin is not an essential complication in the thermal context, once its structure is understood in vacuum.

At the end of each chapter there are in general one or two problems, which are fully worked out. These are of different varieties, like important side results or pieces of calculation needed in the same or later chapters, which, if included in text, could affect the continuity of the presentation. The appendices attempt to derive all the results, mostly elementary, which are needed or mentioned in the text. Numerical evaluations are not reported, except on two occasions; where necessary, they are referred to in the literature.

Though a monograph, the book should be accessible to all graduate students interested in relativistic thermal field theory. The only prerequisite is acquaintance with the conventional perturbative framework using the canonical method of field theory. All other technical details are developed in the text and appendices.

Acknowledgements by S.M.

This book owes a great deal to Professor Heinrich Leutwyler. Through countless discussions with him, I could straighten out my understanding and hence the presentation of this book in many places. I take this opportunity to express my deep gratitude to him. Of course, we authors alone are responsible for any errors that may still remain. It is a pleasure for me to thank Professor Leutwyler's wife, Ursula, for her kind and warm hospitality during my stays with them. At this point I thank Professor A. Ereditato (Albert Einstein Centre for Fundamental Physics) for supporting my visits to the University of Bern. I also thank Professor H. Bebie, Professor J. Gasser and Professor P. Minkowski for their encouragement to complete the book.

I have lectured at several institutions on different parts of the book. For these invitations I thank Professor B. Ananthanarayan (Indian Institute of Science, Bangalore), Professor A. Harindranath (Saha Institute of Nuclear Physics, Kolkata), Professor H. Mishra (Physical Research Laboratory, Ahmedabad) and Professor H.S. Sharatchandra (Institute of Mathematical Sciences, Chennai). For help and suggestions with LaTex I thank Professor P.B. Pal (Saha Institute of Nuclear Physics). I also thank Professor M.K. Sanyal and Professor B.K. Chakrabarti (Saha Institute of Nuclear Physics) for extending academic facilities of the Institute during the period of writing the book. Finally I thank my daughter, Joyita and son-in-law, Saket for help with the computer.

This book was written under a project granted by the Department of Science and Technology, Government of India, during the period from May 2011 to April 2013. I am glad to have corresponded with Mr S.S. Kohli (Scientist F) during this period and thank him for his patience in awaiting the book's publication.

Notation

Greek indices run over t, x, y, z or 0, 1, 2, 3. The indices α, β, γ may also run over the generators of a symmetry group. Latin indices are used to run over components of other quantities. Repeated indices are summed over, unless stated otherwise.

The Lorentz metric is diagonal, $g_{\mu\nu} = g^{\mu\nu} = diag(1, -1, -1, -1)$. The d'Alembertian is $\Box = g^{\mu\nu}\partial^2/\partial x^\mu x^\nu = \partial_t^2 - \nabla^2$, where ∇^2 is the Laplacian $\partial^2/\partial x^i \partial x^i$.

Spatial coordinates and momentum three-vectors are denoted by boldface small letters. Isospin three-vectors are denoted by an arrow over the letters.

The complex conjugate of any quantity A is denoted A^*. The transpose and hermitian adjoint of a matrix or vector R is denoted R^T and $R^\dagger = R^{*T}$ respectively. The hermitian adjoint of an operator $\mathcal{O}$ is denoted $\mathcal{O}^\dagger$. A bar on a Dirac spinor ψ is defined by $\overline{\psi} = \psi^\dagger \gamma_0$. The conventions for Dirac gamma matrices are spelt out in Section 1.3.

Vacuum propagators are denoted by Δ_F for bosonic fields and by S_F for fermionic fields with necessary indices.

Thermal matrices are written in boldface capital letters. In particular, thermal propagators are denoted by $\boldsymbol{D}$ and $\boldsymbol{S}$ for bosonic and fermionic ones respectively, again with necessary indices. Thus, for bosonic propagators we make a notational distinction between the vacuum (Δ_F) and thermal (D) ones, which we do not maintain for fermionic propagators. When the matrices are diagonalised, the diagonal elements are written with a bar on the symbols, such as $\overline{D}$, $\overline{S}$ etc.

The mass of the pion is denoted by M to lowest order in chiral pertubation theory. Masses of other particles will be denoted by m with a subscript for the particle.

Following convention, we use natural units, $\hbar = c = 1$. So the fine structure constant is $\alpha = e^2/(4\pi) = 1/137$.

The equations in the text are numbered as usual, prefixed with chapter and section numbers. The equations in Problems are prefixed with P and chapter number. The appendices are denoted by A, B etc. and their equations are prefixed with the corresponding letter.

1

Free Fields in Vacuum

This chapter introduces notations and obtains results for free fields in vacuum that we shall utilise in the rest of the book. In applications, one generally considers particles of spin 0, $\frac{1}{2}$, 1 and $\frac{3}{2}$, which are conventionally described respectively by scalar, Dirac, vector and Rarita–Schwinger fields. Accordingly, we discuss each of these fields separately and derive their spectral functions and Feynman propagators. But we shall not limit ourselves to these fields only, and we will extend their formal constructions to general fields describing particles of arbitrary spin.

The conventional approach to quantum field theory begins with the Lagrangian, from which a Klein–Gordon equation is derived for each field component. These are then expanded in terms of one-particle annihilation and creation operators. There is also the other approach, expounded by Weinberg [1], that starts from particle states and constructs fields requiring causality and Lorentz covariance. In this brief review we exploit both points of view, starting with the Lagrangian for low-spin fields and with the one-particle states for high-spin fields. In either case, we do not need to construct explicitly the coefficient functions in the expansion of fields. As we shall be interested only in the spectral functions and propagators, all we require are the spin sums over the product of such functions. These sums may be obtained directly by using the supplementary conditions eliminating components in excess of those needed to describe different spin degrees of freedom.

We begin with general fields to present a unified view of some properties of free fields. Then, in the following sections, we treat separately the individual fields of low spin and show how the supplementary conditions on the field suffice to evaluate the spin sums. We come back to general fields in the last section.

1.1 Generalities

Let $|\boldsymbol{q}, \sigma\rangle$ be the state vector of a particle of mass m, momentum $\boldsymbol{q}$ and spin j with z-component σ in its rest frame ($\sigma = j, j - 1, \cdots, -j$). It is normalised in a Lorentz invariant way:

$$\langle \boldsymbol{q}', \sigma' | \boldsymbol{q}, \sigma \rangle = (2\pi)^3 2\omega \delta_{\sigma\sigma'} \delta^3(\boldsymbol{q} - \boldsymbol{q}'), \qquad \omega = +\sqrt{\boldsymbol{q}^2 + m^2}. \tag{1.1.1}$$

Define as usual the creation operator $a^\dagger(\boldsymbol{q}, \sigma)$ to produce this state by its action on the vacuum state

$$|\boldsymbol{q}, \sigma \rangle = a^\dagger(\boldsymbol{q}, \sigma) |0\rangle, \tag{1.1.2}$$

and the annihilation operator $a(\boldsymbol{q}, \sigma)$ by its adjoint with $a(\boldsymbol{q}, \sigma)|0\rangle = 0$. Then (1.1.1) gives the commutation/anticommutation relation

$$[a(\boldsymbol{q}, \sigma), a^\dagger(\boldsymbol{q}', \sigma')]_\mp = (2\pi)^3 2\omega \delta_{\sigma\sigma'} \delta^3(\boldsymbol{q} - \boldsymbol{q}'). \tag{1.1.3}$$

As always, the top and bottom signs refer to bosonic and fermionic degrees of freedom. Thus the subscript $\mp$ indicates commutator or anticommutator according to whether the particles destroyed and created by a and $a^\dagger$ are bosons or fermions respectively. If the field also describes a distinct antiparticle, we denote the corresponding destruction and creation operator by $b(\boldsymbol{q}, \sigma)$ and $b^\dagger(\boldsymbol{q}, \sigma)$ satisfying

$$[b(\boldsymbol{q}, \sigma), b^\dagger(\boldsymbol{q}', \sigma')]_\mp = (2\pi)^3 2\omega \delta_{\sigma\sigma'} \delta^3(\boldsymbol{q} - \boldsymbol{q}'), \tag{1.1.4}$$

all other commutatators/anticommutators being zero.

A multicomponent field $\psi_l(x)$ is needed to describe particles of non-zero spin. In the conventional description, the index l denotes one or more (Lorentz) vector indices for bosonic fields and an additional (Dirac) spinor index for fermionic fields. Such a description introduces extra components, which are then eliminated by imposing supplementary conditions on $\psi_l(x)$. (The index structure and supplementary conditions are reviewed in Appendix A.) We expand the field in terms of annihilation and creation operators [1]:

$$\psi_l(x) = \int \frac{d^3q}{(2\pi)^3 2\omega} \sum_\sigma \left(u_l(\boldsymbol{q}, \sigma) e^{-iq \cdot x} a(\boldsymbol{q}, \sigma) + v_l(\boldsymbol{q}, \sigma) e^{iq \cdot x} b^\dagger(\boldsymbol{q}, \sigma) \right). \tag{1.1.5}$$

Here and below, four-momenta q^μ with space components integrated over, are generally on mass-shell, $q^2 = m^2$, with q^0 given by the positive square root, $q^0 = \sqrt{\boldsymbol{q}^2 + m^2}$. (For real fields describing a self-charge-conjugate particle, $a \equiv b$.) The coefficient functions u_l and v_l are polarisation spin-tensors in the general case. With the normalisation of creation and annihilation operators already fixed by (1.1.3) and (1.1.4), that of the coefficient functions is determined by the normalisation of the field.

In this chapter we are mainly interested in calculating for different fields the commutator/anticommutator

$$\Delta_{ll'}(x, x') \equiv [\psi_l(x), \psi_{l'}^\dagger(x')]_\mp, \tag{1.1.6}$$

and the Feynman propagator

$$\Delta_{ll'}^F(x, x') \equiv i\langle 0 | T\{\psi_l(x) \psi_{l'}^\dagger(x')\} | 0 \rangle, \tag{1.1.7}$$

where T is the time-ordering symbol

$$T\{\psi_l(x)\psi_{l'}^{\dagger}(x')\} = \theta(t-t')\psi_l(x)\psi_{l'}^{\dagger}(x') \pm \theta(t'-t)\psi_{l'}^{\dagger}(x')\psi_l(x). \qquad (1.1.8)$$

Here $\theta(t)$ is a step function, being equal to $+1$ for $t > 0$ and zero for $t < 0$. Inserting the expansion (1.1.5) for ψ in Δ and Δ^F, both reduce to commutators/anticommutators of creation and annihilation operators. Then using (1.1.3) and (1.1.4) we get

$$\Delta_{ll'}(x,x') = \int \frac{d^3q}{(2\pi)^3 2\omega} \left(e^{-iq\cdot(x-x')} \sum_{\sigma} u_l(\boldsymbol{q},\sigma)u_{l'}^*(\boldsymbol{q},\sigma) \right.$$

$$\left. \mp e^{iq\cdot(x-x')} \sum_{\sigma} v_l(\boldsymbol{q},\sigma)v_{l'}^*(\boldsymbol{q},\sigma) \right), \qquad (1.1.9)$$

and

$$-i\Delta_{ll'}^F(x,x') = \int \frac{d^3q}{(2\pi)^3 2\omega} \left(\theta(t-t')e^{-iq\cdot(x-x')} \sum_{\sigma} u_l(\boldsymbol{q},\sigma)u_{l'}^*(\boldsymbol{q},\sigma) \right.$$

$$\left. \pm \theta(t'-t)e^{iq\cdot(x-x')} \sum_{\sigma} v_l(\boldsymbol{q},\sigma)v_{l'}^*(\boldsymbol{q},\sigma) \right). \qquad (1.1.10)$$

Thus the essential quantities are the spin sums over the products of coefficient functions. Once these are found, it remains to combine the two terms in each of Δ and Δ^F in the form of four-dimensional Fourier transforms. While for Δ it can be done simply by introducing a mass shell delta function, it is a bit more complicated for Δ_F, involving derivatives of theta functions. The latter gives rise to non-covariant terms for fields of spin greater than $\frac{1}{2}$, which may be traced to the too-singular behaviour of commutators of such fields at the origin of the light-cone. However, these terms are local and so may be removed by adding a non-covariant piece in the Hamiltonian, making the resulting theory Lorentz covariant [1, 2].[1]

1.2 Scalar Field

The familiar Lagrangian (density) for a free real (hermitian) field

$$\mathscr{L}(\phi) = \frac{1}{2}\partial_\mu\phi\partial^\mu\phi - \frac{1}{2}m^2\phi^2 \qquad (1.2.1)$$

gives the dynamical (Euler–Lagrange) equation

$$(\Box + m^2)\phi(x) = 0. \qquad (1.2.2)$$

---

[1] Another justification for ignoring the non-covariant terms can be obtained from the path integral formalism, where the propagator is obtained directly from the quadratic terms in the Lagrangian (and the vertices are read off from the interaction terms) [1].

The field $\phi(x)$ can then be expanded as

$$\phi(x) = \int \frac{d^3k}{(2\pi)^3 2\omega} \left(a(\mathbf{k})e^{-ik\cdot x} + a^\dagger(\mathbf{k})e^{ik\cdot x} \right), \quad \omega = +\sqrt{\mathbf{k}^2 + m^2}. \quad (1.2.3)$$

Here the coefficient functions are unity, there being no polarisation. So the field commutator (1.1.9) reduces to

$$\Delta(x - x') = \int \frac{d^3k}{(2\pi)^3 2\omega} \left(e^{-ik\cdot(x-x')} - e^{ik\cdot(x-x')} \right) \qquad (1.2.4)$$

$$\equiv \Delta_+(x - x') - \Delta_+(x' - x) \qquad (1.2.5)$$

where Δ_+ is the standard function

$$\Delta_+(x - x') = \int \frac{d^3k}{(2\pi)^3 2\omega} e^{-ik\cdot(x-x')}. \qquad (1.2.6)$$

On using the formula

$$\delta(k^2 - m^2) = \frac{1}{2\omega}\{\delta(k_0 - \omega) + \delta(k_0 + \omega)\}$$

it can be written as a four-dimensional Fourier transform

$$\Delta_+(x - x') = \int \frac{d^4k}{(2\pi)^4} e^{-ik\cdot(x-x')} 2\pi\theta(k_0)\delta(k^2 - m^2). \qquad (1.2.7)$$

With this representation the commutator (1.2.5) takes the form

$$\Delta(x - x') = \int \frac{d^4k}{(2\pi)^4} e^{-ik\cdot(x-x')} \rho_0(k) \qquad (1.2.8)$$

where ρ_0 is the free spectral function given by

$$\rho_0(k) = 2\pi\epsilon(k_0)\delta(k^2 - m^2). \qquad (1.2.9)$$

Here $\epsilon(k_0)$ is another step function, $\epsilon(k_0) \equiv \theta(k_0) - \theta(-k_0)$, which is $+1$ for $k_0 > 0$ and -1 for $k_0 < 0$.

The Feynman propagator formula (1.1.10) gives directly the spatial Fourier transform

$$-i\Delta_F(x, x') = \theta(t - t')\Delta_+(x - x') + \theta(t' - t)\Delta_+(x' - x)$$

$$= -i \int \frac{d^3k}{(2\pi)^3} e^{ik\cdot(x-x')} \Delta_F(\omega, t - t') \qquad (1.2.10)$$

where[2]

$$\Delta_F(\omega, t - t') = \frac{i}{2\omega}\{\theta(t - t')e^{-i\omega(t-t')} + \theta(t' - t)e^{i\omega(t-t')}\} \qquad (1.2.11)$$

---

[2] To prove the second equality, we do the k_0 integral by closing the integration contour by a large semicircle: If $t > t'$, it is clockwise in the lower half-plane, when the integral picks up the residue at $k_0 = \omega - i\eta$; if $t < t'$, it is anticlockwise in the upper half-plane, when the integral picks up the residue at $k_0 = -\omega + i\eta$.

$$= -\int_{-\infty}^{\infty} \frac{dk_0}{2\pi} \frac{e^{-ik_0(t-t')}}{\{k_0 + i\eta\epsilon(k_0)\}^2 - \omega^2}. \tag{1.2.12}$$

Here η is a positive infinitesimal quantity introduced to define the poles of the integrand.[3] As the integration is over all values of k_0, it is not restricted to mass-shell, $k_0^2 = \omega^2$. Inserting (1.2.12) into (1.2.10) we get the four-dimensional Fourier transform

$$\Delta_F(x - x') = \int \frac{d^4k}{(2\pi)^4} e^{-ik\cdot(x-x')} \Delta_F(k), \tag{1.2.13}$$

with $\Delta_F(k)$ given by

$$\Delta_F(k) = \frac{-1}{\{k_0 + i\eta\epsilon(k_0)\}^2 - \omega^2} = \frac{-1}{k^2 - m^2 + i\eta} \tag{1.2.14}$$

on noting that $2k_0\eta\epsilon(k_0)$ may be replaced simply by η. Notice that although both $\Delta(x)$ and $\Delta_F(x)$ are written as integrals over four momenta, only $\Delta_F(x)$ involves off-shell (virtual) momenta, originating from the integral (1.2.12). We can also put the free propagator (1.2.14) in the form of a spectral representation[4]:

$$\Delta_F(k) = \int_{-\infty}^{\infty} dk_0' \frac{\{\delta(k_0' - \omega) - \delta(k_0' + \omega)\}/(2\omega)}{k_0' - k_0 - i\eta\epsilon(k_0)}$$

$$= \int_{-\infty}^{+\infty} \frac{dk_0'}{2\pi} \frac{\rho_0(k_0', \mathbf{k})}{k_0' - k_0 - i\eta\epsilon(k_0)} \tag{1.2.15}$$

in an apparently non-covariant form. As we shall see in Chapter 4, complete (interacting) thermal propagators will arise naturally in this form.

$$* \quad * \quad *$$

Theories with a multiplet of scalar fields are of much interest, as we shall see in the next chapter. Here we mention only the special case of a doublet, having the Lagrangian

$$\mathscr{L}(\phi_1, \phi_2) = \frac{1}{2}(\partial_\mu\phi_1\partial^\mu\phi_1 + \partial_\mu\phi_2\partial^\mu\phi_2) - \frac{m^2}{2}(\phi_1^2 + \phi_2^2). \tag{1.2.16}$$

For a compact notation we introduce a complex (non-hermitian) field defined by $\phi(x) = (\phi_1 + i\phi_2)/\sqrt{2}$, when the Lagrangian becomes

$$\mathscr{L}(\phi) = \partial_\mu\phi^\dagger\partial^\mu\phi - m^2\phi^\dagger\phi. \tag{1.2.17}$$

Clearly the complex field has the same expressions for the spectral function and propagator as the real field, if we define them following (1.1.9) and (1.1.10).

---

[3] Besides the time-ordered (Feynman) propagator, one can also define the retarded/ advanced propagator by replacing the denominator in (1.2.12) with $(k_0 \pm i\eta)^2 - \omega^2$ respectively. Taking the retarded case and integrating over k_0 as before, we get zero for $t < t'$. Similarly for the advanced case, it is zero for $t > t'$.

[4] The argument of ϵ function can also be k_0', as it is only at the pole, $k_0' = k_0$, that its presence matters.

1.3 Dirac Field

The free Lagrangian for the Dirac field is

$$\mathscr{L} = \overline{\psi}(i\gamma^{\mu}\partial_{\mu} - m)\psi, \tag{1.3.1}$$

where γ^{μ} are the so-called Dirac gamma matrices of dimension 4×4. It gives the Euler–Lagrange equation

$$(i\gamma^{\mu}\partial_{\mu} - m)\psi(x) = 0. \tag{1.3.2}$$

Applying $(i\gamma^{\nu}\partial_{\nu} + m)$ from the left and requiring the gamma matrices to satisfy the anticommutation relation

$$\{\gamma^{\mu}, \gamma^{\nu}\} = 2g^{\mu\nu}\mathbf{1}_{4\times 4}, \tag{1.3.3}$$

we get the Klein–Gordon equation

$$(\Box + m^2)\psi(x) = 0, \tag{1.3.4}$$

for each component of the Dirac field. As we discuss in Appendix A, the Dirac equation (1.3.2) itself may be interpreted as a supplementary condition, reducing the number of independent components of the field from four to two, which is the number of physical degrees of freedom of a spin-$\frac{1}{2}$ particle[3]. If an explicit choice of gamma matrices is necessary, we shall use the Weyl representation[5]:

$$\gamma^0 = \begin{pmatrix} 0 & 1 \\ 1 & 0 \end{pmatrix}, \gamma^i = \begin{pmatrix} 0 & \sigma^i \\ -\sigma^i & 0 \end{pmatrix}, \gamma^5 = i\gamma^0\gamma^1\gamma^2\gamma^3 = \begin{pmatrix} -1 & 0 \\ 0 & 1 \end{pmatrix}, \tag{1.3.5}$$

where $\mathbf{1}$ is the unit 2×2 matrix and σ^i are the Pauli matrices

$$\sigma^1 = \begin{pmatrix} 0 & 1 \\ 1 & 0 \end{pmatrix}, \quad \sigma^2 = \begin{pmatrix} 0 & -i \\ i & 0 \end{pmatrix}, \quad \sigma^3 = \begin{pmatrix} 1 & 0 \\ 0 & -1 \end{pmatrix}, \tag{1.3.6}$$

satisfying

$$\sigma^i\sigma^j = \delta^{ij} + i\epsilon^{ijk}\sigma^k. \tag{1.3.7}$$

By inspection, we see that γ^0 is hermitian, while γ^i are antihermitian; the two results may be put together by writing $\gamma^{\mu\dagger} = \gamma^0\gamma^{\mu}\gamma^0$. Also $\gamma^{5\dagger} = \gamma^5$.

The field $\psi(x)$ can be expanded as ($\omega = \sqrt{\boldsymbol{p}^2 + m^2}$):

$$\psi(x) = \int \frac{d^3p}{(2\pi)^3 2\omega} \sum_{\sigma} \left\{u(\boldsymbol{p}, \sigma)e^{-ip\cdot x}a(\boldsymbol{p}, \sigma) + v(\boldsymbol{p}, \sigma)e^{ip\cdot x}b^{\dagger}(\boldsymbol{p}, \sigma)\right\}. \tag{1.3.8}$$

Inserting this expansion in (1.3.2), the spinor coefficient functions u and v are found to satisfy

$$(\not{p} - m)u(\boldsymbol{p}, \sigma) = 0, \qquad (\not{p} + m)v(\boldsymbol{p}, \sigma) = 0. \tag{1.3.9}$$

---

[5] The other useful representation is that of Dirac, obtained by interchanging γ_0 and γ^5 in (1.3.5).

We now evaluate the spin sums

$$M(p) \equiv \sum_\sigma u(\boldsymbol{p},\sigma)\bar{u}(\boldsymbol{p},\sigma), \qquad N(p) \equiv \sum_\sigma v(\boldsymbol{p},\sigma)\bar{v}(\boldsymbol{p},\sigma) \qquad (1.3.10)$$

without solving for u and v. Recall that any 4×4 matrix may be expanded in terms of the 16 covariant matrices $\mathbf{1}$, γ^μ, $\sigma^{\mu\nu} \equiv \frac{i}{2}\{\gamma^\mu,\gamma^\nu\}$, $\gamma^\mu\gamma^5$ and γ^5. Here only one four-vector p^μ is available. So we may expand, for example, $M(p)$ as

$$M(p) = a\!\!\not{p} + b + c\!\!\not{p}\gamma^5 + d\gamma^5,$$

where a, b, c, d are constants. Apply $(\not{p} - m)$ on it from the left and also from the right. Subtracting one from the other, we get $c = d = 0$. Then any one of the equations gives $b = ma$. With the conventional normalisation of the field $\psi(x)$, we get both $M(p)$ and $N(p)$ in this way as

$$M(p) = \not{p} + m, \qquad N(p) = \not{p} - m, \qquad (1.3.11)$$

satisfying $M^2 = 2mM$ and $N^2 = 2mN$. Note also that the two spin sums are related as $N(p) = -M(-p)$.

From (1.1.9) the field anticommutation relation is now given by

$$\{\psi(x), \bar{\psi}(x')\} = \int \frac{d^3p}{(2\pi)^3 2\omega}\left\{(\not{p}+m)e^{-ip\cdot(x-x')} + (\not{p}-m)e^{ip\cdot(x-x')}\right\}$$

$$= \int \frac{d^4p}{(2\pi)^4} e^{-ip\cdot(x-x')}\sigma_0(p) \qquad (1.3.12)$$

where $\sigma_0(p)$ is the free spectral function

$$\sigma_0(p) = 2\pi\epsilon(p_0)(\not{p}+m)\delta(p^2 - m^2). \qquad (1.3.13)$$

The Feynman propagator (1.1.10) becomes

$$-iS_F(x-x') = \theta(t-t')\int \frac{d^3p}{(2\pi)^3 2\omega}(\not{p}+m)e^{-ip\cdot(x-x')}$$

$$-\theta(t'-t)\int \frac{d^3p}{(2\pi)^3 2\omega}(\not{p}-m)e^{ip\cdot(x-x')}$$

$$= \theta(t-t')(i\not{\partial}+m)\Delta_+(x-x') + \theta(t'-t)(i\not{\partial}+m)\Delta_+(x'-x). \qquad (1.3.14)$$

If we could pull the theta functions past the (time) derivatives, the Dirac propagator would be related to the scalar one. But it produces an additional term $-i\gamma^0\delta(t-t')\Delta(x-x')$, which, however, turns out to be zero (Problem 1.1a). We thus get the Dirac propagator:

$$S_F(x-x') = (i\not{\partial}+m)\Delta_F(x-x')$$

$$= \int \frac{d^4p}{(2\pi)^4} e^{-ip\cdot(x-x')}S_F(p), \qquad S_F(p) = \frac{-(\not{p}+m)}{p^2 - m^2 + i\eta}. \qquad (1.3.15)$$

Like (1.2.15) in the scalar case, we have the spectral representation of the free Dirac propagator

$$S_F(p) = \int_{-\infty}^{+\infty} \frac{dp_0'}{2\pi} \frac{\sigma_0(p_0', \boldsymbol{p})}{p_0' - p_0 - i\eta\epsilon(p_0)}. \tag{1.3.16}$$

As we shall see at the end of Section 1.6, Feynman propagators for all higher-spin fields also admit such spectral representations.

1.4 Vector Field

We first derive the free Lagrangian for the vector field B^μ and the supplementary condition on it to describe a spin-one particle. The most general form may be written as [1, 4]:

$$\mathscr{L} = \frac{1}{2}\left\{ a\, \partial_\mu B_\nu \partial^\mu B^\nu + b\, \partial_\mu B_\nu \partial^\nu B^\mu + c\, (\partial_\mu B^\mu)^2 \right\} + \frac{m^2}{2} B_\mu B^\mu, \tag{1.4.1}$$

where a, b, c and m^2 are arbitrary constants. The Euler–Lagrange equation is

$$a\Box B^\mu + (b + c)\partial^\mu(\partial_\nu B^\nu) - m^2 B^\mu = 0. \tag{1.4.2}$$

Taking divergence it gives

$$(a + b + c)\Box \partial_\mu B^\mu - m^2 \partial_\mu B^\mu = 0, \tag{1.4.3}$$

which is the equation of motion for the scalar field $\partial_\mu B^\mu$ with $-m^2/(a + b + c)$ as the squared mass. Since we want here to describe only particles of spin one and not zero, we can avoid $\partial_\mu B^\mu$ as a propagating field by setting $a + b + c = 0$, when we have $\partial_\mu B^\mu = 0$. Eliminating b, the Lagrangian (1.4.1) becomes

$$\mathscr{L} = \frac{a}{2}\partial_\mu B_\nu(\partial^\mu B^\nu - \partial^\nu B^\mu) + \frac{c}{2}(\partial_\mu B^\mu \partial_\nu B^\nu - \partial_\mu B_\nu \partial^\nu B^\mu) + \frac{m^2}{2} B_\mu B^\mu. \tag{1.4.4}$$

The second term turns out to be a total divergence,

$$\partial_\mu B^\mu \partial_\nu B^\nu - \partial_\mu B_\nu \partial^\nu B^\mu = \partial_\mu\{B_\nu(g^{\mu\nu}\partial_\rho B^\rho - \partial^\nu B^\mu)\}. \tag{1.4.5}$$

Omitting this term and absorbing a in the definition of B_μ and m, we get[6]

$$\mathscr{L} = -\frac{1}{4}F_{\mu\nu}F^{\mu\nu} + \frac{m^2}{2} B_\mu B^\mu, \qquad F_{\mu\nu} = \partial_\mu B_\nu - \partial_\nu B_\mu. \tag{1.4.6}$$

The dynamical equation for $B_\mu(x)$ is then simply

$$(\Box + m^2)B^\mu(x) = 0, \tag{1.4.7}$$

with the supplementary condition

$$\partial_\mu B^\mu = 0. \tag{1.4.8}$$

---

[6] We choose a to be negative so that the space-like components, which are the physical ones, can contribute negatively to the Lagrangian density. Then, for these components, the signs of different terms agree with those for the scalar case (1.2.1).

The condition (1.4.8) reduces the independent components of $B^\mu(x)$ from four to three, which is the number of degrees of freedom of a spin-one particle. (This is the subsidiary condition (A.3) of Appendix A.)

The vector field may again be expanded in terms of creation and annihilation operators,

$$B^\mu(x) = \int \frac{d^3k}{(2\pi)^3 2\omega} \sum_\sigma \left\{ e^\mu(\mathbf{k},\sigma)e^{-ik\cdot x}a(\mathbf{k},\sigma) + e^{\mu*}(\mathbf{k},\sigma)e^{ik\cdot x}b^\dagger(\mathbf{k},\sigma) \right\}$$

$$(1.4.9)$$

when the supplementary condition gives

$$k_\mu e^\mu = 0. \tag{1.4.10}$$

The spin sum

$$E^{\mu\nu}(k) = \sum_\sigma e^\mu(\mathbf{k},\sigma)e^{\nu*}(\mathbf{k},\sigma) \tag{1.4.11}$$

must be a linear combination of the two available second rank tensors $g^{\mu\nu}$ and $k^\mu k^\nu$. The condition (1.4.10) fixes the relative coefficient and we choose the normalisation to write

$$E^{\mu\nu}(k) = -g^{\mu\nu} + \frac{k^\mu k^\nu}{m^2}, \tag{1.4.12}$$

satisfying $E^{\mu\nu}E_{\nu\lambda} = -E^\mu_\lambda$. Then the field commutator (1.1.9) becomes

$$\Delta^{\mu\nu} = \int \frac{d^4k}{(2\pi)^4} e^{-ik\cdot(x-x')} \rho_0^{\mu\nu}(k), \tag{1.4.13}$$

where the free spectral function is given by

$$\rho_0^{\mu\nu}(k) = \left(-g^{\mu\nu} + \frac{k^\mu k^\nu}{m^2} \right) 2\pi\epsilon(k^0)\delta(k^2 - m^2). \tag{1.4.14}$$

Also, the Feynman propagator (1.1.10) is given by

$$-i\Delta_F^{\mu\nu}(x-x') = \theta(t-t')\left(-g^{\mu\nu} - \frac{\partial^\mu\partial^\nu}{m^2} \right)\Delta_+(x-x')$$

$$+\theta(t'-t)\left(-g^{\mu\nu} - \frac{\partial^\mu\partial^\nu}{m^2} \right)\Delta_+(x'-x). \tag{1.4.15}$$

In pulling the theta functions past the tensors, the time derivatives may produce additional terms. As we saw in the case of the Dirac propagator, the first derivative does not produce any, but the second derivative produces one. Using the properties of $\Delta(x-x')$ in Problems 1.1(a) and (b), one gets

$$\theta(t-t')\partial_t^2\Delta_+(x-x') + \theta(t'-t)\partial_t^2\Delta_+(x'-x)$$

$$= -i\partial_t^2\Delta_F(x-x') + i\delta^4(x-x'). \tag{1.4.16}$$

The propagator (1.4.15) then becomes

$$\Delta_F^{\mu\nu}(x-x') = \left(-g^{\mu\nu} - \frac{\partial^\mu \partial^\nu}{m^2}\right)\Delta_F(x-x') + \frac{1}{m^2}\delta_0^\mu \delta_0^\nu \delta^4(x-x'). \qquad (1.4.17)$$

Taking a Fourier transform, one gets

$$\Delta_F^{\mu\nu}(k) = \left(-g^{\mu\nu} + \frac{k^\mu k^\nu}{m^2}\right)\frac{-1}{k^2 - m^2 + i\epsilon} + \frac{1}{m^2}\delta_0^\mu \delta_0^\nu. \qquad (1.4.18)$$

As already mentioned, we may drop the non-covariant local term to write it simply as

$$\Delta_F^{\mu\nu}(k) = \left(-g^{\mu\nu} + \frac{k^\mu k^\nu}{m^2}\right)\frac{-1}{k^2 - m^2 + i\epsilon}. \qquad (1.4.19)$$

$$* \quad * \quad *$$

For a massless vector field $A_\mu(x)$, the propagator (1.4.19) may still be used if the interaction is linear in $A_\mu(x)$, $\mathscr{L}_{int} \sim J^\mu(x)A_\mu(x)$ and the current $J_\mu(x)$ is conserved, $(\partial^\mu J_\mu = 0)$, as in electrodynamics. Then in matrix elements, where the photon propagator is sandwiched between currents, the $k^\mu k^\nu/m^2$ term vanishes. So it amounts to taking

$$\Delta^{\mu\nu}(k)|_{m=0} = \frac{g^{\mu\nu}}{k^2 + i\epsilon}. \qquad (1.4.20)$$

1.5 Rarita–Schwinger Field

A spin $\frac{3}{2}$ particle can be described by a Rarita–Schwinger field ψ_A^μ, a vector-spinor having $4 \times 4 = 16$ components [5]. Following Appendix A we reduce the number of independent components to four, needed to describe the spin components. Suppressing the Dirac index as before, we impose

$$\gamma_\mu \psi^\mu(x) = 0 \qquad (1.5.1)$$

to take away four components and then the Dirac equation

$$(i\slashed{\partial} - m)\psi^\mu(x) = 0 \qquad (1.5.2)$$

to remove another eight components, retaining just four of them. A different supplementary condition

$$\partial_\mu \psi^\mu(x) = 0 \qquad (1.5.3)$$

follows from the first two conditions[7].

We shall not try to write the general form of the Lagrangian for this field, which is somewhat involved [6]. As before we expand $\psi^\mu(x)$ in terms of positive and negative frequency modes,

---

[7] To get it, multiply (1.5.2) by γ_μ from the left and use anticommutation relation (1.3.3) of gamma matrices and the other supplementary condition (1.5.1).

$$\psi^\mu(x) = \int \frac{d^3p}{(2\pi)^3 2\omega} \sum_\sigma \left\{ u^\mu(\boldsymbol{p},\sigma)e^{-ip\cdot x}a(\boldsymbol{p},\sigma) + v^\mu(\boldsymbol{p},\sigma)e^{ip\cdot x}b^\dagger(\boldsymbol{p},\sigma) \right\},$$

$$(1.5.4)$$

where u^μ and v^μ satisfy

$$(\slashed{p} - m)u^\mu = 0, \qquad p_\mu u^\mu = 0, \qquad \gamma_\mu u^\mu = 0 \tag{1.5.5}$$

$$(\slashed{p} + m)v^\mu = 0, \qquad p_\mu v^\mu = 0, \qquad \gamma_\mu v^\mu = 0. \tag{1.5.6}$$

Let us find the spin sum over u^μ:

$$\Lambda^{\mu\nu}(p) = \sum_\sigma u^\mu(\boldsymbol{p},\sigma)\overline{u}^\nu(\boldsymbol{p},\sigma). \tag{1.5.7}$$

The two spin-tensors[8]

$$(\slashed{p} + m)\left(-g^{\mu\nu} + \frac{p^\mu p^\nu}{m^2}\right) \quad \text{and} \quad \left(\frac{p^\mu}{m} + \gamma^\mu\right)(\slashed{p} - m)\left(\frac{p^\nu}{m} + \gamma^\nu\right) \tag{1.5.8}$$

separately satisfy the first two conditions in (1.5.5). Assume the spin sum to be a linear combination of the two expressions in (1.5.8),

$$a(\slashed{p} + m)\left(-g^{\mu\nu} + \frac{p^\mu p^\nu}{m^2}\right) + b\left(\frac{p^\mu}{m} + \gamma^\mu\right)(\slashed{p} - m)\left(\frac{p^\nu}{m} + \gamma^\nu\right)$$

and apply on it the third condition in (1.5.5) to get $a + 3b = 0$. After choosing the overall normalisation, we write

$$\Lambda^{\mu\nu}(p) = (\slashed{p} + m)\left\{-g^{\mu\nu} + \frac{p^\mu p^\nu}{m^2} - \frac{1}{3}\left(\frac{p^\mu}{m} - \gamma^\mu\right)\left(\frac{p^\nu}{m} + \gamma^\nu\right)\right\}, \tag{1.5.9}$$

satisfying $\Lambda^{\mu\nu}\Lambda_{\nu\lambda} = -2m\Lambda^\mu_{\ \lambda}$. The other spin sum is (see Problem 1.6)

$$\sum_\sigma v^\mu(\boldsymbol{p},\sigma)\overline{v}^\nu(\boldsymbol{p},\sigma) = -\Lambda^{\mu\nu}(-p). \tag{1.5.10}$$

Knowing the spin sums, we proceed through by now familiar steps. The field anticommutation relation is given by

$$\{\psi^\mu(x), \overline{\psi}^\nu(x')\} = \int \frac{d^4p}{(2\pi)^4}e^{-ip\cdot(x-x')}\sigma_0^{\mu\nu}(p) \tag{1.5.11}$$

with the spectral function

$$\sigma_0^{\mu\nu}(p) = 2\pi\epsilon(p_0)\Lambda^{\mu\nu}(p)\delta(p^2 - m^2). \tag{1.5.12}$$

---

[8] Two identities are helpful in these manipulations [7],

$$\left(\frac{p^\mu}{m} + \gamma^\mu\right)(\slashed{p} - m) = (\slashed{p} + m)\left(\frac{p^\mu}{m} - \gamma^\mu\right)$$

$$(\slashed{p} - m)\left(\frac{p^\nu}{m} + \gamma^\nu\right) = \left(\frac{p^\nu}{m} - \gamma^\nu\right)(\slashed{p} + m).$$

The Feynman propagator is

$$
-iS_F^{\mu\nu}(x,x') = \theta(t-t') \int \frac{d^3p}{(2\pi)^3 2\omega} \Lambda^{\mu\nu}(p) e^{-ip\cdot(x-x')}
$$
$$
+\theta(t'-t) \int \frac{d^3p}{(2\pi)^3 2\omega} \Lambda^{\mu\nu}(-p) e^{ip\cdot(x-x')}. \qquad (1.5.13)
$$

Again we express the result in terms of Δ_F by bringing the spin sums outside the respective integrals as differential operators and commuting theta functions past them. We thus get the Feynman propagator for the Rarita–Schwinger field as

$$
S_F^{\mu\nu}(p) = \frac{-\Lambda^{\mu\nu}(p)}{p^2 - m^2 + i\epsilon} + N^{\mu\nu}, \qquad (1.5.14)
$$

where $N^{\mu\nu}$ collects the non-covariant, local terms, not needed in covariant calculations[9].

1.6 General Fields

We return to our discussion of general fields in Section 1.1 and obtain their spectral functions and propagators in terms of spin sums over coefficient functions [1]. We denote these by (Problem 1.6):

$$
\sum_\sigma u_l(\boldsymbol{q},\sigma) u_{l'}^*(\boldsymbol{q},\sigma) \equiv P_{ll'}(\omega,\boldsymbol{q}), \quad \sum_\sigma v_l(\boldsymbol{q},\sigma) v_{l'}^*(\boldsymbol{q},\sigma) = \pm P_{ll'}(-\omega,-\boldsymbol{q}), \quad (1.6.1)
$$

which, as usual, are defined on mass-shell with $\omega = \sqrt{\boldsymbol{q}^2 + m^2}$. Then the commutator/anticommutator (1.1.9) becomes

$$
\Delta_{ll'}(x,x') = \int \frac{d^3q}{(2\pi)^3 2\omega} \left\{ e^{-iq\cdot(x-x')} P_{ll'}(\omega,\boldsymbol{q}) - e^{iq\cdot(x-x')} P_{ll'}(-\omega,-\boldsymbol{q}) \right\}
$$
$$
= \int \frac{d^3q}{(2\pi)^3} e^{iq\cdot(x-x')} C_{ll'}(\omega,t-t'), \qquad (1.6.2)
$$

where

$$
C_{ll'}(\omega,t-t') = \frac{1}{2\omega} \left\{ e^{-i\omega(t-t')} P_{ll'}(\omega,\boldsymbol{q}) - e^{i\omega(t-t')} P_{ll'}(-\omega,\boldsymbol{q}) \right\}
$$
$$
= \int_{-\infty}^{+\infty} dq_0 e^{-iq_0(t-t')} \delta(q^2 - m^2) \epsilon(q_0) P_{ll'}(q_0,\boldsymbol{q}). \qquad (1.6.3)
$$

Inserting (1.6.3) in (1.6.2) we get the Fourier transform of the commutator/anticommutator

$$
\Delta_{ll'}(x,x') = \int \frac{d^4q}{(2\pi)^4} e^{-iq\cdot(x-x')} \rho_{ll'}^{(0)}(q), \qquad (1.6.4)
$$

---

[9] For completeness, we write the expression for $N^{\mu\nu}$

$$
N^{\mu\nu} = \frac{2}{3m^2}(\not{p}+m)\delta_0^\mu \delta_0^\nu + \frac{2p^i}{3m^2}\gamma^0(\delta_0^\mu \delta_i^\nu + \delta_i^\mu \delta_0^\nu) - \frac{1}{3m}\gamma^0(\delta_0^\mu \gamma^\nu - \delta_0^\nu \gamma^\mu).
$$

giving the free spectral function

$$\rho_{ll'}^{(0)}(q) = 2\pi\delta(q^2 - m^2)\epsilon(q_0)P_{ll'}(q_0, \boldsymbol{q}). \tag{1.6.5}$$

Though we integrate over all q^μ in (1.6.4), the delta function restricts it to mass shell.

Similarly, the Feynman propagator (1.1.10) is given by

$$-i\Delta_{ll'}^F(x, x') = \theta(t - t') \int \frac{d^3q}{(2\pi)^3 2\omega} e^{-iq\cdot(x-x')} P_{ll'}(\omega, \boldsymbol{q})$$
$$+\theta(t' - t) \int \frac{d^3q}{(2\pi)^3 2\omega} e^{iq\cdot(x-x')} P_{ll'}(-\omega, -\boldsymbol{q}). \tag{1.6.6}$$

Now we deal with the non-covariance problem in a slightly different way from our approach to the previous individual cases. Instead of trying to commute the spin sum with the theta function, we first separate its terms into even and odd powers of ω [1]:

$$P_{ll'}(\omega, \boldsymbol{q}) = P_{ll'}^{(1)}(\omega^2, \boldsymbol{q}) + \omega P_{ll'}^{(2)}(\omega^2, \boldsymbol{q}). \tag{1.6.7}$$

(Clearly the polynomials $P_{ll'}^{(1,2)}$ are functions of $\boldsymbol{q}$ only, but we keep the ω^2 dependence explicit for convenience later in continuing off the mass-shell.) Inserting this decomposition in (1.6.6) and changing the sign of $\boldsymbol{q}$ in the second term, we get

$$-i\Delta_{ll'}^F(x, x') =$$
$$\int \frac{d^3q}{(2\pi)^3 2\omega} e^{iq\cdot(x-x')} \left[\theta(t - t') \left\{ P_{ll'}^{(1)}(\omega^2, \boldsymbol{q}) + \omega P_{ll'}^{(2)}(\omega^2, \boldsymbol{q}) \right\} e^{-i\omega(t-t')} \right.$$
$$\left. +\theta(t' - t) \left\{ P_{ll'}^{(1)}(\omega^2, \boldsymbol{q}) - \omega P_{ll'}^{(2)}(\omega^2, \boldsymbol{q}) \right\} e^{i\omega(t-t')} \right]. \tag{1.6.8}$$

Next, replace ω multiplying $P^{(2)}$ as a time derivative acting on the exponential, when the two expressions within the curly brackets above become identical. Also, the theta functions can be brought inside this derivative, producing terms with delta functions, which, however, cancel out. We thus get

$$\Delta_{ll'}^F(x, x') = \int \frac{d^3q}{(2\pi)^3} e^{iq\cdot(x-x')} \left\{ P_{ll'}^{(1)}(\omega^2, \boldsymbol{q}) + P_{ll'}^{(2)}(\omega^2, \boldsymbol{q}) i\frac{\partial}{\partial t} \right\}$$
$$\frac{i}{2\omega} \left\{ \theta(t - t')e^{-i\omega(t-t')} + \theta(t' - t)e^{i\omega(t-t')} \right\}. \tag{1.6.9}$$

The factor with theta functions is $\Delta_F(\omega, t - t')$, defined by (1.2.11). Inserting its integral representation (1.2.12) on which the time derivative can act, we get

$$\Delta_{ll'}^F(x - x') = -\int \frac{d^4q}{(2\pi)^4} \frac{e^{-iq(x-x')}}{q^2 - m^2 + i\eta} \left\{ P_{ll'}^{(1)}(\omega^2, \boldsymbol{q}) + q_0 P_{ll'}^{(2)}(\omega^2, \boldsymbol{q}) \right\}. \tag{1.6.10}$$

The problem of non-covariance of the propagator now surfaces in the polynomial within the curly bracket, which is Lorentz covariant only at the pole, i.e.

on the mass shell. As we integrate over all values of q^μ, not restricted to this mass shell, the polynomial, which is linear in q^0 but not in $\boldsymbol{q}$ in general, cannot be covariant. But we can restore covariance if it is expressed as a function of the four-vector $q^\mu = (q^0, \boldsymbol{q})$, not restricted to mass shell. For this purpose we write $\omega^2 = q_0^2 - (q^2 - m^2)$ in the arguments of $P_{ll'}^{(1,2)}$ and expand them in ω^2 around q_0^2 to get

$$P_{ll'}^{(1)}(\omega^2, \boldsymbol{q}) + q_0 P_{ll'}^{(2)}(\omega^2, \boldsymbol{q}) = P_{ll'}(q) + (q^2 - m^2)N_{ll'}. \tag{1.6.11}$$

Here the first term is

$$P_{ll'}(q) = P_{ll'}^{(1)}(q_0^2, \boldsymbol{q}) + q_0 P_{ll'}^{(2)}(q_0^2, \boldsymbol{q}), \tag{1.6.12}$$

which collects itself into a covariant expression, just as $P_{ll'}(\omega, \boldsymbol{q})$ does on the mass shell, but now without this constraint. The second term is the non-covariant piece, given by the remaining terms of the expansion, if any, with at least one factor of $(q^2 - m^2)$, which we show explicitly in (1.6.11). We thus have the covariant propagator with an additive non-covariant, local piece:

$$\Delta_{ll'}^F(q) = \frac{-P_{ll'}(q)}{q^2 - m^2 + i\eta} + N_{ll'}. \tag{1.6.13}$$

Looking back at the individual cases considered previously, we find that for the scalar and Dirac fields, there is no ω^2 term in $P_{ll'}$, so we did not find any non-covariant piece. But for the vector and Rarita–Schwinger fields, the presence of such terms gave rise to non-covariant pieces in agreement with what is given by $N_{ll'}$. However, as it is local, it can be removed by adding a corresponding term in the Hamiltonian. Accordingly, we shall omit any $N_{ll'}$ term in the expression for the propagator. We close our discussion by noting the spectral representation of the propagator:

$$\Delta_{ll'}^F(q) = \int_{-\infty}^{+\infty} \frac{dq_0'}{2\pi} \frac{\rho_{ll'}^{(0)}(q_0', \boldsymbol{q})}{q_0' - q_0 - i\eta\epsilon(q_0)}, \tag{1.6.14}$$

which may again be verified by integrating over q_0' with $\rho_{ll'}^{(0)}$ given by (1.6.5).

Problems

Problem 1.1: Obtain the following properties of the scalar field commutator defined by (1.2.4).

$$a)\ \Delta(x - x')|_{t=t'} = 0, \qquad\qquad b)\ \partial_t \Delta(x - x')|_{t=t'} = -i\delta^3(\boldsymbol{x} - \boldsymbol{x'}),$$

$$c)\ \Delta(x - x')|_{(x-x')^2 < 0} = 0, \qquad d)\ (\Box_x + m^2)\Delta(x - x') = 0. \tag{P1.1}$$

Solution:

a) Set $t = t'$ in (1.2.4) and put $-\mathbf{k}$ for $\mathbf{k}$ in any of the two terms, when they cancel each other.

b) Carrying out the diferentiation, we get

$$\partial_t \Delta(x - x') = \frac{-i}{2} \int \frac{d^3k}{(2\pi)^3} \left(e^{-ik\cdot(x-x')} + e^{ik\cdot(x-x')} \right). \qquad \text{(P1.2)}$$

Now set $t = t'$, when each term becomes a three-dimensional delta-function, giving the result.

c) The function Δ_+ is manifestly Lorentz invariant:

$$\Delta_+(x - x') = \int \frac{d^4k}{(2\pi)^3} \theta(k_0)\delta(k^2 - m^2)e^{-ik\cdot(x-x')}. \qquad \text{(P1.3)}$$

So, for space-like $(x - x')$, it depends only on $-(x - x')^2 > 0$, when $\Delta_+(x - x')$ and $\Delta_+(x' - x)$ become equal, making $\Delta(x - x')$ vanish. It can also be verified by explicit evaluation. (This result ensures causality in quantum field theory [1, 8].)

d) The function $\Delta(x - x')$ is a commutator of the scalar field $\phi(x)$,

$$\Delta(x - x') = [\phi(x), \phi(x')].$$

Applying $(\Box_x + m^2)$ on it and using the equation of motion (1.2.2), we get zero. The result also follows from (1.2.8), as $(\Box_x + m^2)$ operates on the exponential, giving $(k^2 - m^2)\delta(k^2 - m^2) = 0$ in the integrand.

Problem 1.2: Show that the differential equation satisfied by the scalar propagator is

$$(\Box_x + m^2)\Delta_F(x, x') = \delta^4(x - x'). \qquad \text{(P1.4)}$$

Solution: If $(\Box_x + m^2)$ operates on the propagator (1.2.13), we immediately get (P1.4). But let us derive it from the definition

$$-i\Delta_F(x - x') = \langle 0|T\{\phi(x)\phi^\dagger(x')\}|0\rangle, \qquad \text{(P1.5)}$$

where T is the time-ordering symbol (1.1.8) with the plus sign. The differential operator acts on the field operator and not on the (vacuum) state, so that we consider

$$(\Box_x + m^2)T\{\phi(x)\phi^\dagger(x')\}.$$

If we could pull $(\Box_x + m^2)$ past the T symbol, it would act on $\phi(x)$ giving zero. However, ∂_t^2 present in this operator acts on the theta functions defining T, which give additional terms. We find

$$\begin{aligned}
\partial_t^2 &T\{\phi(x)\phi^\dagger(x')\} \\
&= \partial_t \left[T\{\partial_t\phi(x)\phi^\dagger(x')\} + \delta(t - t')\Delta(x - x') \right] \\
&= T\{\partial_t^2\phi(x)\phi^\dagger(x')\} + \delta(t - t')[\partial_t\phi(x), \phi^\dagger(x')], \qquad \text{(P1.6)}
\end{aligned}$$

where we set the equal time commutator to zero in the second line. Also, the second term in the last line becomes

$$\delta(t - t')[\partial_t \phi(x), \phi^\dagger(x')] = \delta(t - t')\partial_t \Delta(x - x') = -i\delta^4(x - x')$$

by Problem 1.1b). We thus get

$$(\Box_x + m^2)T\{\phi(x)\phi(x')\} = -i\delta^4(x - x'), \tag{P1.7}$$

giving the differential equation (P1.4) for the propagator.

Problem 1.3: Derive the scalar field propagator by solving its differential equation with Feynman boundary conditions. (In Section 4.2 we shall follow a similar procedure to derive the thermal propagator, replacing the Feynman boundary condition with the thermal one.)

Solution: It follows from (1.1.10) that

$$\Delta_F(\boldsymbol{x}, \boldsymbol{x'}, t, t') \to \begin{cases} e^{-i\omega(t-t')} & t > t' \\ e^{i\omega(t-t')} & t < t', \end{cases} \tag{P1.8}$$

which serve as the boundary conditions to solve (P1.4) for the vacuum propagator.

It is a simple exercise in Green's functions [9], which we do in detail now. As the boundary condition does not involve spatial coordinates, it simplifies to work with spatial Fourier transform:

$$\Delta_F(x, x') = \int \frac{d^3k}{(2\pi)^3} e^{ik\cdot(\boldsymbol{x}-\boldsymbol{x'})} \Delta_F(\boldsymbol{k}, t, t'). \tag{P1.9}$$

The transform satisfies

$$\left(\frac{d^2}{dt^2} + \omega^2\right)\Delta_F(\boldsymbol{k}, t, t') = \delta(t - t'), \quad \omega^2 = k^2 + m^2. \tag{P1.10}$$

As with (P1.4), it is written as a differential equation in the unprimed variable (t), keeping the primed one (t') fixed.

For $t \neq t'$, $\Delta_F(\boldsymbol{k}, t, t')$ satisfies a homogeneous equation having two independent solutions, namely $e^{\pm i\omega t}$. So in each of the segments, $t > t'$ and $t < t'$, the general solutions are linear combinations of these solutions. However, the boundary conditions (8) allow only one of these in each segment:

$$\Delta_F(\boldsymbol{k}, t, t') = \begin{cases} A(t')\, e^{-i\omega(t-t')}, & t > t' \\ B(t')\, e^{i\omega(t-t')}, & t < t'. \end{cases} \tag{P1.11}$$

As a Green's function of a second order differential equation, $\Delta_F(\boldsymbol{k}, t, t')$ must be continuous at $t = t'$, while its first derivative has a discontinuity there given by the coefficient of the delta function. These two conditions yield

$$A(t') = B(t')$$
$$-i\omega\{A(t') + B(t')\} = 1, \tag{P1.12}$$

giving $A = B = i/2\omega$. Then (P1.11) gives

$$\Delta_F(\omega, t - t') = \frac{i}{2\omega}\{\theta(t - t')e^{-i\omega(t-t')} + \theta(t' - t)e^{i\omega(t-t')}\}, \qquad \text{(P1.13)}$$

reproducing the Feynman propagator (1.2.11) in vacuum.

Problem 1.4: Find the symmetry transformations keeping the Lagrangian for the scalar field doublet invariant.

Solution: The Lagrangian (1.2.16) is invariant under $SO(2)$ transformations

$$\begin{pmatrix} \phi_1 \\ \phi_2 \end{pmatrix} \rightarrow \begin{pmatrix} \phi_1' \\ \phi_2' \end{pmatrix} = \begin{pmatrix} \cos\theta & -\sin\theta \\ \sin\theta & \cos\theta \end{pmatrix} \begin{pmatrix} \phi_1 \\ \phi_2 \end{pmatrix} \qquad \text{(P1.14)}$$

while the equivalent Lagrangian (1.2.17) is invariant under $U(1)$ transformations

$$\phi \rightarrow \phi' = e^{i\theta}\phi. \qquad \text{(P1.15)}$$

We see the isomorphism of $SO(2)$ and $U(1)$ groups. Either way, we get the symmetry current (see Appendix B):

$$j_\mu = i\phi^\dagger(x) \overleftrightarrow{\partial_\mu} \phi(x) \equiv i\left(\phi^\dagger(x)\partial_\mu\phi(x) - \partial_\mu\phi^\dagger(x)\phi(x)\right). \qquad \text{(P1.16)}$$

Problem 1.5: Find the symmetry transformations keeping the Dirac Lagrangian invariant. Describe the transformations also in terms of two-component (Weyl) fields.

Solution: The Dirac Lagrangian (1.3.1) is invariant under the phase transformation

$$\psi \rightarrow e^{i\theta_V}\psi, \qquad \text{(P1.17)}$$

which form the group $U(1)_V$. Further, if the field is massless, the Lagrangian reduces to

$$\mathscr{L}^{(0)} = \overline{\psi}i\slashed{\partial}\psi \qquad \text{(P1.18)}$$

when it has an additional invariance under the transformation

$$\psi \rightarrow e^{i\theta_A\gamma^5}\psi \qquad \text{(P1.19)}$$

of the group $U(1)_A$. Considering infinitesimal x-dependent transformations, we see from (B.8) of Appendix B that these symmetries lead respectively to the conserved currents, the vector and axial-vector,

$$V^\mu = \overline{\psi}\gamma^\mu\psi, \qquad A^\mu = \overline{\psi}\gamma^\mu\gamma^5\psi. \qquad \text{(P1.20)}$$

This two-fold symmetry of the massless Dirac Lagrangian is related to the fact we already mentioned in Section 1.3, namely that the Dirac field has twice as many components as are needed to describe a particle of spin $\frac{1}{2}$. Indeed, we may split ψ into the right- and left-handed components,

$$\psi_R = \frac{1}{2}(1 + \gamma^5)\psi, \qquad \psi_L = \frac{1}{2}(1 - \gamma^5)\psi \qquad \text{(P1.21)}$$

For a massless field ψ, the components ψ_R and ψ_L describe particles of helicity $+\frac{1}{2}$ and $-\frac{1}{2}$ respectively[10]. But the decomposition is useful, be the field massless or not.

The transformation (P1.17) acts in the same way on ψ_R and ψ_L,

$$\psi_R \to e^{i\theta_V}\psi_R, \quad \psi_L \to e^{i\theta_V}\psi_L, \tag{P1.22}$$

while the transformation (P1.19) acts differently on them,

$$\psi_R \to e^{i\theta_A}\psi_R, \quad \psi_L \to e^{-i\theta_A}\psi_L. \tag{P1.23}$$

When the transformations (P1.22) and (P1.23) are combined, we get the transformation

$$\psi_R \to e^{i\theta_R}\psi_R, \quad \theta_R = \theta_V + \theta_A, \tag{P1.24}$$

$$\psi_L \to e^{i\theta_L}\psi_L, \quad \theta_L = \theta_V - \theta_A, \tag{P1.25}$$

---

[10] This follows immediately in the Weyl representation (1.3.5) of γ-matrices, where the Dirac spinor decomposes into two two-component Weyl spinors:

$$\psi = \begin{pmatrix} \psi_L \\ \psi_R \end{pmatrix}.$$

The massless Dirac equation with plane wave solution of positive energy reads from (1.3.9) as

$$\slashed{p}\,u(p) = 0.$$

Decomposing u into u_L and u_R as in (P1.21), we get

$$(|\boldsymbol{p}| - \boldsymbol{\sigma} \cdot p)u_R = 0, \quad (|\boldsymbol{p}| + \boldsymbol{\sigma} \cdot p)u_L = 0,$$

giving $\boldsymbol{\sigma} \cdot p/|\boldsymbol{p}| = \pm 1$ and hence helicity $\pm\frac{1}{2}$ for u_R and u_L respectively.

To see this result in an arbitrary representation [10], let us write the massless equation for motion along, say the 3-direction,

$$|\boldsymbol{p}|(\gamma^0 - \gamma^3)u(p) = 0$$

giving

$$\gamma^0 u = \gamma^3 u.$$

The matrices $J^{\mu\nu} = \frac{i}{4}[\gamma^\mu, \gamma^\nu]$ are generators of the homogeneous Lorentz group [1, 4]. It gives the component of spin angular momentum in 3-direction as

$$J_3 = \frac{i}{4}[\gamma^1, \gamma^2] = \frac{i}{2}\gamma^1\gamma^2 = \frac{i}{2}\gamma^0\gamma^1\gamma^2\gamma^0.$$

As we just saw, γ^0 acting on u can be replaced by γ^3, giving

$$J_3 = \frac{i}{2}\gamma^0\gamma^1\gamma^2\gamma^3 = \frac{1}{2}\gamma^5.$$

So

$$J_3 u_R = \frac{1}{2}\gamma^5\frac{1}{2}(1+\gamma^5)u = \frac{1}{2}u_R, \quad \text{and} \quad J_3 u_L = \frac{1}{2}\gamma^5\frac{1}{2}(1-\gamma^5)u = -\frac{1}{2}u_L.$$

so that ψ_R and ψ_L transform independently of each other. If we rewrite the Lagrangian (P1.18) in terms of ψ_R and ψ_L,

$$\mathscr{L}^{(0)} = \overline{\psi}_R i\partial\!\!\!/ \, \psi_R + \overline{\psi}_L i\partial\!\!\!/ \, \psi_L, \tag{P1.26}$$

we see that it does not contain any term which connects the two components. The massless Lagrangian, therefore, remains invariant under 'chiral rotations'; that is, under independent transformations of the right- and left-handed fermion fields. (Clearly a mass term $\sim \overline{\psi}_R\psi_L + \overline{\psi}_L\psi_R$ spoils its invariance under both transformations, unless they are applied simultaneously with $\theta_R = \theta_L$.) Later, in Chapter 3 we shall deal with two species (flavours) of Dirac fields, when the Lagrangian will have additional internal symmetries.

Problem 1.6: If the spin sums over positive and negative energy coefficient functions for general fields are denoted by

$$P_{ll'}(\omega, \boldsymbol{q}) = \sum_\sigma u_l(\boldsymbol{q}, \sigma) u_{l'}^*(\boldsymbol{q}, \sigma), \qquad Q_{ll'}(\omega, \boldsymbol{q}) = \sum_\sigma v_l(\boldsymbol{q}, \sigma) v_{l'}^*(\boldsymbol{q}, \sigma),$$

show that they are related as

$$Q_{ll'}(\omega, \boldsymbol{q}) = \pm P_{ll'}(-\omega, -\boldsymbol{q}).$$

Solution: Using the notation for the spin sums, the commutator/anticommutator (1.1.9) becomes

$$\begin{aligned}
\Delta_{ll'}(x, x') &= \int \frac{d^3q}{(2\pi)^3 2\omega} \left(e^{-iq\cdot(x-x')} P_{ll'}(q) \mp e^{iq\cdot(x-x')} Q_{ll'}(q) \right) \\
&= P_{ll'}(i\partial_x)\Delta_+(x - x') \mp Q_{ll'}(-i\partial_x)\Delta_+(x' - x). \tag{P1.27}
\end{aligned}$$

Specialising to space-like separation, $\Delta_{ll'}(x - x')$ must be zero and the two Δ_+ functions equal. So, for such separations (P1.27) becomes

$$\begin{aligned}
0 &= \{P_{ll'}(i\partial_x) \mp Q_{ll'}(-i\partial_x)\} \Delta_+(x - x') \\
&= \int \frac{d^3q}{(2\pi)^3 2\omega} e^{-iq\cdot(x-x')} \{P_{ll'}(q) \mp Q_{ll'}(-q)\}, \tag{P1.28}
\end{aligned}$$

whence the result follows.

References

[1] S. Weinberg, *The Quantum Theory of Fields, Vol. 1: Foundations*, Cambridge, UK: Cambridge University Press (1995).

[2] S. Weinberg, Phys. Rev. **133**, B1318 (1964).

[3] A. Zee, *Quantum Field Theory in a Nutshell*, Princeton University Press (2003).

[4] C. Itzykson and J.-B. Zuber, *Quantum Field Theory*, McGraw-Hill (1980).

[5] W. Rarita and J. Schwinger, Phys. Rev. **60**, 61 (1941).

[6] K. Johnson and E.C.G. Sudarshan, Annals Phys. **13**, 126 (1961).

[7] J. Schwinger, *Particles, Sources, and Fields*, volume 1, Reading, Mass. (1970).

[8] M.E. Peskin and D.V. Schroeder, *An Introduction to Quantum Field Theory*, Westview Press (1995).

[9] K.F. Riley, M.P. Hobson and S.J. Bence, *Mathematical Methods for Physics and Engineering*, Cambridge University Press (2006).

[10] H. Georgi, *Weak Interactions and Modern Particle Theory*, Benjamin/Cummings Publishing Co. (1984).

2

Spontaneous Symmetry Breaking

In this chapter and the following one we attempt an elementary and self-contained review of the theory of strong interactions at low energy. As is well known, the fundamental theory of strong interactions, called *quantum chromodynamics (QCD)*, is not immediately applicable to processes at low energy. An effective field theory is called for, based on the (approximate) flavour symmetry of the QCD Lagrangian. This chapter develops the idea of spontaneously broken global symmetry, which is central to the construction of such an effective theory. The next chapter uses this framework to write the effective Lagrangian of this theory.

The main result of spontaneous symmetry breaking is the Goldstone theorem that proves the existence of massless, spinless particles known as Nambu–Goldstone bosons (or Goldstone bosons, for short). After illustrating this result with linear σ-models in classical field theory, we describe two classic proofs of the Goldstone theorem in quantum theory, using *effective potential* and *current algebra* [1]. Some results relating to interaction of the Goldstone bosons can be derived immediately. We then go on to construct the non-linear version of the σ-model as the prototype of effective theories and generalise the construction to find transformation rules for fields under the action of an arbitrary symmetry group. Symmetry broken spontaneously in the presence of explicit breaking, as is the case with QCD, is also discussed.

2.1 Linear σ-Models

We first investigate spontaneously broken symmetry in quantum field theory in tree approximation (without loops), which is essentially the classical field theory. The examples of field theory we are going to discuss are called linear σ-models [2–5]. They are given by the Lagrangian (density) of N real (hermitian) scalar fields $\phi_n(x)$,

$$\mathscr{L}(\phi) = \frac{1}{2}\partial_\mu\phi_n(x)\partial^\mu\phi_n(x) - U(\phi), \qquad (2.1.1)$$

where the potential function $U(\phi)$ has the form

$$U(\phi) = \frac{\mu^2}{2}\phi_n\phi_n + \frac{\lambda}{4}(\phi_n\phi_n)^2 . \qquad (2.1.2)$$

A sum is understood over n taking values $1, 2, \cdots N$. The coupling parameter λ must be positive for U to remain bounded from below, but μ^2 in the mass-like term may take either sign. For $N \geq 2$, the internal symmetry group of the Lagrangian is $O(N)$, the group of orthogonal linear transformation in N dimensions. That is, it is invariant under global (space–time-independent) transformation of the field components

$$\phi_n(x) \to \phi'_n(x) = R_{nm}\phi_m(x), \qquad R \in O(N) . \qquad (2.1.3)$$

Let us now discuss the invariance of the vacuum state. Here we have to find the state (or states) of minimum energy. Now, the energy of a state is given by the Hamiltonian (density),

$$\mathcal{H} = \frac{1}{2}\partial_0\phi_n\partial_0\phi_n + \frac{1}{2}\boldsymbol{\nabla}\phi_n \cdot \boldsymbol{\nabla}\phi_n + U(\phi).$$

So for the state(s) of lowest energy, the field must be constant (space–time-independent) and its values be determined by the minimum (or minima) of the potential $U(\phi)$, $\partial U/\partial\phi_n = 0$. If there is a single such state, we have the unique vacuum. But if there is *degeneracy*, any one of the degenerate states is a candidate for the vacuum state. However, to build a quantum field theory we have to choose one of these as *the* vacuum state. Then those transformations of $O(N)$ that do not leave this vacuum state invariant are said to be broken spontaneously. The remaining ones keeping it invariant constitute the symmetry group of the vacuum state[1]. Clearly this breaking reduces in general the symmetry of the vacuum state and multiplets built on it; but the Lagrangian is still symmetric under the full group. Before taking up the general case of N fields, we first discuss the cases of a single field and of two fields.

(A) The simplest σ-model is given by the Lagrangian of a single scalar field $\phi(x)$,

$$\mathscr{L}(\phi) = \frac{1}{2}\partial_\mu\phi(x)\partial^\mu\phi(x) - U(\phi) \qquad (2.1.4)$$

with

$$U(\phi) = \frac{\mu^2}{2}\phi^2 + \frac{\lambda}{4}\phi^4 . \qquad (2.1.5)$$

---

[1] A question arises as to the possibility of defining the vacuum as a linear superposition of all the degenerate states, which would be invariant under the full symmetry group. While it is possible to do so in quantum mechanics, it is not possible in quantum field theory. See [2].

This Lagrangian has reflection symmetry under

$$\phi \to -\phi. \tag{2.1.6}$$

If $\mu^2 > 0$, the potential is as shown in Figure 2.1(a). The ground state or the vacuum is at $\phi = 0$; the symmetry is manifest and the scalar particle has mass μ.

If, however, $\mu^2 < 0$, the situation becomes different. The potential no longer has a minimum at the origin of the field, but is shifted to $\overline{\phi}$ given by

$$\overline{\phi}^2 \equiv -\frac{\mu^2}{\lambda} > 0. \tag{2.1.7}$$

We can now rewrite the potential as

$$U(\phi) = \frac{\lambda}{4}(\phi^2 - \overline{\phi}^2)^2 - \frac{\lambda}{4}\overline{\phi}^4. \tag{2.1.8}$$

(A field-independent term in the potential, such as the last one in (2.1.8), appears as a cosmological constant, which is irrelevant in the absence of gravity.) Clearly the potential has two minima at $\phi = \pm\overline{\phi}$, as shown in Figure 2.1(b). As already pointed out, we must choose one of these as the vacuum of the theory. Because of the symmetry (2.1.6) of the Lagrangian, this choice does not affect the resulting physics. However, whichever one we choose, the vacuum state will not have this symmetry: the symmetry is said to be spontaneously broken. Let us choose $\phi = +\overline{\phi}$ for the vacuum and consider fluctuations around it, defining the shifted field ϕ' by $\phi = \overline{\phi} + \phi'$. Then from (2.1.4) and (2.1.8) we get

$$\mathcal{L}(\phi') = \frac{1}{2}\partial_\mu\phi'\partial^\mu\phi' - \lambda(\overline{\phi}^2\phi'^2 + \overline{\phi}\phi'^3 + \frac{1}{4}\phi'^4). \tag{2.1.9}$$

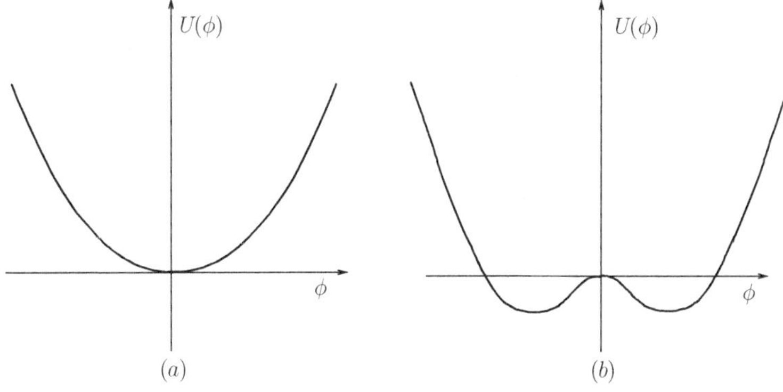

Figure 2.1 The potential function $U(\phi)$ for (a) $\mu^2 > 0$ and (b) for $\mu^2 < 0$.

We see that the true mass of the particle as given by the ϕ'^2 term is $\sqrt{2}|\mu|$. Due to the appearance of ϕ'^3, the symmetry (2.1.6) of $\mathscr{L}$ is hidden.

(B) Next, consider the Lagrangian with two real scalar fields,

$$\mathscr{L} = \frac{1}{2}\partial_\mu\phi_1\partial^\mu\phi_1 + \frac{1}{2}\partial_\mu\phi_2\partial^\mu\phi_2 - U(\phi_1,\phi_2) \tag{2.1.10}$$

and

$$U = \frac{\mu^2}{2}(\phi_1^2 + \phi_2^2) + \frac{\lambda}{4}(\phi_1^2 + \phi_2^2)^2 \tag{2.1.11}$$

having $O(2)$ symmetry. As before, if $\mu^2 > 0$, the minimum of U giving the vacuum state is at $\phi_1 = \phi_2 = 0$, which is invariant under $O(2)$ and there is no spontaneous symmetry breaking.

But if $\mu^2 < 0$, the potential exhibits minima on a circle in the (ϕ_1,ϕ_2) plane,

$$\bar{\phi}_1^2 + \bar{\phi}_2^2 = -\frac{\mu^2}{\lambda} > 0, \tag{2.1.12}$$

giving an infinitely degenerate vacuum state, as compared to a twofold degeneracy in the previous case. The choice of a definite $(\bar{\phi}_1, \bar{\phi}_2)$ as the vacuum state for the quantum theory breaks the $O(2)$ symmetry spontaneously. To get the resulting mass spectrum, we rewrite (2.1.11) as

$$U(\phi_1, \phi_2) = \frac{\lambda}{4}(\phi_1^2 + \phi_2^2 - \bar{\phi}_1^2 - \bar{\phi}_2^2)^2, \tag{2.1.13}$$

dropping an inessential constant. Using the $SO(2)$ symmetry of the Lagrangian, we may rotate an arbitrary vacuum state to $\bar{\phi}_1 = +\sqrt{-\mu^2/\lambda}$, $\bar{\phi}_2 = 0$. Again shifting the field, $\phi_1 = \bar{\phi}_1 + \phi_1'$, the potential (2.1.13) is seen to have a quadratic term only in ϕ_1', namely $\lambda\bar{\phi}_1^2\phi_1'^2$. Thus the field $\phi_1'(x)$ has mass $\sqrt{2}|\mu|$, while $\phi_2(x)$ is massless.

We note here the difference between the cases $N = 1$ and $N = 2$. In the former case it is a *discrete* symmetry of $\mathscr{L}$, which is broken by the vacuum state and no massless particle results. In the latter case the vacuum breaks the $O(2)$ symmetry of $\mathscr{L}$, which is *continuous*, leading to a massless particle. This result can also be seen geometrically. As a function of $\sqrt{\phi_1^2 + \phi_2^2}$, the potential given by (2.1.13) has the same shape as of Figure 2.1(b) for a single real field. But ϕ_1 and ϕ_2 are independent variables and the potential develops into a Mexican hat in three dimensions. Referring to vacuum states considered above, excitations in ϕ_1' are in the radial direction, having a curvature; so we get a massive particle. On the other hand, ϕ_2 describes excitations in the orthogonal direction, which is flat, giving a massless particle.

(C) We now consider the general case of N scalar fields, with the Lagrangian given by (2.1.1–2). It follows essentially the same steps as for two fields. Again, the case $\mu^2 > 0$ is not interesting for us. For $\mu^2 < 0$, the minima of the potential are given by

$$\overline{\phi}_n \overline{\phi}_n = -\frac{\mu^2}{\lambda} > 0, \tag{2.1.14}$$

when the potential (2.1.2) may be written as

$$U(\phi_n) = \frac{\lambda}{4}(\phi_n\phi_n - \overline{\phi}_m\overline{\phi}_m)^2 \,, \tag{2.1.15}$$

again dropping a constant. Let us now choose the vacuum state not along the ϕ_1 axis, as we did for the $O(2)$ case above, but along an arbitrary direction specified by the components $\overline{\phi}_n$. Introducing vector notation, we write the N dimensional field as

$$\vec{\phi}(x) = (\phi_1(x), \phi_2(x), \cdots, \phi_N(x))$$

when the vacuum state can be represented as

$$\vec{\phi}_0 = (\overline{\phi}_1, \overline{\phi}_2, \cdots, \overline{\phi}_N), \qquad |\vec{\phi}_0|^2 = -\frac{\mu^2}{\lambda}.$$

In this notation the potential (2.1.15) becomes

$$U(\vec{\phi}) = \frac{\lambda}{4}(|\vec{\phi}|^2 - |\vec{\phi}_0|^2)^2. \tag{2.1.16}$$

The vacuum state breaks the $O(N)$ symmetry of the Lagrangian to $O(N-1)$, because the subset of rotations in N dimensions that leave $\vec{\phi}_0$ invariant form a subgroup $O(N-1)$. Decompose the field as

$$\vec{\phi}(x) = \vec{\phi}_0 + \vec{\chi}(x) + \vec{\phi}_\perp(x) \,, \tag{2.1.17}$$

where $\vec{\chi}$ is along $\vec{\phi}_0$, while $\vec{\phi}_\perp$, containing the remaining $N-1$ fields, is orthogonal to $\vec{\phi}_0$ and $\vec{\chi}$. Then the potential (2.1.16) takes the form

$$U(\vec{\phi}) = \frac{\lambda}{4}(2|\vec{\phi}_0||\vec{\chi}| + |\vec{\chi}|^2 + |\vec{\phi}_\perp|^2)^2 \,, \tag{2.1.18}$$

showing the absence of quadratic term in $\vec{\phi}_\perp$. Thus only the field $\vec{\chi}$ has mass $\sqrt{2}|\mu|$ and the other $N-1$ independent fields remain massless. Like the $N = 2$ case, the massless modes may also be seen geometrically: on the N-dimensional sphere (2.1.14) containing the degenerate vacua, there are $(N-1)$ linearly independent directions to leave a particular point and still stay on the sphere [5].

*   *   *

We introduce here some notations relating to an arbitrary symmetry group G breaking spontaneously to a subgroup H. As we saw above, this phenomenon implies that elements $g \in G$ acting on fields $\phi(x)$ keep the Lagrangian invariant,

$$\mathscr{L}(g\phi) = \mathscr{L}(\phi), \tag{2.1.19}$$

but only a subset of elements $h \in H$ keep the vacuum field $\bar{\phi}$ invariant, while the remaining ones, call them k, break this vacuum symmetry,

$$h\bar{\phi} = \bar{\phi}, \qquad k\bar{\phi} \neq \bar{\phi} \tag{2.1.20}$$

(The elements h form a subgroup: if two elements h_1 and h_2 are such that $h_1\bar{\phi} = \bar{\phi}$, $h_2\bar{\phi} = \bar{\phi}$, then so does their product, $h_1 h_2 \bar{\phi} = \bar{\phi}$.)

We now adapt our notation for the (matrix) generators of G to the above pattern of symmetry breaking. Let t_i denote the independent generators of H. As H is a subgroup, t_i form a subalgebra

$$[t_i, t_j] = i c_{ijk} t_k. \tag{2.1.21}$$

Denote the *other* independent generators of G by s_a in a basis with totally antisymmetric structure constants. As s_a does not appear on the right side of (2.1.21), all $c_{ija} = 0$. From total antisymmetry, it follows that $c_{iaj} = 0$, so we have

$$[t_i, s_a] = i c_{iab} s_b. \tag{2.1.22}$$

Finally, we have in general

$$[s_a, s_b] = i c_{abi} t_i + i c_{abc} s_c. \tag{2.1.23}$$

Introducing coordinates θ_i and θ_a in the group space, the finite elements in the canonical form are

$$h(\theta_i) = \exp(i\theta_i t_i), \qquad k(\theta_a) = \exp(i\theta_a s_a).$$

Then (2.1.20) become

$$t_i \bar{\phi} = 0, \qquad s_a \bar{\phi} \neq 0. \tag{2.1.24}$$

A general matrix generator will be called l_α. So the canonical form of a finite element $g \in G$ is

$$g(\theta_\alpha) = \exp(i\theta_\alpha l_\alpha). \tag{2.1.25}$$

Returning to the linear σ model, where the $O(N)$ symmetry is broken spontaneously to $O(N-1)$, we can count the broken generators. The number of generators of $O(N)$ is $\frac{1}{2}N(N-1)$. Then the number of broken generators is $\frac{1}{2}N(N-1) - \frac{1}{2}(N-1)(N-2) = N-1$, which is the number of Goldstone bosons we have found. This is a general result, as we shall see below.

2.2 Goldstone Theorem (First Proof)

In this section and the following one we describe two proofs of the Goldstone theorem [1]. The first proof uses the *effective potential*, which includes quantum corrections to the ordinary potential [2, 3]. To define this potential we consider quantum field theory in the presence of external classical fields. We begin with a brief introduction to this method.

External Field Method

The conventional field theory begins with the Lagrangian as a sum of free and interaction parts. Taking a multiplet of scalar fields $\phi_n(x)$, it is

$$\mathscr{L}(\phi_n) = \mathscr{L}_0(\phi_n) + \mathscr{L}_{\text{int}}(\phi_n), \tag{2.2.1}$$

which may contain other fields as well. It produces S-matrix elements for the processes $\alpha \longrightarrow \beta$, given by

$$S_{\beta\alpha} = \langle\beta|T\exp\left(i\int d^4x\mathscr{L}_{\text{int}}(\phi_n)\right)|\alpha\rangle. \tag{2.2.2}$$

Here the fields in the interaction Lagrangian are in the interaction representation, carrying the time dependence given by the free Hamiltonian.

We now study this theory in the presence of arbitrary classical (external) fields $h_n(x)$ by adding linear couplings of $\phi_n(x)$ with $h_n(x)$ to the original Lagrangian,

$$\mathscr{L}(\phi_n) \longrightarrow \mathscr{L}(\phi_n) + h_n(x)\phi_n(x), \tag{2.2.3}$$

where a sum over n is implied as usual. Consider the functional

$$Z[h] = \langle 0|T\exp\left(i\int d^4xh_n(x)\phi_n(x))\right)|0\rangle. \tag{2.2.4}$$

Suppose the external fields vanish smoothly as $t \longrightarrow \pm\infty$. Then comparing with the perturbation theory formula (2.2.2) we see that $Z[h]$ is the vacuum-to-vacuum amplitude in presence of external fields,

$$Z[h] = \langle 0, \text{out}|0, \text{in}\rangle_h. \tag{2.2.5}$$

Expanding the exponential in (2.2.4),

$$Z[h] = \sum_{r=0}^{\infty} \frac{1}{r!}\int d^4x_1\cdots d^4x_r h_{n_1}(x_1)\cdots h_{n_r}(x_r)\langle 0|T\phi_{n_1}(x_1)\cdots\phi_{n_r}(x_r)|0\rangle, \tag{2.2.6}$$

we see that the functional collects all the (complete) Green's functions. Equivalently we can generate all the Green's functions by differentiating the functional with respect to $h_n(x)$; hence the name-generating functional for $Z[h]$.

Unlike in (2.2.2) the fields in (2.2.4) and (2.2.6) are in the Heisenberg representation, the entire original Hamiltonian giving their time dependence. The Green's functions in (2.2.6) can be expanded perturbatively as usual in the interaction representation

$$\langle 0|T\phi_{n_1}(x_1)\cdots\phi_{n_r}(x_r)|0\rangle = \langle 0|T\phi_{n_1}^{in}(x_1)\cdots\phi_{n_r}^{in}(x_r)\exp\left(i\int d^4x\mathscr{L}_{\text{int}}(x))\right)|0\rangle. \tag{2.2.7}$$

The graphs contributing to this Green's function have r vertices (in addition to those from $\mathscr{L}_{int}$), to each of which is attached a single internal line. Removing the propagators from these lines, we get essentially the S-matrix elements (2.2.2). These graphs include both connected and disconnected ones. However, if we write the functional $Z[h]$ as

$$Z[h] = e^{iW[h]} \qquad (2.2.8)$$

and expand $W[h]$ in powers of h as before, then the corresponding coefficients represent only connected graphs,

$$W[h] = \sum_{r=0}^{\infty} \frac{1}{r!} \int d^4x_1 \cdots d^4x_r h_{n_1}(x_1) \cdots h_{n_r}(x_r) \langle 0|T\phi_{n_1}(x_1) \cdots \phi_{n_r}(x_r)|0\rangle_{\mathrm{conn}}.$$

$$(2.2.9)$$

Effective Potential

We return to the definition of the effective potential. First define the classical field $\phi_n^c(x)$ as the vacuum expectation value of the quantum field $\phi_n(x)$ in presence of the external fields,

$$\phi_n^c(y) = \frac{\langle 0, \mathrm{out}|\phi_n(y)|0, \mathrm{in}\rangle_h}{\langle 0, \mathrm{out}|0, \mathrm{in}\rangle_h}$$

$$= \frac{-i}{Z[h]} \frac{\delta Z[h]}{\delta h_n(y)} = \frac{\delta W[h]}{\delta h_n(y)}. \qquad (2.2.10)$$

The quantum effective action $\Gamma[\phi^c]$ is defined by a functional Legendre transform,

$$\Gamma[\phi^c] = W[h] - \int d^4x \phi_n^c(x) h_n(x). \qquad (2.2.11)$$

From this definition we get

$$\frac{\delta \Gamma}{\delta \phi_m^c(y)} = \int d^4x \frac{\delta W[h]}{\delta h_n(x)} \frac{\delta h_n(x)}{\delta \phi_m^c(y)} - \int d^4x \, \phi_n^c(x) \frac{\delta h_n(x)}{\delta \phi_m^c(y)} - h_m(y) = -h_m(y)$$

$$(2.2.12)$$

on using (2.2.10). We shall presently come back to this equation.

Just like $Z[h]$ and $W[h]$, we can also expand $\Gamma[\phi^c]$ in powers of ϕ_n^c,

$$\Gamma[\phi^c] = \sum_r \frac{1}{r!} \int d^4x_1 \cdots d^4x_r \Gamma_{n_1 \cdots n_r}^{(r)}(x_1, \cdots x_r) \phi_{n_1}^c(x_1) \cdots \phi_{n_r}^c(x_r). \qquad (2.2.13)$$

Compared to $W[h]$ the effective action $\Gamma[\phi^c]$ goes one step further: $\Gamma^{(r)}$ gives the sum of all one-particle-irreducible (not just connected) graphs with r external lines, from which the propagators are removed. In particular, $\Gamma^{(2)}(x_1, x_2)$ is the inverse propagator.

An alternative way to expand $\Gamma[\phi^c]$ is in powers of external momenta. In coordinate space it will assume the form

$$\Gamma[\phi^c] = \int d^4x\{-V(\phi^c) + Z_{n_1 n_2}(\phi^c)\partial_\mu \phi_{n_1}^c \partial_\mu \phi_{n_2}^c + \cdots\}, \qquad (2.2.14)$$

where $V(\phi^c)$ is an ordinary function, called *effective potential*. In tree approximation V is just the ordinary potential we called U in Section 2.1. To relate $V(\phi^c)$ to $\Gamma^{(r)}$, we insert the Fourier transform $\widetilde{\Gamma}^{(r)}(p_1, \cdots, p_r)$ of $\Gamma^{(r)}(x_1, \cdots, x_r)$ in (2.2.13) and specialise to constant (position independent) classical fields. Then the x-integrals give rise to δ-functions in momentum, which remove the momentum integrals, giving

$$\Gamma[\phi^c] = \sum \frac{1}{r!} \widetilde{\Gamma}^{(r)}_{n_1 \cdots n_r}(p_1 = 0, \cdots, p_r = 0) \phi^c_{n_1}(0) \cdots \phi^c_{n_r}(0). \qquad (2.2.15)$$

Comparing the two expansions (2.2.14) and (2.2.15) for $\Gamma(\phi^c)$, we see that the n-th derivative of V is the sum of all one-particle-irreducible graphs with n vanishing external momenta, up to a multiplicative four-dimensional volume. In particular, the second derivative of $V(\phi^c)$ is the mass matrix that we need below.

Goldstone Modes

We can now study spontaneous symmetry breaking. Let G be the internal symmetry group of the Lagrangian. It is broken spontaneously if the quantum field $\phi_n(x)$ develops a non-zero expectation value in the absence of external fields. From (2.2.10) and (2.2.12) this occurs if

$$\frac{\delta \Gamma}{\phi^c_n} = 0 \qquad (2.2.16)$$

for some non-zero values of $\phi^c_n(x)$. Further restricting to expectation values of ϕ that are translationally invariant (x-independent), it becomes

$$\frac{\partial V}{\partial \phi^c_n} = 0\,, \qquad (2.2.17)$$

again for some non-zero values of ϕ^c_n. To simplify notation we drop the index c on ϕ^c_n from now on, so that ϕ_n will denote constant classical fields in vacuum in the rest of this section. We then rewrite the condition (2.2.17) for spontaneous symmetry breaking as

$$\frac{\partial V}{\partial \phi_n}\bigg|_{\phi=\bar{\phi}\neq 0} = 0\,. \qquad (2.2.18)$$

It is now simple to prove the Goldstone theorem [2]. Under the action of G, the fields undergo linear infinitesimal transformation

$$\delta \phi_n = i\epsilon_\alpha [l_\alpha]_{nm} \phi_m \qquad (2.2.19)$$

in the notation defined at the end of Section 2.1. The group G is also the symmetry group of the effective potential, so that

$$\frac{\partial V}{\partial \phi_n} \epsilon_\alpha [l_\alpha]_{nm} \phi_m = 0. \qquad (2.2.20)$$

As ϵ_α are arbitrary, it holds separately for each generator

$$\frac{\partial V}{\partial \phi_n}[l_\alpha]_{np}\phi_p = 0. \tag{2.2.21}$$

Differentiating the symmetry requirement (2.2.21) again with respect to ϕ_m, we get

$$\frac{\partial V}{\partial \phi_n}[l_\alpha]_{nm} + \frac{\partial^2 V}{\partial \phi_n \partial \phi_m}[l_\alpha]_{np}\phi_p = 0.$$

Evaluating it at the minimum of $V(\phi)$ given by (2.2.18), it becomes

$$\left.\frac{\partial^2 V}{\partial \phi_n \partial \phi_m}\right|_{\phi=\bar\phi}(l_\alpha\bar\phi)_n = 0, \qquad \bar\phi \equiv \langle 0|\phi|0\rangle. \tag{2.2.22}$$

If l_α is a generator belonging to H, then (2.2.22) is trivially satisfied. But if l_α is outside H, we have $(s_a\bar\phi)$ as an eigenvector of the mass matrix with zero eigenvalue. We thus arrive at the existence of *Goldstone bosons*: for every independent broken generator of a symmetry group of the Lagrangian, there is one massless boson in the theory.

2.3 Goldstone Theorem (Second Proof)

The second proof uses operator methods [1, 2]. If the Lagrangian is invariant under a Lie group G, it leads to the existence of symmetry currents $J_\alpha^\mu(x)$ that are conserved,

$$\partial_\mu J_\alpha^\mu(x) = 0, \tag{2.3.1}$$

giving charges

$$Q_\alpha = \int d^3x\, J_\alpha^0(x), \tag{2.3.2}$$

which are independent of time,

$$[Q_\alpha, H] = 0. \tag{2.3.3}$$

A simple proof of the Goldstone theorem suggests itself. Suppose a subset of the generators Q_a say, are broken; that is, do not annihilate the vacuum,

$$Q_a|0\rangle \neq 0. \tag{2.3.4}$$

Then (2.3.3) shows that $Q_a|0\rangle$ describes a state with the same energy as that of the vacuum. In a relativistic theory this implies that the spectrum of the theory contains as many massless particles as the number of broken generators. This proof is, however, mathematically sloppy, as the norm of the state $Q_a|0\rangle$ is not defined.[2] Nevertheless, we shall use (2.3.4) to arrive at physical situations in which spontaneous symmetry breaking may occur.

The rigorous proof is based on commutators of symmetry generators with scalar field components $\phi_n(x)$, which may be elementary (appearing in the Lagrangian) or composite thereof [2]. To find these commutators, we recall that the symmetries act on the space of state vectors. An element $g \in G$ is represented by an unitary transformation

$$|\psi\rangle \longrightarrow |\psi'\rangle = \mathscr{U}(g)|\psi\rangle.$$

Let the matrix elements of the field multiplet $\phi_n(x)$ transform according to a representation $D(g)$ of G

$$\langle\psi|\phi_n(x)|\chi\rangle \longrightarrow \langle\psi'|\phi_n(x)|\chi'\rangle = D(g)_{nm}\langle\psi|\phi_m(x)|\chi\rangle$$

giving[3]

$$\mathscr{U}^\dagger(g)\phi_n(x)\mathscr{U}(g) = D(g)_{nm}\phi_m(x). \tag{2.3.5}$$

If $\omega_\alpha, \alpha = 1, 2\cdots$ are the infinitesimal coordinates of g in the neighbourhood of the unit element, the operator $\mathscr{U}(g)$ and the representative $D(g)$ admit expansions

$$\mathscr{U}(1 + \omega) = 1 + i\omega_\alpha Q_\alpha + O(\omega^2) \tag{2.3.6}$$

$$D(1 + \omega) = 1 + i\omega_\alpha l_\alpha + O(\omega^2) \tag{2.3.7}$$

where the charges (2.3.2) appear as quantum generators and the matrices in (2.1.25) as group generators. Inserting these in (2.3.5) we get the desired commutators

$$[Q_\alpha, \phi_n(x)] = -[l_\alpha]_{nm}\phi_m(x). \tag{2.3.8}$$

Unlike the charges, which are formal objects, the fields $\phi_n(x)$ are local operators. So the commutators (2.3.8) are well defined.[4] Also it should be noted here that operator relations like those written above remain unaffected by spontaneous symmetry breaking; the latter only reduces the symmetry of the vacuum state and those built on it.

---

[2] To see this, note that the charges are formal objects, being integrals over the whole space. For the norm we have the double integral

$$\langle 0|Q_a Q_b|0\rangle = \int d^3x\, d^3y \langle 0|J_a^0(x) J_b^0(y)|0\rangle$$

which is clearly unbounded, as the vacuum matrix element above is a function of the single variable $|\boldsymbol{x} - \boldsymbol{y}|$.

[3] Let us take G to be $O(N)$ and compare the transformation rules (2.1.3) with (2.3.5). In writing (2.1.3) we infer the symmetry group from the Lagrangian, where the fields are in the defining or vector representation. But in writing (2.3.5) we are given the symmetry group and the fields may transform according to any of its representations $D(g)$, including of course the defining one.

[4] As the commutator $[J_\alpha^0(y), \phi_n(x)]$ vanishes for space-like separations, its spatial integral over $\boldsymbol{y}$ receives contribution only from a finite region around $\boldsymbol{x}$ given by $|\boldsymbol{y} - \boldsymbol{x}| \leq |y^0 - x^0|$.

We shall use the commutation relation (2.3.8) later. Let us first derive the Källén–Lehmann spectral representation [6, 7] for the vacuum expectation value of the commutator, $\langle 0|[J_\alpha^\mu(y), \phi_n(x)]|0\rangle$. Taking the first term of this commutator, we insert a complete set of states $|N\rangle$ between the operators

$$\langle 0|J_\alpha^\mu(y)\phi_n(x)|0\rangle = \sum_N \langle 0|J_\alpha^\mu(y)|N\rangle\langle N|\phi_n(x)|0\rangle \tag{2.3.9}$$

where the sum over N includes sums over discrete indices as well as integrals over momenta. Choosing these states to be eigenstates of four-momentum operator P_μ, we can extract the space–time dependence of the operators by using translation invariance, which for an operator $\mathcal{O}(x)$ is

$$\mathcal{O}(x) = e^{iP\cdot x}\mathcal{O}(0)e^{-iP\cdot x}. \tag{2.3.10}$$

Then (2.3.9) gives

$$\langle 0|J_\alpha^\mu(y)\phi_n(x)|0\rangle = \sum_N e^{-ip_N\cdot(y-x)}\langle 0|J_\alpha^\mu(0)|N\rangle\langle N|\phi_n(0)|0\rangle \tag{2.3.11}$$

where p_N is the total four-momentum of the state N. Inserting the resolution of unity, $1 = \int d^4p\,\delta^4(p - p_N)$ within the sum, it becomes

$$\langle 0|J_\alpha^\mu(y)\phi_n(x)|0\rangle = \int \frac{d^4p}{(2\pi)^3} e^{-ip\cdot(y-x)} i\rho_{\alpha,n}^\mu(p) \tag{2.3.12}$$

where the sum over states defines the spectral density ρ,

$$(2\pi)^3 \sum_N \langle 0|J_\alpha^\mu(0)|N\rangle\langle N|\phi_n(0)|0\rangle\delta^4(p - p_N)$$
$$\equiv i\rho_{\alpha,n}^\mu(p)$$
$$= ip^\mu\rho_{\alpha,n}(p^2)\theta(p_0)\,, \tag{2.3.13}$$

the last line following from Lorentz invariance and the fact that physical states have positive energy. Again introduce $1 = \int ds\,\delta(s - p^2)$ in the integrand of (2.3.12) to get

$$\langle 0|J_\alpha^\mu(y)\phi_n(x)|0\rangle = -\frac{\partial}{\partial y_\mu}\int ds\,\rho_{\alpha,n}(s)\Delta_+(y - x; s) \tag{2.3.14}$$

where Δ_+ is the function (1.2.7) with squared mass s.

In just the same way, we write the second term of the commutator as

$$\langle 0|\phi_n(x)J_\alpha^\mu(y)|0\rangle = +\frac{\partial}{\partial y_\mu}\int ds\,\widetilde{\rho}_{\alpha,n}(s)\Delta_+(x - y; s) \tag{2.3.15}$$

where again the sum over states defines a second spectral density $\widetilde{\rho}$,

$$(2\pi)^3 \sum_N \langle 0|\phi_n(0)|N\rangle\langle N|J_\alpha^\mu(0)|0\rangle\delta^4(p - p_N)$$
$$\equiv i\widetilde{\rho}_{\alpha,n}^\mu(p)$$
$$= ip^\mu\widetilde{\rho}_{\alpha,n}(p^2)\theta(p_0). \tag{2.3.16}$$

With (2.3.14) and (2.3.15) we get the commutator as

$$\langle 0|[J_\alpha^\mu(y), \phi_n(x)]|0\rangle$$
$$= -\frac{\partial}{\partial y_\mu} \int ds \{\rho_{\alpha,n}(s)\Delta_+(y - x; s) + \widetilde{\rho}_{\alpha,n}(s)\Delta_+(x - y; s)\}. \quad (2.3.17)$$

The two spectral functions are not independent. As shown in Problem 1.1(c), if we restrict $(x - y)$ to space-like separations, $\Delta_+(x - y; s)$ becomes an even function of $(x - y)$, so that $\Delta_+(y - x; s)$ and $\Delta_+(x - y; s)$ are equal. Also, the commutator on the left side must vanish in this region, giving

$$\rho_{\alpha,n}(s) + \widetilde{\rho}_{\alpha,n}(s) = 0. \quad (2.3.18)$$

Thus, for general x and y, (2.3.17) gives the desired spectral representation

$$\langle 0|[J_\alpha^\mu(y), \phi_n(x)]|0\rangle = -\frac{\partial}{\partial y_\mu} \int ds\, \rho_{\alpha,n}(s)\Delta(y - x; s), \quad (2.3.19)$$

where the function Δ is defined by (1.2.5).

We now use the fact that $J_\alpha^\mu(y)$ is conserved. Applying the derivative $\partial/\partial y^\mu$ on both sides and noting the equation, $(\Box_y + s)\Delta(y - x; s) = 0$, the commutator (2.3.19) gives for all x and y,

$$0 = \int ds\, s\, \rho_{\alpha,n}(s)\Delta(y - x; s) \quad (2.3.20)$$

leading to $s\, \rho_{\alpha,n}(s) = 0$, whose general solution is

$$\rho_{\alpha,n}(s) = c_{\alpha,n}\delta(s) \quad (2.3.21)$$

where $c_{\alpha,n}$ are constants. Thus the spectral density vanishes everywhere, except possibly at $s = 0$. Next, we are going to use the commutator (2.3.8) to show that $c_{\alpha,n}$ are indeed non-zero in theories with spontaneous symmetry breaking. Set $\mu = 0$ and $x^0 = y^0 = t$ in (2.3.19) and use (2.3.21) to get

$$\langle 0|[J_\alpha^0(\boldsymbol{y}, t), \phi_n(\boldsymbol{x}, t)]|0\rangle = -c_{\alpha,n}\frac{\partial}{\partial y^0}\Delta(y - x)|_{y^0=x^0} = ic_{\alpha,n}\delta^3(\boldsymbol{x} - \boldsymbol{y}) \quad (2.3.22)$$

by the result of Problem 1.1(b). Integrating this equation over $\boldsymbol{y}$, we get the charge (2.3.2) on the left side, when (2.3.8) gives the value of the resulting commutator. We then get

$$-[l_\alpha]_{nm}\langle 0|\phi_m|0\rangle = ic_{\alpha,n}$$

giving finally the spectral density as

$$\rho_{\alpha,n}(s) = i\delta(s)(l_\alpha\overline{\phi})_n, \quad (2.3.23)$$

where $\overline{\phi}$ denotes the column vector with the vacuum expectation values $\overline{\phi}_n = \langle 0|\phi_n|0\rangle$ as components.

So far l_α is any of the generators of the symmetry group G. If l_α is one of the unbroken generators t_i, then $t_i\overline{\phi} = 0$. In this case the entire spectral density vanishes and nothing interesting happens. If, however, l_α is one of the broken generators, $s_a\overline{\phi} \neq 0$, the spectral density is non-zero:

$$\rho_{a,n}(s) = i\delta(s)(s_a\overline{\phi})_n \neq 0 \,. \tag{2.3.24}$$

The delta function present in it signals the existence of massless particles in the theory. The singularity is exactly at $s = p_N^2 = 0$, showing that it arises from single particle (and not multiparticle) state of zero mass. To get their quantum numbers, we look at the matrix elements comprising the spectral function (2.3.13): $\langle N|\phi_n(0)|0\rangle$ vanishes for any state $|N\rangle$ of non-zero angular momentum and $\langle 0|J_a^0(0)|N\rangle$ vanishes for any state $|N\rangle$ differing in parity and other internal quantum numbers from that of $J_a^0(x)$. We thus arrive at the *Goldstone theorem*: For each spontaneously broken symmetry $s_a\overline{\phi} \neq 0$, we have a massless, spinless particle (the Goldstone boson) having the same parity and internal quantum numbers as $J_a^0(x)$.

* * *

As only a single Goldstone boson (B_a) contributes to the spectral function (2.3.13), it can be worked out immediately. From Lorentz invariance we get

$$\langle 0|J_a^\mu(x)|B_b(k)\rangle = iF_{ab}k^\mu e^{-ik\cdot x}\,. \tag{2.3.25}$$

It also satisfies current conservation, as $k^2 = 0$. The other matrix element needed is

$$\langle B_b(k)|\phi_n(y)|0\rangle = Z_{bn}e^{ik\cdot y} \tag{2.3.26}$$

where Z_{bn} are constants. Inserting these matrix elements in (2.3.13) we get

$$
\begin{aligned}
ip^\mu \rho_{a,n}(p^2)\theta(p_0) &= \int \frac{d^3k}{2\omega_k}\langle 0|J_a^\mu(0)|B_b(k)\rangle\langle B_b(k)|\phi_n(0)|0\rangle\delta^4(k-p) \\
&= ip^\mu F_{ab}Z_{bn}\frac{1}{2|\boldsymbol{p}|}\delta(p_0 - |\boldsymbol{p}|) \\
&= ip^\mu F_{ab}Z_{bn}\delta(p^2)\theta(p_0)
\end{aligned}
\tag{2.3.27}
$$

so that

$$\rho_{a,n}(p^2) = F_{ab}Z_{bn}\delta(p^2)\,. \tag{2.3.28}$$

Inserting this result in (2.3.24) we get

$$F_{ab}Z_{bn} = i(s_a\overline{\phi})_n\,, \tag{2.3.29}$$

which we shall evaluate in the $O(4)$ model in Section 2.5.

2.4 Other Matrix Elements of Current

The proof of the Goldstone theorem utilises only the matrix element (2.3.25) of the broken symmetry current between the vacuum and one Goldstone boson

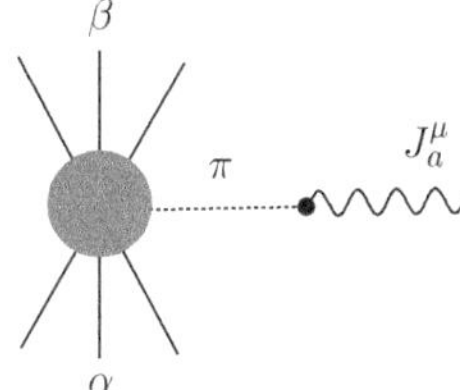

Figure 2.2 Feynman graph showing Goldstone boson pole in the matrix element of $J^\mu(x)$ between general states α and β.

state. By taking more general matrix elements and applying the condition of current conservation on them, we can learn about the interaction of Goldstone bosons. Let us begin with the Fourier transform of the matrix element of $J_a^\mu(x)$ between arbitrary states α and β,

$$I_a^\mu(k) \equiv \int d^4x\, e^{ik\cdot x}\langle\beta|J_a^\mu(x)|\alpha\rangle = (2\pi)^4\delta^4(k + p_\beta - p_\alpha)\langle\beta|J_a^\mu(0)|\alpha\rangle, \quad (2.4.1)$$

where $p_{\alpha,\beta}$ denote total momenta of particles in states α and β. Because of the non-zero matrix element (2.3.25), the matrix element (2.4.1) is expected to have a pole as $k^2 \to 0$ (see Figure 2.2).

To see how the pole arises, consider the integral as a function of k_0. As the integrand is regular in k_0, the source of singularity can only be the infinite range of integration of x_0. There are two such intervals, extending to $\pm\infty$ from finite values. The two intervals can be separated formally by inserting $1 = \theta(x_0) + \theta(-x_0)$ in the integrand.[5] We then insert a complete set of intermediate states in (2.4.1), from which we isolate the contribution of the single Goldstone boson. If we choose to have it in the *out* state, we get

$$I_a^\mu(k) \to \int \frac{d^3l}{(2\pi)^3 2\omega_l} \int d^4x\, e^{ik\cdot x}\theta(x_0)\langle 0|J_a^\mu(x)|B_b(l)\rangle\langle B_b(l),\beta|\alpha\rangle, \quad (2.4.2)$$

where a summation over the index b of the Goldstone boson is implied. Extracting the x-dependence of the matrix element in (2.4.2) the x-integral gives

$$\int d^4x\, e^{i(k-l)\cdot x}\theta(x_0) = \frac{i(2\pi)^3\delta(\mathbf{k}-\mathbf{l})}{k_0 - l_0 + i\epsilon} \to \frac{i(2\pi)^3 2\omega_l\delta(\mathbf{k}-\mathbf{l})}{k^2 + i\epsilon}. \quad (2.4.3)$$

If the Goldstone bosons form an irreducible multiplet of the unbroken subgroup H, we can use (2.3.25) with $F_{ab} = \delta_{ab}F$. Then (2.4.2) yields

$$I_a^\mu(k) \to -\frac{Fk^\mu\langle B_a(k),\beta|\alpha\rangle}{k^2 + i\epsilon}. \quad (2.4.4)$$

It is now clear why $\theta(x_0)$ puts the Goldstone boson in the *out* state in the matrix element $\langle B_a,\beta|\alpha\rangle$: the pole in (2.4.3) with proper $i\epsilon$ prescription arises from the

---

[5] For two or more operators their time ordering would provide such theta functions.

integral which diverges in the limit $x_0 \to +\infty$. Similarly $\theta(-x_0)$ will bring in the matrix element $\langle \beta | B_a, \alpha \rangle$ with the Goldstone boson in the *in* state.

The S-matrix element in (2.4.4) is related to the invariant amplitude by

$$\langle B_a(k), \beta | \alpha \rangle = i(2\pi)^4 \delta^4(k + p_\beta - p_\alpha) M_{\alpha \to \beta + B}. \tag{2.4.5}$$

From (2.4.1), (2.4.4) and (2.4.5), we get the pole contribution as

$$\langle \beta | J_a^\mu(0) | \alpha \rangle|_{\text{Goldstone pole}} = -\frac{iF k^\mu M_{\alpha \to \beta + B}}{k^2 + i\epsilon}. \tag{2.4.6}$$

This result is an elementary example of the general reduction formula, relating correlation functions and S-matrix elements, first derived by Lehmann, Symanzik and Zimmermann [8]. The method used here [9–11] is somewhat different and simpler from that of the original derivation. Moreover it applies to any sort of operators, not those restricted to fields appearing in the Lagrangian.

We shall now derive some results on Goldstone boson interactions.

Goldstone Boson Emission Amplitude

In general, the Goldstone pole contribution (2.4.6) is not the only singularity of $\langle \beta | J^\mu | \alpha \rangle$ as $k_\mu \to 0$. The current J_a^μ can also attach directly to a line in the graphs for the process, which may be external or internal [12, 13]. When it attaches externally to a line carrying momentum p as in Figure 2.3, an extra propagator is generated that is close to the mass-shell,

$$\frac{1}{(p \pm k)^2 - m^2} = \frac{1}{\pm 2p \cdot k + k^2} \to \pm \frac{1}{2p \cdot k}, \tag{2.4.7}$$

where we omit the numerator of the propagator, which depends on the spin of the particle. The corresponding contribution to (2.4.1) from each of the external lines has the form

$$I_a^\mu(k) \to \pm \frac{C_a^\mu}{2p \cdot k} \langle \beta | \alpha \rangle \tag{2.4.8}$$

where C_a^μ denotes the coupling of J_a^μ with the external line times the numerator of the propagator left out in (2.4.7). On the other hand, if the current attaches

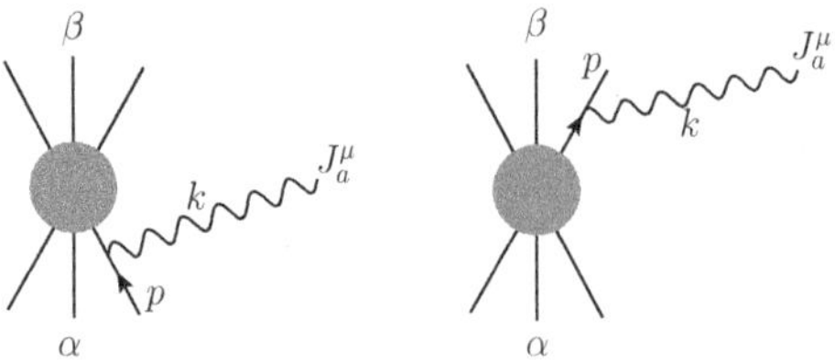

Figure 2.3 Feynman graph showing heavy particle poles in the matrix element of $J^\mu(x)$ between states α and β, when the current attaches directly to an external line.

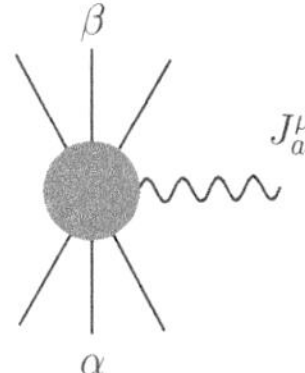

Figure 2.4 Feynman graph for $J^\mu(x)$ between states α and β showing the current attaching directly to an internal line.

to an internal line as in Figure 2.4, the extra propagator does not produce any singular behaviour, as internal lines in a graph are always off their mass shells.

Adding the contributions (2.4.4) and (2.4.8) from the two sources, we get the complete singular behaviour of $I(k)$ as

$$\int d^4x\, e^{ik\cdot x} \langle\beta|J_a^\mu(x)|\alpha\rangle \xrightarrow{k\to 0} -\frac{Fk^\mu \langle B_a(k),\beta|\alpha\rangle}{k^2 + i\epsilon} + \sum_{\text{ext. lines}} (\pm)\frac{C_a^\mu}{2p\cdot k}\langle\beta|\alpha\rangle. \tag{2.4.9}$$

Multiply (2.4.9) by k_μ and integrate the left-hand side partially, when it vanishes by current conservation. Then we get

$$\langle B_a(k),\beta|\alpha\rangle \longrightarrow \frac{1}{2F}\sum_{\text{ext. lines}} \pm\frac{C_a\cdot k}{p\cdot k}\langle\beta|\alpha\rangle \qquad \text{as } k_\mu \to 0. \tag{2.4.10}$$

We thus have the result that the matrix element for one soft Goldstone boson emission in the process $\alpha \to \beta$ can be expressed in terms of the process $\alpha \to \beta$ itself. If the current cannot attach to an external line ($C_a^\mu = 0$), (2.4.10) tells us that the matrix element $\langle B_a(k),\beta|\alpha\rangle$ for emitting a Goldstone boson in the transition $\alpha \to \beta$ vanishes as $k \to 0$, getting an 'Adler zero' in the amplitude [13].

Goldstone Boson Interactions

From the above results we can learn about the interaction of Goldstone bosons at low energy. It is interesting here to take the case of massless QCD (to be covered in the next chapter) and go through the above steps once again. In this theory the axial charges Q_a, ($a = 1, 2, 3$) are assumed to be spontaneously broken, leading to a triplet of massless pions as Goldstone bosons. Like (2.3.25) the axial-vector current has the matrix element

$$\langle 0|A_a^\mu(x)|\pi_b(k)\rangle = ik^\mu \delta_{ab} F, \tag{2.4.11}$$

where a, b are pion isospin indices. Now consider its matrix element as in (2.4.6); if the states α and β consists of pions only, the current cannot attach to an external line, because of parity or G-parity. Thus the entire singular contribution

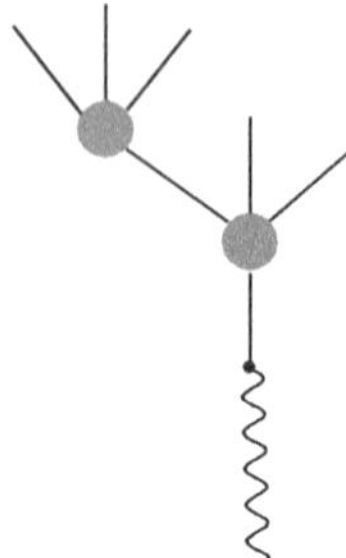

Figure 2.5 One-particle reducible poles in the axial-vector matrix element.

in (2.4.6) is from the pion pole only. Taking all the pions in the out state, for example, it gives

$$\langle \pi_{a_1}(k_1), \pi_{a_2}(k_2) \cdots | A^\mu_{a_n}(0) | 0 \rangle = \frac{iF k^\mu_n M_{a_1,a_2 \cdots a_n}(k_1, k_2 \cdots k_n)}{k^2 + i\epsilon} + O(k^0).$$

(2.4.12)

Applying current conservation we get for the multi-pion amplitude

$$M_{a_1,a_2 \cdots a_n}(k_1, k_2 \cdots k_n) \to 0,$$

(2.4.13)

as k_n and hence by Bose symmetry any of the momenta, tends to zero. Thus to leading order at low momenta, the pion amplitudes are quadratic in momenta, by Lorentz invariance [14]. (It is easy to check that graphs with two or more poles follow the same behaviour. Consider, for example, the graph of Figure 2.5 having two poles. Here each of the two four-point vertices is suppressed by two powers of momenta, cancelling the associated pole. So this amplitude is again quadratic at low momenta.) We conclude that the strength of interaction among pions is weak at low energy – pions of zero energy do not interact at all.

2.5 Linear σ-Model Revisited

Here we work out explicitly the infinitesimal symmetry transformation rules for fields in the $O(4)$ linear σ-model. In the next section we shall see how these rules give rise to those for fields in the non-linear version. The group $O(4)$ is especially interesting, as it is isomorphic to $SU(2) \times SU(2)$, the symmetry group of the QCD Lagrangian, to be studied in the next chapter. From (2.1.1) and (2.1.2) we get the Lagrangian

$$\mathscr{L}(\phi) = \frac{1}{2} \partial_\mu \phi_n \partial^\mu \phi_n - \frac{\mu^2}{2} \phi_n \phi_n - \frac{\lambda}{4} (\phi_n \phi_n)^2 , \quad n = 1, 2, 3, 4, \quad (2.5.1)$$

for which the symmetry group G is $O(4)$, that is, it is invariant under

$$\phi_n \to \phi'_n = R_{nm} \phi_m, \qquad R \in O(4) . \qquad (2.5.2)$$

For infinitesimal transformation, $R = 1 + \omega$, $(\omega \ll 1)$, this transformation rule becomes

$$\delta\phi_n(x) = \omega_{nm}\phi_m(x) \tag{2.5.3}$$

and the orthogonality condition, $R^T R = 1$, reduces to the antisymmetry of ω, $\omega_{nm} + \omega_{mn} = 0$.

As we discussed in Section 2.1, for $\mu^2 < 0$, the $O(4)$ symmetry of the Lagrangian is broken partially by the vacuum state. Let us choose this state as $\overline{\phi} = (0, 0, 0, v)$, where $v = \sqrt{-\mu^2/\lambda}$. Then we shift the fourth component of the field to $\phi_4'(x) = \phi_4(x) - v$ and introduce the three-vector $\vec{\phi} = (\phi_1, \phi_2, \phi_3)$ to write the Lagrangian (2.5.1) as

$$\mathcal{L} = \frac{1}{2}\partial^\mu\vec{\phi}\cdot\partial_\mu\vec{\phi} + \frac{1}{2}\partial^\mu\phi_4'\partial_\mu\phi_4' - |\mu^2|\phi_4'^2$$
$$- \frac{\lambda}{4}\{4v\phi_4'(\vec{\phi}\cdot\vec{\phi} + \phi_4'^2) + (\vec{\phi}\cdot\vec{\phi} + \phi_4'^2)^2\}, \tag{2.5.4}$$

which shows that the quanta of ϕ_4' have mass $\sqrt{2}|\mu|$, while those of ϕ_1, ϕ_2 and ϕ_3 are massless.

The rotations in i-j planes $(i, j = 1, 2, 3)$ leave the vacuum state $\overline{\phi}$ invariant, giving $SO(3)$ as the unbroken subgroup H. The remaining rotations in i-4 planes do not leave the vacuum state invariant. We call these isospin and chiral rotations respectively and denote their respective parameters as θ_i and ϵ_i, which are related to ω_{mn} by $\omega_{ij} = \epsilon_{ijk}\theta_k$, $\omega_{a4} = -\omega_{4a} = \epsilon_a$. The transformation rule (2.5.3) then splits into (unbroken) isospin rotations

$$\delta\vec{\phi} = \vec{\phi}\times\vec{\theta}, \quad \delta\phi_4 = 0 \tag{2.5.5}$$

and (broken) chiral rotations

$$\delta\vec{\phi} = \vec{\epsilon}\phi_4, \quad \delta\phi_4 = -\vec{\epsilon}\cdot\vec{\phi} \tag{2.5.6}$$

We can also write the infinitesimal transformation rule for field components in terms of the matrix generators. In the notation described at the end of Section 2.1, it is

$$\delta\phi_n = i(\theta_i[T_i]_{nm} + \epsilon_a[S_a]_{nm})\phi_m \tag{2.5.7}$$

where we use capital letters to denote generators of this particular representation. From (2.5.5) we get the unbroken generators

$$T_1 = -i\begin{pmatrix} 0 & 0 & 0 & 0 \\ 0 & 0 & 1 & 0 \\ 0 & -1 & 0 & 0 \\ 0 & 0 & 0 & 0 \end{pmatrix}, T_2 = -i\begin{pmatrix} 0 & 0 & -1 & 0 \\ 0 & 0 & 0 & 0 \\ 1 & 0 & 0 & 0 \\ 0 & 0 & 0 & 0 \end{pmatrix}, T_3 = -i\begin{pmatrix} 0 & 1 & 0 & 0 \\ -1 & 0 & 0 & 0 \\ 0 & 0 & 0 & 0 \\ 0 & 0 & 0 & 0 \end{pmatrix}$$

while (2.5.6) gives the broken ones

$$S_1 = -i \begin{pmatrix} 0 & 0 & 0 & 1 \\ 0 & 0 & 0 & 0 \\ 0 & 0 & 0 & 0 \\ -1 & 0 & 0 & 0 \end{pmatrix}, S_2 = -i \begin{pmatrix} 0 & 0 & 0 & 0 \\ 0 & 0 & 0 & 1 \\ 0 & 0 & 0 & 0 \\ 0 & -1 & 0 & 0 \end{pmatrix}, S_3 = -i \begin{pmatrix} 0 & 0 & 0 & 0 \\ 0 & 0 & 0 & 0 \\ 0 & 0 & 0 & 1 \\ 0 & 0 & -1 & 0 \end{pmatrix}$$

We now work out in $O(4)$ model the relation (2.3.29) arising from the derivation of the Goldstone theorem. From (2.5.5) we see that the three Goldstone bosons transform according to the vector representation of H, forming a single irreducible multiplet, so that

$$F_{ab} = F\delta_{ab}, \qquad Z_{ab} = Z\delta_{ab}, \qquad Z_{a4} = 0. \tag{2.5.8}$$

Also

$$(S_a\overline{\phi})_b = -iv\delta_{ab}, \qquad (S_a\overline{\phi})_4 = 0. \tag{2.5.9}$$

Then (2.3.29) becomes

$$FZ = v. \tag{2.5.10}$$

As ϕ is an elementary field, it is conventional to set the field renormalisation constant $Z = 1$, when we have $F = v$. So F measures the strength of spontaneous symmetry breaking. As we shall see in Chapter 3, Goldstone fields scale with $1/F$. Thus the stronger this symmetry breaking, the weaker is the interaction involving Goldstone bosons.

2.6 Non-Linear σ-Model

In this section and the following one, we follow Weinberg to construct the $O(4)$ non-linear σ-model and generalise it to arbitrary symmetry groups [2].

The Lagrangian (2.5.1) or its shifted version (2.5.4) does not allow, in general, a perturbative calculation of scattering amplitudes. But as we show below, the Goldstone fields can be so redefined that they interact through derivative couplings only [15]. Then the Feynman graphs can be ordered as an expansion in powers of Goldstone boson momenta. In this process of redefinition, however, the broken symmetries act non-linearly on the new fields.

To transform the Lagrangian in this way, we define a field $\widetilde{\phi}(x)$ from which the Goldstone components are removed, $\widetilde{\phi}(x) = (0, 0, 0, \sigma(x))$. Then if we put

$$\phi_n(x) = R_{nm}\widetilde{\phi}(x) = R_{n4}\sigma(x),$$

where R is a symmetry of $\mathscr{L}$, it reduces $\mathscr{L}$ to that of $\sigma(x)$,

$$\mathscr{L}(\sigma) = \frac{1}{2}\partial_\mu\sigma\partial^\mu\sigma - \frac{\mu^2}{2}\sigma^2 - \frac{\lambda}{4}\sigma^4, \tag{2.6.1}$$

missing the Goldstone fields altogether. But if we put

$$\phi_n(x) = R_{nm}(x)\tilde{\phi}_m(x) = R_{n4}(x)\sigma(x)\,, \tag{2.6.2}$$

where $R(x)$, with its parameters space–time dependent, is no longer a symmetry of $\mathscr{L}$, the Goldstone fields reappear as these parameters. The advantage of reinstating these fields through $R(x)$ is that the non-derivative terms in the original Lagrangian remain free from $R(x)$; only those with derivatives (here the kinetic term) involve $R(x)$. In other words, every term in the transformed Lagrangian containing the new Goldstone fields must have at least one space–time derivative on them.

Accordingly, substituting (2.6.2) in (2.5.1) we get

$$\mathscr{L} = \frac{1}{2}(\sigma\partial_\mu R_{n4} + R_{n4}\partial_\mu\sigma)(\sigma\partial^\mu R_{n4} + R_{n4}\partial^\mu\sigma) - \frac{\mu^2}{2}\sigma^2 - \frac{\lambda}{4}\sigma^4. \tag{2.6.3}$$

From $R^T(x)R(x) = 1$, we get $R_{n4}R_{n4} = 1$. Then the coefficient of $\frac{1}{2}\partial_\mu\sigma\partial^\mu\sigma$ in (2.6.3) is unity, while that of the crossed terms is zero, as

$$R_{n4}\partial_\mu R_{n4} = \frac{1}{2}\partial_\mu(R_{n4}R_{n4}) = 0.$$

We thus arrive at the Lagrangian

$$\mathscr{L} = \mathscr{L}(\sigma) + \frac{\sigma^2}{2}(\partial_\mu\vec{R}\cdot\partial^\mu\vec{R} + \partial_\mu R_{44}\partial^\mu R_{44})\,, \tag{2.6.4}$$

where we introduce the vector triplet $\vec{R} = (R_{14}, R_{24}, R_{34})$ and so $R_{44} = \sqrt{1 - \vec{R}\cdot\vec{R}}$. The defining equation (2.6.2) relates the new set of fields $(\vec{R}, \sigma)$ to the old set $(\vec{\phi}, \phi_4)$ by

$$\sigma = \sqrt{\vec{\phi}\cdot\vec{\phi} + \phi_4^2}, \qquad \vec{R} = \frac{\vec{\phi}}{\sigma}. \tag{2.6.5}$$

Instead of taking $\vec{R}$ themselves as field variables, one can choose other parametrizations to get simpler forms for the transformed Lagrangian [15]. But the form (2.6.4) suffices to illustrate the main properties of a non-linear Lagrangian.

Though not apparent, the Lagrangian (2.6.4) *is* invariant under $SO(4)$. To see this we use the relations (2.6.5) to find the transformation rules of the new fields from those of the old ones given in (2.5.5) and (2.5.6). Thus under isospin rotation we get

$$\delta\sigma = 0\,, \quad \delta\vec{R} = \frac{\delta\vec{\phi}}{\sigma} = \vec{R}\times\vec{\theta} \tag{2.6.6}$$

showing that σ is an isoscalar and $\vec{R}$ an isovector, so that $\mathscr{L}$ is manifestly isospin invariant. Next, under chiral rotations,

$$\delta\sigma = \frac{\vec{\phi}\cdot\delta\vec{\phi} + \phi_4\delta\phi_4}{\sqrt{\phi_n\phi_n}} = 0\,, \quad \delta\vec{R} = \frac{\delta\vec{\phi}}{\sigma} = \vec{\epsilon}R_{44} \tag{2.6.7}$$

and hence

$$\delta R_{44} = \delta(\sqrt{1 - \vec{R}\cdot\vec{R}}) = -\vec{\epsilon}\cdot\vec{R}. \tag{2.6.8}$$

Thus the two pieces in the second term of (2.6.4) taken together are also invariant under chiral transformations. While the isospin transformation (2.6.6) is realised linearly, i.e. as a representation, the chiral transformation (2.6.7) is realised in a *non-linear* way. Thus we have here a non-linear realisation of $O(4)$ giving the so-called non-linear σ-model [16].

The Goldstone fields $\vec{R}(x)$ by themselves provide an $O(4)$ invariant Lagrangian. The massive field $\sigma(x)$ is an $O(4)$ scalar and may simply be treated as a constant, c. Expanding R_{44} and normalising $\vec{R}$ so as to get the standard kinetic term, the Lagrangian (2.6.4) becomes

$$\mathcal{L} = \frac{1}{2}\partial_\mu\vec{R}\cdot\partial^\mu\vec{R} + \frac{1}{2c^2}\vec{R}\cdot\partial_\mu\vec{R}\,\vec{R}\cdot\partial^\mu\vec{R} + O(R^6). \tag{2.6.9}$$

As expected, the fields $\vec{R}$ describe massless particles, the Goldstone bosons, whose interaction terms involve derivatives. Also note the generation of terms containing six and higher Goldstone fields. The associated question of renormalisability will be discussed in Section 3.5.

2.7 Effective Field Theories

In deriving the Lagrangian of the non-linear σ-model in the last section, we did not really need to start with the Lagrangian of any specific theory, like the linear σ-model. All that was necessary to know was the symmetry group $O(4)$ of the Lagrangian, which is broken spontaneously to $O(3)$. This gives us the transformation rules for the Goldstone and non-Goldstone fields, using which we can construct the effective Lagrangian. We now generalise this construction to arbitrary groups, leading to effective field theories that are free from assumptions of particular models, relying entirely on the assumed symmetry of the Lagrangian [2, 17].

Consider a quantum field theory, whose Lagrangian is symmetric under an arbitrary compact Lie group G: that is, it is invariant under global (space–time-independent) linear transformations of the field components $\psi_n(x)$

$$\psi_n(x) \longrightarrow \psi'_n(x) = g_{nm}\psi_m(x), \quad g \in G. \tag{2.7.1}$$

Further, let G be broken spontaneously to some subgroup H, so that only the elements of H leave the vacuum expectation values of ψ invariant. As only the scalar components of ψ are involved here, we denote them by $\overline{\phi}_m = \langle 0|\psi_m|0\rangle$, getting

$$h_{nm}\overline{\phi}_m = \overline{\phi}_n, \quad h \in H. \tag{2.7.2}$$

Following the example of $SO(4)$ σ-model in the preceding section, we express the fields $\psi(x)$ as a space–time-dependent G transformation $\gamma(x)$ acting on fields $\widetilde{\psi}(x)$, from which the Goldstone fields are removed

$$\psi_n(x) = \gamma_{nm}(x)\widetilde{\psi}_m(x), \qquad \gamma \in G. \tag{2.7.3}$$

According to the Goldstone theorem the massless Goldstone fields are represented by vectors $s_a\overline{\phi}$, where s_a are the broken generators. So the condition that $\widetilde{\psi}$ does not contain Goldstone modes is that $\widetilde{\psi}$ be orthogonal to $s_a\overline{\phi}$

$$\widetilde{\psi}_n(s_a)_{nm}\overline{\phi}_m = 0. \tag{2.7.4}$$

The number of such independent conditions is clearly equal to the number of broken generators. If γ has as many parameters, they will bring back the Goldstone fields.

We now show that the transformation matrices γ may always be chosen to satisfy (2.7.4). Working with a real representation[6] of the compact group G, we define

$$V_\psi(g) = \psi_n(x)g_{nm}\overline{\phi}_m \tag{2.7.5}$$

which is a real, continuous and therefore bounded function of $g \in G$. So at each space–time point x, $V(g)$ reaches a maximum for some g, say $\gamma(x)$. At $g = \gamma(x)$, $V(g)$ is then stationary with respect to arbitrary variation in g. Using the notation at the end of Section 2.1, we write a general element g as $g = \exp{(il_\alpha\theta_\alpha)}$. An infinitesimal change in g is $\delta g = i\epsilon_\alpha g l_\alpha$, where il_α are real and ϵ_α arbitrary infinitesimal parameters. So we have at $g = \gamma(x)$

$$0 = \delta V(\gamma(x)) = \psi_n(x)(\delta g)_{nm}\overline{\phi}_m = i\epsilon_a\psi_n(x)\gamma(x)_{np}(s_a)_{pm}\overline{\phi}_m , \tag{2.7.6}$$

where we replace l_α by s_a, as the generators belonging to H give zero, $t_i\overline{\phi} = 0$. It must be satisfied for all variations and so for all ϵ_a

$$0 = i\psi_n(x)\gamma_{np}(x)(s_a)_{pm}\overline{\phi}_m. \tag{2.7.7}$$

For a compact group, we can choose a unitary representation, which for real representation is orthogonal, $\gamma\gamma^T = 1$, giving $\gamma_{np} = (\gamma^{-1})_{pn}$. Then (2.7.7) becomes

$$0 = (\gamma^{-1}(x))_{pn}\psi_n(x)i(s_a)_{pm}\overline{\phi}_m = (\gamma^{-1}(x)\psi(x))_p i(s_a\overline{\phi})_p \tag{2.7.8}$$

showing that (2.7.4) is indeed satisfied by the transformation (2.7.3).

As in the case of $O(4)$ non-linear σ-model, the transformation (2.7.3) has an important effect on the Goldstone boson couplings. Since the Lagrangian is invariant under constant γ, it has the effect that all dependence on γ will drop out except where a derivative acts on γ. We can then construct a low-energy

---

[6] If the representation is complex, one can work with the real and imaginary parts of ψ_n, getting a real, though reducible, representation. [2]

perturbative framework to calculate amplitudes for Goldstone bosons, possibly including heavy particles also.

It should be clear that the element γ in (2.7.3) is generally not unique. As (2.7.2) shows, the elements $h \in H$ keep $\overline{\phi}$ invariant. So $V_\psi(g)$ defined by (2.7.5) remains invariant under right multiplication of g by h. Thus if γ maximises $V_\psi(g)$, so does γh. Hence γ is unique up to right multiplication of h. This non-uniqueness is evident in the case of (bosonic) σ-models considered earlier, where the non-Goldstone field $\widetilde{\psi}_n$ and the vacuum expectation value $\overline{\phi}$ point in the same direction. We now have to find a convenient parametrization of $\gamma(x)$.

The set of all elements of the form γh with a particular γ and h running over H defines an equivalence class, called a right coset of G with respect to H. All elements of the group G can be sorted into such right cosets. If we choose one representative group element γ from each right coset, we get a parametrisation of the space of right cosets or quotient space G/H of the two groups. A standard parametrization follows from writing a general element $g \in G$ in the product form

$$g = \exp\left(i\xi_a s_a\right) \exp\left(i\theta_i t_i\right) \equiv \gamma h. \tag{2.7.9}$$

We now define $\gamma(x)$ as

$$\gamma(x) \equiv \gamma(\xi(x)) = \exp\left(i\xi_a(x)s_a\right) \tag{2.7.10}$$

where we take the x-dependent parameters $\xi_a(x)$ as proportional to the Goldstone fields.

The transformation rules for $\xi_a(x)$ and $\widetilde{\psi}(x)$ may now be found. Under the action of a general element $g \in G$, the field $\psi(x)$ transforms as

$$\psi(x) \longrightarrow \psi'(x) = g\,\psi(x) = g\,\gamma(\xi(x))\,\widetilde{\psi}(x). \tag{2.7.11}$$

Here the product $g\gamma(\xi)$ of two elements is again an element of G, so it must be in the right coset of some $\gamma(\xi')$. Then from (2.7.9) we get

$$g\,\gamma(\xi(x)) = \gamma(\xi'(x))\,h(\xi, g) \tag{2.7.12}$$

which gives the transformation rule $\xi(x) \to \xi'(x)$ for the Goldstone fields under the action of g. Inserting this decomposition in (2.7.11) we get

$$\psi'(x) = \gamma(\xi'(x))\,\widetilde{\psi}'(x) \tag{2.7.13}$$

where

$$\widetilde{\psi}'(x) = h(\xi, g)\,\widetilde{\psi}(x) \tag{2.7.14}$$

gives the transformation rule for non-Goldstone fields. The rule for $\xi(x)$ is non-linear and space–time dependent, requiring us to construct covariant derivatives for building terms in the effective Lagrangian [2]. However, for the chiral symmetry group, in which we are interested, a much simpler transformation rule emerges, as we shall see in the next chapter.

2.8 Approximate Symmetry

So far we considered the spontaneous breakdown of an exact symmetry, where the symmetry group of the (complete) Lagrangian is G, while that of the vacuum is a subgroup H of G. We now discuss the effect of adding small symmetry breaking terms in the Lagrangian itself. As we shall see in the next chapter, the spontaneous breaking of such an approximate symmetry does not give rise to massless Goldstone bosons, but to low-mass spinless particles, called pseudo-Goldstone bosons [2].

It is possible to relate the explicit and spontaneous symmetry breakings, following steps similar to those in the proof of the Goldstone theorem using the effective potential in Section 2.2. Again let ϕ_n denote components of a space–time-independent expectation value of a scalar field, elementary or composite. Suppose that the complete effective potential is given by

$$V(\phi) = V_0(\phi) + V_1(\phi) \tag{2.8.1}$$

where $V_0(\phi)$ is invariant under the symmetry group G and and $V_1(\phi)$ is a small perturbation breaking this symmetry explicitly. As before ϕ_n transforms as (2.2.19), so that we have for each α

$$\frac{\partial V_0}{\partial \phi_n}[l_\alpha]_{nm}\phi_m = 0. \tag{2.8.2}$$

Let the potential $V_0(\phi)$ attain minimum at ϕ_0,

$$\left.\frac{\partial V_0(\phi)}{\partial \phi_n}\right|_{\phi=\phi_0} = 0 \tag{2.8.3}$$

which is shifted to $\overline{\phi}$ by the perturbing potential,

$$\left.\frac{\partial V(\phi)}{\partial \phi_n}\right|_{\phi=\overline{\phi}} = 0 . \tag{2.8.4}$$

We write $\overline{\phi} = \phi_0 + \phi_1$, where ϕ_1 is small. Expanding (2.8.4) to first order in the neighbourhood of ϕ_0, we get

$$\left.\frac{\partial V_0(\phi)}{\partial \phi_n}\right|_{\phi_0} + \left.\frac{\partial^2 V_0(\phi)}{\partial \phi_n \partial \phi_m}\right|_{\phi_0}\phi_{1m} + \left.\frac{\partial V_1(\phi)}{\partial \phi_n}\right|_{\phi=\phi_0} = 0 \tag{2.8.5}$$

where the first term is zero by (2.8.3). Earlier we derived (2.2.22) for the symmetric potential $V_0(\phi)$, namely

$$\left.\frac{\partial^2 V_0(\phi)}{\partial \phi_n \partial \phi_m}\right|_{\phi=\phi_0}[l_\alpha]_{np}\phi_{0p} = 0. \tag{2.8.6}$$

So multiplying (2.8.5) by $(l_\alpha)_{np}\phi_{0p}$, its second term vanishes, leaving

$$(l_\alpha\phi_0)_n \left.\frac{\partial V_1(\phi)}{\partial \phi_n}\right|_{\phi=\phi_0} = 0. \tag{2.8.7}$$

Equation (2.8.7) is known as a *vacuum alignment* condition, because by relating ϕ_0 to V_1 it brings the direction of the spontaneously broken vacuum state into some sort of alignment with the symmetry breaking potential. If the perturbation has the form

$$V_1(\phi) = u_n \phi_n \tag{2.8.8}$$

then it reduces to

$$u_n (l_\alpha \phi_0)_n = 0. \tag{2.8.9}$$

As an example, take the case of $O(N)$ broken to a subgroup $O(N-1)$ by the vacuum state, with $\phi_{0n}\phi_{0n} = -\mu^2/\lambda$. The latter condition specifies no particular direction of ϕ_0 and hence no particular subgroup to which the symmetry is broken. But in presence of perturbation V_1, this is no longer true. The generators l_α of $O(N)$ are antisymmetric: replacing the index α by $r\,s$ with $r,s = 1, \cdots N$, these are

$$[l_{rs}]_{pq} = -i(\delta_{rp}\delta_{sq} - \delta_{rq}\delta_{sp})$$

whence (2.8.9) becomes

$$\frac{\phi_{0r}}{\phi_{0s}} = \frac{u_r}{u_s} \tag{2.8.10}$$

so that ϕ_0 must be in the same direction as u.

We can put the vacuum alignment condition (2.8.9) in operator form. Suppose the symmetry breaking perturbation in the Hamiltonian to be of the form

$$H_1 = \sum u_n \Phi_n \tag{2.8.11}$$

where the operators Φ_n belong to a representation of the symmetry group G. Then we have from (2.3.8)

$$[Q_\alpha, \Phi_n(x)] = -(l_\alpha)_{nm}\Phi_m.$$

The symmetry breaking potential V_1 is given by the expectation value of H_1 in the state in which Φ_n has expectation value ϕ_n. Now multiply it by u_n, sum over n and take vacuum expectation value. Then using the vacuum alignment condition (2.8.9) we get

$$\langle 0|[Q_\alpha, H_1]|0\rangle = 0. \tag{2.8.12}$$

The vacuum alignment condition in this form has been used to study the explicit symmetry breaking of QCD theory [2].

Problem

Problem 2.1: Derive the Goldberger–Treiman relation. (Historically this relation played a significant role in identifying pions as the Goldstone bosons arising from the spontaneous breaking of flavour symmetry.)

Solution: Originally obtained in a model [18], Nambu [19] derived it in a model-independent way, which we follow here. Consider again the axial-vector current $A_a^\mu(x)$, as we did at the end of Section 2.4, now taking the matrix element between one nucleon states. By Lorentz invariance and parity conservation, it takes the form

$$\langle N(p')|A_a^\mu(x)|N(p)\rangle$$
$$= \bar{u}(p')\left[\gamma^\mu\gamma^5 g_A(k^2) + k^\mu\gamma^5 h_A(k^2) + \frac{i\sigma^{\mu\nu}k_\nu}{2m_N}\gamma^5 f_A(k^2)\right]\frac{\tau_a}{2}u(p)e^{ik\cdot x} \quad \text{(P2.1)}$$

where $k = p' - p$ and g_A, h_A and f_A are form factors. Requiring current conservation we get a relation between the first two form factors,

$$2m_N g_A(k^2) + k^2 h_A(k^2) = 0. \quad \text{(P2.2)}$$

Here the quantity $g_A(0)$ is non-zero, as can be measured in low-energy nuclear β-decays, $g_A(0) \equiv g_A = 1.26$, Thus as $k^2 \to 0$, $h_A(k^2)$ must have a pole,

$$h_A(k^2) \to -\frac{2m_N g_A}{k^2}. \quad \text{(P2.3)}$$

Fourier transforming (P2.1) and retaining only this pole contribution, we get

$$\int d^4x e^{ik\cdot x}\langle N(p')|A_a^\mu(x)|N(p)\rangle \to (2\pi)^4\delta^4(k + p' - p)h_A(k^2)k^\mu\bar{u}(p')\gamma^5\frac{\tau_a}{2}u(p). \quad \text{(P2.4)}$$

The same pole is also given by the graph of Figure 2.2, when both α and β are states of a nucleon: Formula (2.4.4) gives

$$\int d^4x e^{ik\cdot x}\langle N(p')|A_a^\mu(x)|N(p)\rangle \to -\frac{Fk^\mu\langle\pi_a(k), N(p')|N(p)\rangle}{k^2}. \quad \text{(P2.5)}$$

Assuming a phenomenological πN Lagrangian, $L_{\pi N} = -iG_{\pi N}\overline{N}\vec{\tau}\cdot\vec{\phi}\gamma^5 N$, the S-matrix element in (P2.5) is given by

$$\langle N(p'), \pi_a(k)|N(p)\rangle = (2\pi)^4\delta^4(p' - p - k)G_{\pi N}\bar{u}(p')\gamma^5\tau_a u(p). \quad \text{(P2.6)}$$

Inserting (P2.6) in (P2.5) and then comparing it with (P2.4) we get

$$h_A(k^2) \to -\frac{2FG_{\pi N}}{k^2}. \quad \text{(P2.7)}$$

Putting together (P2.3) and (P2.7) we get the Goldberger–Treiman relation

$$m_N g_A = FG_{\pi N}. \quad \text{(P2.8)}$$

References

[1] J. Goldstone, A. Salam and S. Weinberg, Phys. Rev. **127**, 965 (1962).

[2] S. Weinberg, *The Quantum Theory of Fields*, vol. 2, Cambridge University Press (1995).

[3] S. Coleman, *Aspects of Symmetry*, Cambridge University Press (1988).

[4] A. Zee, *Quantum Field Theory in a Nutshell*, Princeton University Press (2003).

[5] R.E. Marshak, *Conceptual Foundations of Modern Particle Physics*, World Scientific (1993).

[6] G. Källen, Helv. Phys. Acta, **25**, 417 (1952).

[7] H. Lehmann, Nuovo Cim. **11**, 342 (1954).

[8] H. Lehmann, K. Symanzik and W. Zimmermann, Nuovo Cimento, 1, 205 (1955).

[9] S. Weinberg, in *Lectures on Elementary Particles and Quantum Field Theory*, vol. 1, edited by S. Deser et al., MIT Press (1970).

[10] S. Weinberg, *The Quantum Theory of Fields*, vol. 1, Cambridge University Press (1995).

[11] M.E. Peskin and D.V. Schroeder, *An Introduction to Quantum Field Theory*, Westview (1995).

[12] Y. Nambu and D. Lurie, Phys. Rev. **125**, 1429 (1961).

[13] S.L. Adler, Phys. Rev. **137**, B 1022, (1965).

[14] H. Leutwyler, *Principles of Chiral Perturbation Theory*, lectures given at the workshop, 'Hadrons 1994', Gramado, RS, Brazil (1994).

[15] S. Weinberg, Phys. Rev. Lett., **18**, 188 (1967).

[16] S. Weinberg, Phys. Rev. **166**, 1568 (1968).

[17] C.G. Callan, S. Coleman, J. Wess and B. Zumino, Phys. Rev. **177**, 2247 (1969).

[18] M.L. Goldberger and S.B. Treiman, Phys. Rev. **110**, 1478 (1958).

[19] Y. Nambu, Phys. Rev. Lett. **4**, 380 (1960).

3

Chiral Perturbation Theory

It is interesting to note that much of the effective theory of strong interactions was developed well before the existence of QCD [1]. The isotopic spin symmetry of strong interactions was established long ago by works in nuclear physics. Modern developments began in 1960, when Nambu recognised the existence of nearly massless pions as a symptom of a spontaneously broken symmetry [2]. (This topic is reviewed in the previous chapter.) He and co-workers also showed how to calculate amplitudes involving a single pion [3]. In 1964 Gell-Mann introduced current algebra [4], which allowed the treating of processes involving more than a single pion. In particular, exact sum rules could be written for pion–hadron amplitudes, taking the pion to be massless [5, 6]. But for processes involving three or more pions, the calculations became complicated. Further, there was no way to treat approximate symmetries.

The numerical success of the sum rules did show that strong interactions must possess an underlying $SU(2) \times SU(2)$ symmetry broken to the $SU(2)$ symmetry of isospin. In an attempt to find a simpler and more physical method of calculation, Weinberg [7] introduced in 1967 the effective Lagrangian incorporating this symmetry. Originally it was justified on the basis of current algebra: an effective Lagrangian with the symmetry of the underlying theory would produce conserved Noether currents in terms of the effective fields. These currents in turn would yield the equal-time commutators of current algebra. So if one calculates low energy amplitudes directly with this Lagrangian, it must yield the same results as from current algebra. The essential uniqueness of the effective Lagrangian was also established [8].

The QCD theory was proposed in 1972 by Fritzsch and Gell-Mann [1] as a theory of quarks and gluons. Soon after, Leutwyler [9] found that not only the mass difference of u and d quarks, but also the masses themselves, were small compared to the scale of strong interactions, which immediately explained the $SU(2) \times SU(2)$ symmetry of strong interactions. (Including the heavier s quark leads to a less accurate $SU(3) \times SU(3)$ symmetry.) The mass terms provide a definite pattern of symmetry breaking.

Meanwhile, Weinberg [10] explored the field theory of chiral Lagrangian, obtaining corrections to results from current algebra. He also gave a remarkably simple, yet far-reaching, justification of all effective field theories, invoking the foundations of relativistic quantum field theory. It is the realisation 'that when we calculate a physical amplitude from Feynman diagrams using the most general Lagrangian that involves the relevant degrees of freedom and satisfies the assumed symmetries of the theory, we are simply constructing the most general amplitude that is consistent with general principles of relativity, quantum mechanics, and the assumed symmetries' [11].

Finally, Gasser and Leutwyler [12, 13] worked out the complete effective Lagrangian to fourth order to study systematically the mesonic sector of strong interactions. An assumption implicitly made in these derivations is that the symmetries of the underlying theory imply a symmetric effective Lagrangian. This assumption has since also been addressed conclusively [14]. Here the starting point is the generating functional, which collects all the Green's functions of symmetry currents by coupling these to external gauge fields. The Ward identities for the Green's functions then follow from the gauge invariance of the generating functional. Using this framework it is possible to show that the effective Lagrangian can indeed be put in a manifestly gauge-invariant form, provided we use the freedom to add total derivatives and to perform point transformations on the pion field variable.

Though chiral perturbation theory describes strong interactions among hadrons, it does not depend on the colour gauge symmetry of QCD. It merely abstracts the flavour symmety of QCD and its explicit violation due to the quark mass term. Nevertheless, one can understand intuitively how the strong nuclear force among colour-neutral particles, like pions and nucleons, originates from the strong force among coloured quarks and gluons. It is similar to the van der Waals force among neutral atoms or molecules [15]. Thus if two hadrons approach each other, the colour polarisation leads to residual forces producing the observed short-range interaction between them.

Our thermal applications will be for temperatures $T \leq 150$ MeV, low compared to the masses of observed strange particles. Thus it suffices to consider the symmetry group $SU(2) \times SU(2)$, rather than $SU(3) \times SU(3)$. This simplifies the theory: the effective Lagrangian involves fewer terms and is free from anomalies.

In this chapter we construct the effective pion Lagrangian to fourth order and illustrate its use with simple examples. In the pion–nucleon sector this Lagrangian is worked out to first non-leading order, while for vector and axial-vector mesons it is given only to leading order. The examples of nucleon and spin-one mesons are considered only to provide a basis for later evaluations at finite temperature.

3.1 Symmetries of Massless QCD

We consider QCD with the quark field $q(x)$ in two flavours,

$$q = \begin{pmatrix} u \\ d \end{pmatrix},$$

where each of $u(x)$ and $d(x)$ is a Dirac field. They also carry a colour quantum number, which is not of interest to us here. These quarks happen to have relatively small masses. In the approximation that they are massless, the Lagrangian is

$$\mathscr{L}_{QCD}^{(0)} = \bar{q}\, i\slashed{D} q, \tag{3.1.1}$$

where the superscript (0) on $\mathscr{L}$ indicates also omission of terms with only gluon fields and/or other heavier quark flavours. Here D_μ is the colour-gauge-covariant derivative involving the gluon field $G_\mu(x)$. Acting on the quark field, $D_\mu q = (\partial_\mu - iG_\mu)\,q$, it gives the interaction of quarks with gluons.

As in the case of single flavour treated in Problem 1.5, the full symmetry of the Lagrangian (3.1.1) becomes manifest, if we rewrite it in terms of the right- and left-handed quark doublet fields,

$$q_R = \begin{pmatrix} u_R \\ d_R \end{pmatrix}, \qquad q_L = \begin{pmatrix} u_L \\ d_L \end{pmatrix},$$

with R and L components defined by

$$q_R = \frac{1}{2}(1 + \gamma^5)q, \qquad q_L = \frac{1}{2}(1 - \gamma^5)q$$

when it becomes

$$\mathscr{L}_{QCD}^{(0)} = \bar{q}_R i\slashed{D} q_R + \bar{q}_L i\slashed{D} q_L. \tag{3.1.2}$$

Each of the terms in this Lagrangian is separately invariant under transformations of the quark fields by 2×2 matrices V, which are unitary ($V^\dagger V = 1$) and unimodular ($\det V = 1$) forming the group $SU(2)$[1]. In other words, the full Lagrangian is invariant under

$$\begin{aligned} q_R \to q_R' = V_R q_R, \qquad & V_R \in SU(2)_R \\ q_L \to q_L' = V_L q_L, \qquad & V_L \in SU(2)_L. \end{aligned} \tag{3.1.3}$$

It is convenient in this section to define $\vec{t} = \vec{\tau}/2$, where $\vec{\tau}$ is the three-vector of Pauli matrices, denoted by $\vec{\sigma}$ in Section 1.3. Then the elements of $SU(2)_R$ and $SU(2)_L$ may be written in the canonical form as

$$V_R = \exp\left(i\vec{\theta}_R \cdot \vec{t}\right), \qquad V_L = \exp\left(i\vec{\theta}_L \cdot \vec{t}\right). \tag{3.1.4}$$

---

[1] The terms are also invariant under separate phase transformations on q_R and q_L, giving rise to two independent $U(1)$ groups, which commute with $SU(2)_R \times SU(2)_L$.

The symmetry group of the massless QCD Lagrangian of two flavours is therefore $SU(2)_R \times SU(2)_L$. It is useful to combine the two groups of transformations by letting them act on the Dirac field [11],

$$q \equiv q_R + q_L \to \left\{ \exp\left(i\vec{\theta}_R \cdot \vec{t}\right) \frac{1}{2}(1 + \gamma^5) + \exp\left(i\vec{\theta}_L \cdot \vec{t}\right) \frac{1}{2}(1 - \gamma^5) \right\} q$$

$$= \exp\left(i\vec{\theta}_R \cdot \vec{t}_R + i\vec{\theta}_L \cdot \vec{t}_L\right) q \qquad (3.1.5)$$

where

$$\vec{t}_R = \frac{1}{2}(1 + \gamma^5)\vec{t}, \quad \vec{t}_L = \frac{1}{2}(1 - \gamma^5)\vec{t}. \qquad (3.1.6)$$

These generators satisfy the commutation relations

$$[t_{Ri}, t_{Rj}] = i\epsilon_{ijk} t_{Rk},$$
$$[t_{Li}, t_{Lj}] = i\epsilon_{ijk} t_{Lk},$$
$$[t_{Ri}, t_{Lj}] = 0 \qquad (3.1.7)$$

giving two commuting subalgebras for $SU(2)_R$ and $SU(2)_L$. If we split the group parameters as $\vec{\theta}_R = \vec{\theta}_V + \vec{\theta}_A$, $\vec{\theta}_L = \vec{\theta}_V - \vec{\theta}_A$, the transformation rule (3.1.5) becomes

$$q \to q' = \exp\left(i\vec{\theta}_V \cdot \vec{t} + i\vec{\theta}_A \cdot \vec{s}\right) q \qquad (3.1.8)$$

with generators $\vec{t}$ and $\vec{s}$,

$$\vec{t} = \vec{t}_R + \vec{t}_L, \quad \vec{s} = \vec{t}_R - \vec{t}_L = \gamma^5 \vec{t} \qquad (3.1.9)$$

satisfying commutation relations

$$[t_i, t_j] = i\epsilon_{ijk} t_k,$$
$$[t_i, s_j] = i\epsilon_{ijk} s_k,$$
$$[s_i, s_j] = i\epsilon_{ijk} t_k . \qquad (3.1.10)$$

We see the presence of another subgroup with generators t_i. It transforms the right- and the left-handed components of the quark field in the same way, $\theta_A = 0$, forming the ordinary isospin subgroup $SU(2)_V$. As we argue below, phenomenology suggests strongly that the symmetry $SU(2)_R \times SU(2)_L$ of the massless QCD Lagrangian is broken spontaneously to $SU(2)_V$. So, in accordance with our notation at the end of Section 2.1, we shall denote from now on the components of $\vec{t}$ and $\vec{s}$ by indices $i, j, \cdots$ and $a, b, \cdots$ respectively, even though they run over the same set, namely, 1, 2 and 3.

Applying infinitesimal form of transformations (3.1.8) to the Lagrangian (3.1.1) we find the Noether currents, the vector and axial-vector as

$$V_i^\mu(x) = \bar{q}\gamma^\mu t_i q, \quad A_a^\mu(x) = \bar{q}\gamma^\mu \gamma^5 t_a q \qquad (3.1.11)$$

which are conserved, $\partial_\mu V_i^\mu = \partial_\mu A_a^\mu = 0$, giving the charges

$$Q_i = \int d^3x\, q^\dagger(x) t_i q(x), \quad Q_a^5 = \int d^3x\, q^\dagger(x) \gamma^5 t_a q(x). \tag{3.1.12}$$

As already pointed out in Section 2.3, these charges are formal quantities and so are their commutators with each other. But with quark *fields*, they are well-defined

$$[Q_i, q] = -t_i q, \qquad [Q_a^5, q] = -\gamma^5 t_a q \tag{3.1.13}$$

which may be derived in the same way as we got (2.3.8) for scalar fields[2].

3.2 Spontaneous Symmetry Breaking

In constructing an effective theory of strong interactions, it is important to know how the above symmetries of massless QCD are realised in nature. Theoretically, this is a difficult problem, as it involves all the complications of strong interactions at low energy. Fortunately, however, phenomenology gives a clear indication on this point. If the symmetry were realised in the Wigner–Weyl way, that is, the full symmetry of $\mathscr{L}$ were also the symmetry of the vacuum state

$$Q_i|0\rangle = 0, \qquad Q_a^5|0\rangle = 0, \tag{?}$$

it would strongly disagree with the observed hadron spectrum [11]. As the generators Q_a^5 have negative parity, a one-hadron state $|h\rangle$ would require another degenerate state $Q_a^5|h\rangle$ of opposite parity but same spin, baryon number and strangeness. While isospin multiplets are known with nearly degenerate masses, there are no corresponding multiplets of opposite parity. On the other hand, in the Nambu–Goldstone realisation, where the symmetry of the vacuum state is partially broken,

$$Q_i|0\rangle = 0, \qquad Q_a^5|0\rangle \neq 0, \tag{3.2.1}$$

the Goldstone theorem predicts three massless particles with quantum numbers of Q_a^5, which may be identified with the pion triplet. (As we shall see, their observed masses can be accounted for by the quark mass term in the complete QCD Lagrangian.) In this case, $Q_a^5|0\rangle$ is a pion state (see Problem 3.1). So $Q^5|h\rangle$ is a state of the hadron h and a pion, avoiding the need for parity doubling of hadron multiplets.

Having concluded phenomenologically that the symmetry group $G \equiv SU(2)_R \times SU(2)_L$ breaks spontaneously to $H = SU(2)_V$, we turn to the Goldstone theorem to learn more about the asymmetric vacuum [18]. In the linear

---

[2] We can also check (3.1.13) directly by noting the equal time anti-commutator for the quark field, $\{q_{Au}(\boldsymbol{x}, t), q^\dagger_{Bv}(\boldsymbol{x}', t)\} = \delta_{AB}\,\delta_{uv}\,\delta^3(\boldsymbol{x} - \boldsymbol{x}')$, where $A, B \cdots$ are Dirac spinor indices and $u, v \cdots$ are flavour indices running over $1, 2, 3, 4$ and $1, 2$ respectively. Then using the identity, $[XY, Z] = X\{Y, Z\} - \{X, Z\}Y$, the commutators follow.

σ-model (Section 2.1), the elementary scalar field (multiplet) itself develops vacuum expectation value leading to symmetry breaking. In QCD we look for bosonic composite operators, built out of quark fields, which may develop vacuum expectation values. In this context two four-vectors are relevant, which will arise in Section 3.8 from an analysis of the mass term in QCD Lagrangian in presence of external fields [12]. These are

$$\Phi^+ = (-\bar{q}i\gamma^5\vec{\tau}q, \bar{q}q), \qquad \Phi^- = (\bar{q}\vec{\tau}q, \bar{q}i\gamma^5q), \qquad (3.2.2)$$

transforming as independent four-vectors of $SO(4)$. (Here we label the components of four-vectors by indices 1, 2, 3, 4 as in Section 2.5, while later in Section 3.8 the fourth component will be labelled by index 0.) Their commutation rules with the charges can be written in the same way as (2.3.8) for the scalar multiplet,

$$[Q_i, \Phi_n^\pm] = -(T_i)_{nm}\Phi_m^\pm \qquad (3.2.3)$$

$$[Q_a^5, \Phi_n^\pm] = -(S_a)_{nm}\Phi_m^\pm, \qquad (3.2.4)$$

where T_i and S_a are the matrix generators of the algebra of SO(4), isomorphic to $SU(2) \times SU(2)$. (The generators are obtained explicitly in Section 2.5.). These commutation rules may also be checked directly by applying (3.1.13).

Let us next find which components of $\Phi^\pm$ may develop vacuum expectation values. Since we are assuming that the vacuum is invariant under both Lorentz transformations and space reflections, only Lorentz scalar operators can give rise to vacuum expectation values. To see how such an operator $\mathcal{O}$ must behave under the symmetry group, acting as $\mathcal{U}$ on states, we write the identity,

$$\langle 0|\mathcal{O}|0\rangle = \langle 0|\mathcal{U}^\dagger(g)\mathcal{U}(g)\mathcal{O}\mathcal{U}^\dagger(g)\mathcal{U}(g)|0\rangle, \qquad (3.2.5)$$

by inserting $1 = \mathcal{U}^\dagger\mathcal{U}$ twice in the vacuum matrix element. So if $\mathcal{U}$ is a symmetry of the vacuum state $\mathcal{U}|0\rangle = |0\rangle$, then only those operators can develop vacuum expectation values, which are invariant under $\mathcal{U}$, $\mathcal{U}\mathcal{O}\mathcal{U}^\dagger = \mathcal{O}$. Here we *assume* that the vacuum state is invariant only under the subgroup H, $Q_i|0\rangle = 0$, so only H-invariant operators can acquire vacuum expectation values. Thus both the criteria single out $\Phi_4^+ = \bar{q}q$ as the only component of $\Phi^\pm$ that can have $\langle 0|\bar{q}q|0\rangle \neq 0$. We are thus led to choose the non-trivial commutator as

$$[Q_a^5, \Phi_b^+] = -(S_a)_{b4}\Phi_4^+ = i\delta_{ab}\Phi_4^+ \qquad (3.2.6)$$

on noting the matrix generators S_a from Section 2.5.

We now take the vacuum expectation value of (3.2.6) and go through the steps in Section 2.3 to prove the Goldstone theorem. For the spectral function

$$ip^\mu\rho_{ab}(p^2)\theta(p_0) = (2\pi)^3 \sum_N \langle 0|A_a^\mu(0)|N\rangle\langle N|\Phi_b^+(0)|0\rangle\delta^4(p - p_N) \qquad (3.2.7)$$

it gives

$$\rho_{ab}(p^2) = \delta_{ab}\delta(p^2)\langle 0|\bar{q}q|0\rangle. \qquad (3.2.8)$$

With one pion as the only state contributing, the spectral function (3.2.7) can also be worked out directly, as we did for the linear $O(4)$ model at the end of Sections 2.3 and 2.5. Inserting the matrix elements

$$\langle 0|A_a^\mu(0)|\pi_c(k)\rangle = iFk^\mu\delta_{ac}, \tag{3.2.9}$$

$$\langle\pi_c(k)|\Phi_b^+(0)|0\rangle = -\langle\pi_c(k)|\bar{q}i\gamma^5\tau_b q|0\rangle = -G\delta_{cb} \tag{3.2.10}$$

we get

$$\rho_{ab}(p^2) = -\delta_{ab}\delta(p^2)FG. \tag{3.2.11}$$

Putting together (3.2.8) and (3.2.11), we get from the Goldstone theorem

$$FG = -\langle 0|\bar{q}q|0\rangle, \tag{3.2.12}$$

as an exact result in the two-flavour massless QCD theory. The QCD vacuum thus acquires a condensate $\langle 0|\bar{q}q|0\rangle$ as a result of the spontaneous symmetry breaking. As Φ_a^+ is not an elementary field, it is not possible to put $G = 1$. We shall come back to this relation after introducing quark masses in Section 3.7.

$$* \quad * \quad *$$

Note that though $\bar{q}q = \bar{q}_R q_L + \bar{q}_L q_R$ is invariant under $SU(2)_V$, it transforms non-trivially under $SU(2)_R \times SU(2)_L$. One refers to the vacuum expectation value of such operators that are invariant under the symmetry group H of the vacuum, but not under the full group G as *order parameters*. Thus $\langle 0|\bar{q}q|0\rangle$ is an order parameter, called the quark condensate and its non-zero value signals spontaneous symmetry breaking.

The operator $\bar{q}q$ is of the lowest dimension that can acquire vacuum expectation value. There are of course other operators of higher dimension that are H-invariant, but transform non-trivially under G, like $(\bar{q}q)^2$, $(\bar{q}\gamma^5 q)^2$, $(\bar{q}\tau_i q)^2$, $(\bar{q}\gamma_\mu\tau_i q)^2$, etc. Not all of these give rise to independent order parameters. Also there are operators involving the gluon field; under G $\mathrm{tr}\,_cG_{\mu\nu}G^{\mu\nu}$ is a singlet, but $\bar{q}\gamma_{\mu\nu}G^{\mu\nu}q$ is not and so can be a condensate of higher dimension.

The magnitude of the quark condensate depends on the scale of the theory. Nevertheless it can be altered by external conditions, say by exposing the system to a heat bath. As we shall see in Sections 6.1 and 7.1, the condensate gradually 'melts' as temperature rises and probably disappears, when the temperature reaches a certain critical value, restoring the G symmetry of the system.

3.3 Transformation Rule

In QCD theory with two flavours, the chiral symmetry group $G = SU(2)_R \times SU(2)_L$ acts on the quark fields according to (3.1.3). In the effective theory, we have to find the corresponding action on the pion fields that arise as Goldstone bosons from the spontaneous breakdown of G to its subgroup $H = SU(2)_V$. We have seen in Section 2.7 that these fields may be viewed as coordinates $\xi_a(x)$ in the coset or quotient space G/H. Their transformation rules (2.7.12) are in general non-linear and coordinate dependent. But for *chiral symmetries* it is possible to choose the field variable such that the transformation rule becomes linear and coordinate independent [16–18].

As the group G is the direct product of $SU(2)_R$ and $SU(2)_L$, an element $g \in G$ has the form of a pair, $g = \{V_R, V_L\}$, where as before $V_R \in SU(2)_R$ and $V_L \in SU(2)_L$. The unbroken subgroup H is $SU(2)_V$, generated by the charges of the vector current, and so consists of those elements of G for which $V_R = V_L$. That is, we can represent an element $h \in H$ by the pair $h = \{V, V\}$. Then we can decompose g as

$$g = \{V_R, V_L\} = \{(V_R V_L^\dagger)V_L, V_L\} = \{V_R V_L^\dagger, 1\}\{V_L, V_L\} \equiv \gamma h \qquad (3.3.1)$$

where the quotient space element is

$$\gamma = \{U, 1\}, \qquad U = V_R V_L^\dagger \equiv \exp(i\vec{\xi} \cdot \vec{\tau}) \qquad (3.3.2)$$

As in Section 2.7 we now define

$$\gamma(x) = \{U(x), 1\}, \qquad U(x) = \exp(i\vec{\xi}(x) \cdot \vec{\tau}) \qquad (3.3.3)$$

where the x-dependent parameters $\vec{\xi}(x)$ represent the Goldstone fields up to normalisation. To find the action of g on γ, we note that $g\gamma$ must belong to some coset element $\gamma' = \{U', 1\}$ in this space,

$$g\gamma = \{V_R, V_L\}\{U, 1\} = \{V_R U, V_L\} = \{V_R U V_L^\dagger, 1\}\{V_L, V_L\} \equiv \gamma' h, \qquad (3.3.4)$$

so that[3] under g

$$U(x) \to U'(x) = V_R U(x) V_L^\dagger \qquad (3.3.5)$$

Another way to represent the quotient space follows from the decomposition

$$g = \{V_R, V_L\} \equiv \{u, u^\dagger\}\{V, V\}, \qquad V \in SU(2)_V \qquad (3.3.6)$$

A group element g acts on this coset space as

$$g\{u, u^\dagger\} = \{V_R, V_L\}\{u, u^\dagger\} = \{V_R u, V_L u^\dagger\} = \{V_R u V^\dagger, V_L u^\dagger V^\dagger\}\{V, V\} \qquad (3.3.7)$$

---

[3] Another choice of representatives for the quotient space results in a different transformation rule. Decompose the elements of G as

$$g = \{V_R, V_L\} = \{1, U\}\{V_R, V_R\} \equiv \gamma h, \qquad U = V_L V_R^\dagger,$$

and g acts on it as

$$g\{1, U\} = \{V_R, V_L U\} = \{1, V_L U V_R^\dagger\}\{V_R, V_R\}$$

giving the transformation rule

$$U(x) \to U'(x) = V_L U(x) V_R^\dagger.$$

Compared to (3.3.5) it amounts to replacing U by $U^\dagger$.

giving the transformation rules for u,

$$u(x) \to u'(x) = V_R u(x) V^\dagger, \quad u(x) \to u'(x) = V u(x) V_L^\dagger. \tag{3.3.8}$$

As these rules also involve the elements of H, they will be useful in writing the transformation rules for the non-Goldstone (heavy) fields in Section 3.11. Note the transformation for u^2 following from (3.3.8),

$$u^2 \to u'^2 = V_R u^2 V_L^\dagger \tag{3.3.9}$$

which is the same as for U, giving $U = u^2$.

In the rule (3.3.5) the matrix field $U(x)$ transforms linearly under $SU(2)_R \times SU(2)_L$. But the components of $U(x)$ are not independent; they are subject to non-linear constraints, $U^\dagger U = 1$ and $\det U = 1$. (If we write U as $U = U_0 1 + iU_a \tau_a$, both the constraints lead to $(U_0)^2 + (U_a)^2 = 1$.) So the transformation (3.3.5) is a non-linear realisation of $SU(2) \times SU(2)$.

Here we find an effective theory in full glory. Unlike the types of σ-model, where the underlying and effective theories have the same field degrees of freedom, we deal here with two entirely different sets of fields in the two versions: in QCD we have quarks (and gluons), while the effective theory deals with the observed particles, namely pions (and other heavy particles). We abstract the flavour symmetry of the underlying theory and use it to find the transformation rules for the observed particles.

3.4 Effective Lagrangian

Given the transformation rule (3.3.5) we can construct the effective Lagrangian with the matrix field $U(x)$,

$$\mathscr{L}_{\text{eff}} = \mathscr{L}_{\text{eff}}(U, \partial_\mu U, \partial_\mu \partial_\nu U, \cdots), \tag{3.4.1}$$

whose terms would be invariant under this transformation. We arrange them in a series in powers of derivatives. Beginning with one without derivative, such a term can be invariant only if it is a constant, independent of U. As stated earlier, such a constant can be ignored in a non-gravitational context.

Thus the leading term in the series, by Lorentz invariance, must be quadratic in the derivative. A priori, there are two kinds of possible terms,

$$f_1(U) \times \Box U \quad \text{and} \quad f_2(U) \times \partial_\mu U \times \partial^\mu U,$$

where crosses indicate that the coefficients $f_{1,2}$ carry indices, to be contracted with those of $\Box U$ and $\partial_\mu U$. By partial integration, the first term can be transformed into the second one, which we now find. At this point it is useful to replace the derivative $\partial_\mu U$ by a related one, $U^\dagger \partial_\mu U$ or $\partial_\mu U U^\dagger$ [17]. The advantage of working with such variables is that they are invariant under one of the factor groups, $SU(2)_R$ or $SU(2)_L$, and are also traceless[4].

---

[4] The tracelessness of $U^\dagger \partial_\mu U$ and $\partial_\mu U U^\dagger$ follows from formula (C.8) of Appendix C.

For definiteness let us choose $U^\dagger \partial_\mu U \equiv \Delta_\mu$, when the second term in (3.4.2) may be written as

$$\tilde{f}_2(U) \times \Delta_\mu \times \Delta^\mu .$$

Under $SU(2)_R$, $U \to V_R U$, and Δ_μ is invariant. Since the term in the Lagrangian must remain invariant, it follows that $\tilde{f}_2(U)$ is independent of U, as we argued above. Next, under $SU(2)_L$, $U \to UV_L^\dagger$, and $\Delta_\mu \to V_L \Delta_\mu V_L^\dagger$, showing that Δ_μ transforms as the adjoint representation $D^{(1)}$ of $SU(2)_L$. As the Clebsch-Gordan decomposition

$$D^{(1)} \times D^{(1)} = D^{(2)} + D^{(1)} + D^{(0)},$$

contains the singlet only once, there is a single invariant of order p^2,

$$\mathrm{tr}\,(\Delta_\mu \Delta^\mu) = \mathrm{tr}\,(U^\dagger \partial_\mu U U^\dagger \partial^\mu U) = -\,\mathrm{tr}\,(\partial_\mu U \partial^\mu U^\dagger).$$

Here and below $\mathrm{tr}\,(A)$ denotes trace of the 2×2 matrix A. We thus get

$$\mathscr{L}_{\mathrm{eff}}^{(2)} = g\,\mathrm{tr}\,(\partial_\mu U \partial^\mu U^\dagger) . \tag{3.4.2}$$

We note that this way of forming group singlets is quite general. If we have variables like U, all transforming as (3.3.5), then a group singlet is obtained by multiplying consecutively a variable and an adjoint variable and taking the trace over the resulting product. It will be extensively used in the following.

The leading term (3.4.2) in the derivative expansion of the effective Lagrangian contains two free parameters, one being the coupling constant g and the other the scale of the fields in U, $\vec{\xi}(x) = \vec{\phi}(x)/F$, where $\vec{\phi}(x)$ is the pion field triplet,

$$U = \exp\,(i\vec{\tau} \cdot \vec{\phi}/F) . \tag{3.4.3}$$

To identify these constants let us get the quadratic term in $\vec{\phi}$ in the effective Lagrangian,

$$\mathscr{L}_{\mathrm{eff}}^{(2)} = \frac{2g}{F^2} \partial_\mu \vec{\phi} \cdot \partial^\mu \vec{\phi} + O(\phi^4). \tag{3.4.4}$$

The standard normalisation of the kinetic term requires $4g = F^2$, giving

$$\mathscr{L}_{\mathrm{eff}}^{(2)} = \frac{F^2}{4}\,\mathrm{tr}\,(\partial_\mu U \partial^\mu U^\dagger) . \tag{3.4.5}$$

The physical meaning of F is seen from the axial-vector current following from this Lagrangian (see B.19),

$$\mathscr{A}_a^\mu(x) = i\frac{F^2}{4}\,\mathrm{tr}\,(\tau_a\{\partial^\mu U, U^\dagger\}) = -F\partial^\mu \phi_a(x) + \cdots \tag{3.4.6}$$

Comparing its matrix element between the vacuum and pion states with (3.2.9), we see that F is indeed the pion decay constant appearing there.

The full effective Lagrangian is a series

$$\mathscr{L}_{\mathrm{eff}} = \mathscr{L}_{\mathrm{eff}}^{(2)} + \mathscr{L}_{\mathrm{eff}}^{(4)} + \cdots \tag{3.4.7}$$

Later we shall include explicit symmetry breaking in $\mathscr{L}_{\text{eff}}$ and find out the additional pieces in $\mathscr{L}_{\text{eff}}^{(2)}$ as well as the complete $\mathscr{L}_{\text{eff}}^{(4)}$.

3.5 Power Counting

We see above that the interaction Lagrangian of the effective theory consists of a series of terms,

$$\mathscr{L}_{\text{eff}}^{\text{int}} = \sum_i \mathscr{L}^{(i)}, \qquad \mathscr{L}^{(i)} = g_i \partial^{d_i} \phi^{n_i}, \tag{3.5.1}$$

with increasing number of (scalar) fields and/or derivatives acting on these fields, denoted respectively by n_i and d_i in the above symbolic form of the ith term. Its coupling constant g_i has dimension Δ_i given by

$$\Delta_i = 4 - d_i - n_i. \tag{3.5.2}$$

We now describe briefly the renormalisation of terms of the perturbation series for amplitudes generated by such a Lagrangian [29]. As we shall see, it will lead to a power counting rule for amplitudes at low momenta.

A perturbation series is analysed in terms of connected, one-particle irreducible Feynman amplitudes of the form

$$F(\{q\}) = \int \{d^4k\} G(\{q\}, \{k\}). \tag{3.5.3}$$

Here $\{q\}$ and $\{k\}$ denote sets of independent external and internal momenta. The integrand G consists of propagators and vertices. It will be useful to calculate the superficial degree of divergence D of such an amplitude, which is defined as the actual degree of divergence of the integral arising from the integration region, where all the internal momenta $\{k\}$ tend to infinity together. Values of $D = 0, 1, 2, \cdots$ correspond respectively to divergences of type logarithmic, linear, quadratic, $\cdots$.

Consider a graph with E external lines, I internal lines, L loops and V_i vertices of interaction type i. Contributions to D arise from three sources: each internal line gives -2, each vertex of type i gives d_i and each (independent) loop integration gives 4. We thus get

$$D = -2I + \sum_i V_i d_i + 4L. \tag{3.5.4}$$

There are topological identities, which can be used to eliminate from D some internal details of the graph. One such identity is obtained by counting the number of independent internal four-momenta, which is equal to L. Each internal line can be labelled with a four-momentum, but they are related linearly by a delta function at each vertex. Excluding the delta function conserving the total external four-momentum, we get

$$L = I - \left(\sum_i V_i - 1 \right). \tag{3.5.5}$$

Using it to eliminate I from (3.5.4), we get

$$D = \sum_i V_i(d_i - 2) + 2L + 2. \tag{3.5.6}$$

We need this form of expression for D below. But here we eliminate L from (3.5.4) to get

$$D = 2I + \sum_i V_i(d_i - 4) + 4. \tag{3.5.7}$$

Another identity follows from counting the total number of fields in a graph. Each internal line results from two fields and an external line from a single field. Thus we get the identity

$$2I + E = \sum_i V_i n_i \tag{3.5.8}$$

which we use to eliminate I from (3.5.7),

$$D = 4 - E - \sum_i V_i \Delta_i, \tag{3.5.9}$$

where Δ_i is given by (3.5.2).

We now analyse the ultraviolet divergence of graphs. Suppose a graph with E external lines is divergent, with the degree of divergence $D \geq 0$. We can split its amplitude (3.5.3) into different divergent pieces and a finite part by expanding the integrand in a finite Taylor series in the variables $\{q\}$ around $\{0\}$, say [19, 20]. We write this series symbolically as

$$G(\{q\}, \{k\}) = G(\{k\}) + \{q\}\frac{\partial G}{\partial \{q\}}(\{k\}) + \cdots + \frac{1}{D!}\{q\}^D \frac{\partial^D G}{\partial \{q\}^D}(\{k\})$$
$$+ G_c(\{q\}, \{k\}) \tag{3.5.10}$$

where G_c is the remainder after $(D+1)$ terms, obtained by subtracting the latter terms from the full integrand G. Inserting this expression in (3.5.3), we get

$$F(\{q\}) = F^{(0)} + \{q\}F^{(1)} + \cdots + \{q\}^D F^{(D)} + F_c(\{q\}), \tag{3.5.11}$$

where the constants $F^{(0)}, F^{(1)}$, etc. are divergent integrals and the function F_c is in terms of convergent integrals. We have thus expressed the amplitude F as a polynomial of degree D in external momenta with (ultraviolet) divergent coefficients plus a (non-polynomial) finite piece.

The crucial point to note here is that the terms in the polynomial with divergent coefficients in (3.5.11) are of the same form as would be produced by different interactions i with $n_i = E$ fields and $d_i \leq D$ derivatives. So the divergent pieces in the Feynman amplitude simply add 'corrections' to the (bare) coupling constants g_i in (3.5.1). By including a suitable infinite term already in g_is, we can make the sum finite and set them at their measured values. (Divergence of graphs with two external lines are, however, not absorbed by any g_is, but by the bare mass and bare field.) The removal of divergences in this way leads to a *renormalisation* of the bare parameters and fields appearing in the original Lagrangian.

The terms in the Lagrangian are restricted by symmetry principles, such as Lorentz invariance, gauge invariance, etc. But these principles also limit in the same way the appearance of divergent parts of Feynman amplitudes. Thus, provided the interaction Lagrangian includes *all* the allowed terms, there will be a term in it to absorb every divergence, assuring the success of the renormalisation programme.

Two types of theories are distinguished. If all the interactions have $\Delta_i \geq 0$, then (3.5.9) gives an upper bound on D,

$$D \leq 4 - E\,. \tag{3.5.12}$$

Then the divergences arise from graphs with few enough external lines, whose elimination requires only a finite number of interaction terms to be present in the original Lagrangian. These theories are called *renormalisable*. If, on the other hand, any of the interactions has $\Delta_i < 0$, D becomes larger and larger as we insert more and more of such vertices, eventually becoming positive, however large we take E. So one is in need here of an infinite number of coupling constants to absorb all the infinities appearing in the perturbation series. These theories are called *non-renormalisable*. But the discussion above shows that these theories can be renormalised in the same way as the renormalisable ones. In either case, the renormalized amplitude is again given by (3.5.11), but with constants $F^{(0)}, F^{(1)}, \cdots$ now rendered finite. The values of these constants depend generally on the chosen scheme of renormalisation. Clearly the non-analytic pieces in $F_c(\{q\})$ do not depend on such schemes.

So far we have discussed (scalar) field theories in general. Returning to the effective theory at hand, we see that all interaction terms are non-renormalisable, starting from the lowest one in the series (3.4.4) with a coupling constant of dimension $\Delta = -2$. However, as discussed above, renormalisation renders all momentum integrals finite, like $F_c(\{q\})$ in (3.5.11), which are free from any scale, except the external momenta. We can now readily obtain the low energy behaviour for a general process involving an arbitrary number of incoming and outgoing pions. Consider a connected graph with degree of divergence D contributing to the process. Let the external momenta be at most of order Q. As D is also the dimension of the corresponding finite momentum integral with no other scales involved, its contribution is of order Q^ν, where ν just equals D given by (3.5.6) [10, 11]

$$\nu = \sum_i V_i(d_i - 2) + 2L + 2\,. \tag{3.5.13}$$

We see that each term in ν is positive: every interaction has at least two derivatives, $d_i \geq 2$ and, of course, $L \geq 0$. We now make the following observations:

i) The leading term of each process is of order Q^2 and arises solely from tree graphs ($L = 0$) constructed solely from the Lagrangian with two derivatives, $\mathcal{L}_{\text{eff}}^{(2)}$.

ii) The occurrence of loops in a graph reduces its magnitude at low energy, each loop suppressing it by two powers of momentum. This property is crucial for the effective Lagrangian approach to be a coherent framework. It implies that the low energy expansion we are dealing with is, in fact, a standard perturbation series, ordered according to the number of loops.

iii) How small should Q be for the perturbation expansion to be meaningful? It must be small compared with a typical energy scale, like the mass of the nucleon or the ρ. Such an estimate is also given by the factor originating from each additional loop, namely $Q^2/(16\pi^2 F^2)$. Thus the expansion should be useful as long as the pion energies are small compared to $4\pi F$.

3.6 Approximate Chiral Symmetry

Up till now we have considered the QCD Lagrangian with u and d quarks in the limit of vanishing masses, when it has the symmetry group $SU(2)_R \times SU(2)_L$, broken spontaneously to $SU(2)_V$ by the vacuum. We now introduce the quark mass term in the Lagrangian,

$$\mathcal{L}_{QCD} = \mathcal{L}_{QCD}^{(0)} - \bar{q}\mathcal{M}q, \qquad \mathcal{M} = \begin{pmatrix} m_u & 0 \\ 0 & m_d \end{pmatrix}. \qquad (3.6.1)$$

The mass term breaks explicitly the chiral symmetry of $\mathcal{L}_{QCD}^{(0)}$ by connecting the right- and left- handed quark fields,

$$\bar{q}\mathcal{M}q = \bar{q}_R\mathcal{M}q_L + \bar{q}_L\mathcal{M}^{\dagger}q_R. \qquad (3.6.2)$$

So the chiral symmetry of massless QCD is broken not only spontaneously by the vacuum, but also explicitly by the mass term. Writing it as

$$\bar{q}\mathcal{M}q = \frac{1}{2}(m_u + m_d)\bar{q}q + \frac{1}{2}(m_u - m_d)\bar{q}\tau_3 q, \qquad (3.6.3)$$

we see that it also breaks explicitly the isospin $SU(2)_V$ symmetry. This mass term can be treated as a symmetry breaking perturbation in the theory.

It is instructive here to compare the QCD Hamiltonian

$$H_{\text{qcd}} = H_0 + H_1, \qquad H_1 = \int d^3x (m_u \bar{u}u + m_d \bar{d}d), \qquad (3.6.4)$$

with that of a ferromagnet,

$$H = H'_0 - \sum_a \mu \vec{s}_a \cdot \vec{H},$$

where $\vec{s}_a$ denotes the spin associated with lattice a, μ the magnetic moment and $\vec{H}$ an external magnetic field [16]. The term H'_0 is invariant under the simultaneous $O(3)$ rotations of all spin variables, while the other term involving the external magnetic field breaks this symmetry. Thus the quark masses in QCD play a role analogous to the magnetic field and the quark condensate is analogous to magnetisation. In particular, spontaneous magnetisation at zero external field corresponds to a non-zero value of the quark condensate in the chiral limit m_u, $m_d \to 0$.

Now to the construction of the effective Lagrangian [12, 13]. The quark masses representing explicit symmetry breaking in QCD must be considered as external parameters and so should appear in terms for the effective Lagrangian. There is a simple technique to build such terms [17]. Suppose, for the moment, the external matrix-valued field $\mathcal{M}$ to transform as

$$\mathcal{M} \to V_R \mathcal{M} V_L^{\dagger}, \qquad (3.6.5)$$

along with

$$q_R \to V_R q_R, \quad q_L \to V_L q_L,$$

when the full QCD Lagrangian will remain invariant under $SU(2)_R \times SU(2)_L$. So in constructing the individual terms of the effective Lagrangian,

$$\mathscr{L}_{\mathrm{eff}} = \mathscr{L}_{\mathrm{eff}}(U, \partial U, \partial^2 U, \cdots, \mathcal{M}), \qquad (3.6.6)$$

we demand their invariance under the simultaneous transformations

$$U \to V_R U V_L^{\dagger}, \qquad \mathcal{M} \to V_R \mathcal{M} V_L^{\dagger}. \qquad (3.6.7)$$

After the construction is over, we restore $\mathcal{M}$ to its constant physical value.

The effect of the quark masses can now be analysed by expanding this $\mathscr{L}_{\mathrm{eff}}$ in powers of $\mathcal{M}$. The first term in this expansion is the piece (3.4.5) we found earlier. The next term in the expansion is linear in $\mathcal{M}$,

$$\mathscr{L}_{sb} = f(U, \partial U, \cdots) \times \mathcal{M}$$

where f may again be analysed in powers of derivatives. Thus in presence of explicit symmetry breaking, we are dealing with a double Taylor series, involving simultaneous expansion in powers of derivatives or momenta as well as of the symmetry breaking parameters $\mathcal{M}$. This is referred to as *chiral expansion* of the effective Lagrangian.

The leading piece in $\mathscr{L}_{sb}$ must be of the form $f(U) \times \mathcal{M}$ without derivatives. There are only two such independent invariants, namely $\mathrm{tr}\,(\mathcal{M} U^{\dagger})$ and its complex conjugate $\mathrm{tr}\,(U \mathcal{M}^{\dagger})$, giving

$$\mathscr{L}_{sb} = \frac{F^2}{2} \operatorname{tr}\left(B\mathscr{M}U^\dagger + B^* U \mathscr{M}^\dagger\right), \tag{3.6.8}$$

where we extract a factor of $F^2/2$ for later convenience. The symmetry breaking thus introduces a new low energy constant B. The QCD Lagrangian (3.6.1) is invariant under space reflection and we assume the same is true in the effective theory. Under this transformation $q_R(\boldsymbol{x},t) \leftrightarrow \gamma^0 q_L(-\boldsymbol{x},t)$ and $\phi(\boldsymbol{x},t) \to -\phi(-\boldsymbol{x},t)$, giving

$$U(\boldsymbol{x},t) \leftrightarrow U^\dagger(-\boldsymbol{x},t), \qquad \mathscr{M} \to \mathscr{M}^\dagger. \tag{3.6.9}$$

So the parity invariance of $\mathscr{L}_{sb}$ requires the constant B to be real. Thus working with a general quark mass matrix, we get

$$\mathscr{L}_{sb} = \frac{F^2 B}{2} \operatorname{tr}\left(\mathscr{M}U^\dagger + \mathscr{M}^\dagger U\right), \qquad \text{general } \mathscr{M}. \tag{3.6.10}$$

Specialising $\mathscr{M}$ to be real and diagonal, it reduces to

$$\begin{aligned}
\mathscr{L}_{sb} &= \frac{F^2 B}{2} \operatorname{tr}\left\{\mathscr{M}(U + U^\dagger)\right\}, \qquad \text{physical } \mathscr{M} \\
&= \frac{F^2 B}{4}(m_u + m_d)\operatorname{tr}\left(U + U^\dagger\right),
\end{aligned} \tag{3.6.11}$$

where the second equality follows from the fact that, since U is an element of $SU(2)$, $(U + U^\dagger)$ is proportional to the unit matrix. Thus the above symmetry breaking Lagrangian involves only the sum $(m_u + m_d)$ of the two quark masses and therefore conserves isospin. The isospin symmetry is of course broken by pieces beyond the leading one in the effective Lagrangian.

3.7 Results from Approximate Symmetry

We now derive some simple results following from the explicit symmetry breaking piece in the Lagrangian. Expanding U in powers of the pion field $\phi(x)$, we get

$$\mathscr{L}_{sb} = (m_u + m_d)B\left(F^2 - \frac{1}{2}\vec{\phi}\cdot\vec{\phi} + \frac{1}{24F^2}(\vec{\phi}\cdot\vec{\phi})^2 + \cdots\right). \tag{3.7.1}$$

We already ignored a constant in writing the effective Lagrangian (3.4.6), thereby defining the vacuum energy of the symmetric theory to be zero. Now the constant first term above represents the vacuum energy density due to the symmetry breaking perturbation. The second term, quadratic in the pion fields, is the pion mass term. The remaining terms go to modify the interaction among pions. We now derive the following results.

Quark Condensate and Pion Mass

The derivative of the QCD Hamiltonian with respect to $m_u(m_d)$ is the operator $\bar{u}u\,(\bar{d}d)$. So the corresponding derivative of the vacuum energy gives

$\langle 0|\bar{u}u|0\rangle$ ($\langle 0|\bar{d}d|0\rangle$). Equating this derivative to that of the effective Hamiltonian from (3.7.1), we get to leading order

$$\langle 0|\bar{u}u|0\rangle = \langle 0|\bar{d}d|0\rangle = -F^2 B. \qquad (3.7.2)$$

We see that the low energy constant B is related to the quark condensate.

The $\vec{\phi}\cdot\vec{\phi}$ term in (3.7.1) gives equal mass to all the components of the pion triplet,

$$m_{\pi^+}^2 = m_{\pi^-}^2 = m_{\pi^0}^2 = B(m_u + m_d) \equiv M^2 \qquad (3.7.3)$$

as we already remarked earlier. Eliminating B between (3.7.2) and (3.7.3), we get the Gell-Mann–Oakes–Renner formula [21],

$$F^2 M^2 = -(m_u + m_d)\langle 0|\bar{u}u|0\rangle. \qquad (3.7.4)$$

It shows that the pion mass is determined by the product of the quark mass and quark condensate, which measure respectively the explicit and spontaneous symmetry breakings.

The same relation is also obtained from (3.2.12), which follows from QCD and the assumed vacuum symmetry. To relate G appearing in it to the pion decay constant F, we find the divergence of the axial-vector current

$$\partial_\mu A_a^\mu(x) = (m_u + m_d)\bar{q}i\gamma^5 \frac{\tau_a}{2} q \qquad (3.7.5)$$

and take its matrix element between pion and vacuum. Using (3.2.9) and $(3.2.10)^5$ we get

$$M^2 F = \frac{1}{2}(m_u + m_d)G. \qquad (3.7.6)$$

Eliminating G between (3.2.12) and (3.7.6) we again get (3.7.4).

For an on-shell pion momentum p_μ, we have

$$p^2 = M^2 = (m_u + m_d)B. \qquad (3.7.7)$$

Thus in the simultaneous expansion of the effective Lagrangian in powers of the derivative and the quark mass, the latter should be counted like two powers of momentum. So the leading term (to order p^2) is given by the sum of (3.4.5) and (3.6.11),

$$\mathcal{L}_{\text{eff}}^{(2)} = \frac{F^2}{4}\,\text{tr}\,\{\partial_\mu U \partial^\mu U^\dagger + M^2(U + U^\dagger)\}. \qquad (3.7.8)$$

It involves two constants F and M, which we identified as the decay parameter and mass of pion to zeroth order in chiral perturbation theory. (The first-order correction to these quantities will be calculated in Section 3.10.) Inclusion of pion mass in the theory calls for an extension of the power counting rule (3.5.13). If

---

[5] We must distinguish physical values of F and G (quarks with masses) from those in the chiral limit (massless quarks). In the present formula, however, it does not matter, unless we work out pion mass to next order.

m_i factors of M^2 (or quark mass) appear in the interaction of type i, it must be rewritten as

$$\nu = \sum_i V_i(d_i + 2m_i - 2) + 2L + 2. \tag{3.7.9}$$

$$* \quad * \quad *$$

As isolated quarks do not exist, it is not possible to define their masses as pole positions in propagators. It is treated like a coupling constant, when it becomes a running quark mass $m_q(\mu)$, whose dependence on the scale μ is governed by a renormalisation group equation. Reliable values of $m_q(\mu)$ are obtained from calculations on the lattice. At a scale of 2 GeV, the values obtained are [22]

$$m_u(2\,\mathrm{GeV}) = 2.2\,\mathrm{MeV}, \qquad m_d(2\,\mathrm{GeV}) = 4.7\,\mathrm{MeV} \tag{3.7.10}$$

With $M = 140\,\mathrm{MeV}$ and $F = 92\,\mathrm{MeV}$ (see Appendix F), (3.7.4) gives for the quark condensate

$$\langle 0|\bar{u}u|0\rangle(2\,\mathrm{GeV}) = \langle 0|\bar{d}d|0\rangle(2\,\mathrm{GeV}) = -(290\,\mathrm{MeV})^3. \tag{3.7.11}$$

We see that the masses of u and d quarks are not at all degenerate. But as we pointed out at the end of Section 3.6, the leading term (3.6.11) in the effective Lagrangian with chiral symmetry breaking *is* isospin symmetric; it is only the non-leading terms which break it. So $SU(2)_V$ is a good symmetry of strong interaction, not because the masses of u and d quarks are equal, but because they are small.

Pion–Pion Scattering

Given the effective Lagrangian (3.7.8) we can work out the $\pi\pi$ scattering amplitude to leading order. As usual the S matrix operator is

$$S = Te^{i\int d^4x\,\mathscr{L}_{\mathrm{int}}(x)} = 1 + iT\,.$$

The required matrix element is (Figure 3.1)

$$\langle \pi_c(k'), \pi_d(q') | S - 1 | \pi_a(k), \pi_b(q)\rangle = i(2\pi)^4\delta^4(k' + q' - k - q)M_{cd,ab}(k'q' \to kq), \tag{3.7.12}$$

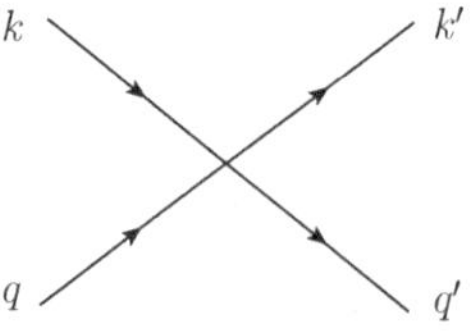

Figure 3.1 $\pi\pi$ scattering to leading order.

giving the invariant amplitude $M_{cd,ab}$, where a, b, c, d are the isospin indices of pions[6].

We expand the effective Lagrangian (3.7.8) to $O(\phi^4)$ to find this amplitude. We write

$$\text{tr}\,(\partial_\mu U \partial^\mu U^\dagger) = \text{tr}\,(\partial_\mu U U^\dagger U \partial^\mu U^\dagger) = -\,\text{tr}\,(U^\dagger \partial_\mu U U^\dagger \partial^\mu U). \qquad (3.7.13)$$

Let $U = \exp(i\psi)$, where $\psi = \vec{\tau} \cdot \vec{\phi}/F$. Then (C.10) of Appendix C gives

$$U^\dagger \partial_\mu U = i\partial_\mu \psi + \frac{1}{2}[\psi, \partial_\mu \psi] - \frac{i}{3!}[\psi, [\psi, \partial_\mu \psi]] + \cdots \qquad (3.7.14)$$

Multiplying two such factors and taking the trace over Pauli matrices, we get the expansion of the first term of (3.7.8). The second term is already evaluated in (3.7.1). We thus get

$$\mathscr{L}_{\text{eff}}^{(2)} = \mathscr{L}_0 + \mathscr{L}_{\text{int}}^{(2)}$$

where

$$\mathscr{L}_0 = \frac{1}{2}\partial_\mu \vec{\phi} \cdot \partial^\mu \vec{\phi} - \frac{M^2}{2}\vec{\phi} \cdot \vec{\phi}$$

$$\mathscr{L}_{\text{int}}^{(2)} = \frac{1}{6F^2}\left\{ \vec{\phi} \cdot \partial_\mu \vec{\phi}\, \vec{\phi} \cdot \partial^\mu \vec{\phi} - \vec{\phi} \cdot \vec{\phi}\, \partial_\mu \vec{\phi} \cdot \partial^\mu \vec{\phi} + \frac{M^2}{4}(\vec{\phi} \cdot \vec{\phi})^2 \right\} + O(\phi^6).$$

$$(3.7.15)$$

The scattering amplitude in (3.7.12) is then given by

$$i\int d^4x \langle \pi^c(k'), \pi^d(q')|T\mathscr{L}_{\text{int}}^{(2)}(x)|\pi^a(k), \pi^b(q)\rangle. \qquad (3.7.16)$$

It depends on two independent variables[7], which we choose as the Mandelstam variables, $s = (k+q)^2$, $t = (k-k')^2$, $u = (k-q')^2$; these are linearly related, $s+t+u = 4M^2$, as can be shown by using the energy-momentum conservation, $k' + q' = k + q$. The scattering amplitude following from (3.7.16) may now be written as

$$M_{cd,ab} = \frac{1}{F^2}\left\{ \delta_{ab}\delta_{cd}(s - M^2) + \delta_{ac}\delta_{bd}(t - M^2) + \delta_{ad}\delta_{bc}(u - M^2) \right\} \qquad (3.7.17)$$

---

[6] There are three sources of factors of i in a Feynman graph, namely the vertices (from $i\int \mathscr{L}_{\text{int}}(x)d^4x$), internal lines (from field contraction) and loop integrations (from Wick rotation of loop momenta). Thus in the notation of Section 3.5 a Feynman graph amplitude has factors of i as

$$i^V \cdot i^I \cdot i^L = i^{2I} i^{L-I+V} = \pm i$$

the last equality following from the topological identity (3.5.5). It is this i which appears in equations like (3.7.12). After removing it we get a real amplitude as long as all propagators are off their mass shells. The amplitude will develop imaginary parts, when some or all the propagators are put on mass shells, as we shall see in Chapter 5.

[7] This is true for any scattering amplitude of two particles going to two particles. To begin with the amplitude depends on 16 variables, namely the components of four four-momenta of the particles. The ten generators of the inhomogeneous Lorentz group can be used to set ten of them to zero. Also the four-momentum conservation gives four relations among the components. We are thus finally left with $16 - 10 - 4 = 2$ independent momentum components, which are conveniently chosen as Lorentz scalars.

displaying isospin and crossing symmetries. Thus the leading terms in the amplitude are quadratic in momenta and pion mass, as follows from the power counting rule (3.7.9). The scattering lengths given by this amplitude are consistent with experiment [11].

3.8 External Fields

Having derived the effective Lagrangian, we can calculate scattering amplitudes involving any number of pions by evaluating the corresponding Feynman graphs. However, Green (correlation) functions of quark currents and densities are also of much importance in extracting physical consequences from QCD theory. One may think of first expressing these composite quark operators in terms of fields in the effective theory and then working with them. But the correlation functions may involve more than one such operator at the same space–time point, thus also requiring their own effective field representations. An elegant way to avoid these problems is to use the method of external fields [12].

We already described the use of external fields in Section 2.2. There the classical external fields were coupled to elementary quantum fields present in the Lagrangian. Now we introduce external fields by coupling them to composite fields, namely the quark currents and densities. Thus we extend the QCD Lagrangian to

$$\mathscr{L}_{QCD} \longrightarrow \mathscr{L}_{ext} = \mathscr{L}_{QCD}^{(0)} + \bar{q}\gamma^{\mu}(v_{\mu} + a_{\mu}\gamma^{5})q - \bar{q}(s - ip\gamma^{5})q \qquad (3.8.1)$$

where $\mathscr{L}_{QCD}^{(0)}$ is given by (3.1.1) and $v_{\mu}(x)$, $a_{\mu}(x)$, $s(x)$, $p(x)$ are hermitian matrices of external fields in flavour space; the first two

$$v_{\mu} = \vec{v}_{\mu} \cdot \frac{\vec{\tau}}{2}, \qquad a_{\mu} = \vec{a}_{\mu} \cdot \frac{\vec{\tau}}{2}, \qquad (3.8.2)$$

couple to the vector and axial-vector currents given by (3.1.11), while the last two

$$s = s^{0}\mathbf{1} + \vec{s} \cdot \vec{\tau}, \qquad p = p^{0}\mathbf{1} + \vec{p} \cdot \vec{\tau}, \qquad (3.8.3)$$

couple to the scalar and pseudo-scalar densities

$$S^{0} = \bar{q}q, \quad \vec{S} = \bar{q}\vec{\tau}q; \qquad P^{0} = \bar{q}i\gamma^{5}q, \quad \vec{P} = \bar{q}i\gamma^{5}\vec{\tau}q. \qquad (3.8.4)$$

(As we restrict the symmetry group G to $SU(2)_{R} \times SU(2)_{L}$, the isoscalar vector and axial-vector currents are left out in (3.8.2).) Written out in components, (8.3.1) becomes

$$\mathscr{L}_{ext} = \mathscr{L}_{QCD}^{(0)} + v_{\mu\,i}V_{i}^{\mu} + a_{\mu\,i}A_{i}^{\mu} - s^{0}S^{0} - \vec{s} \cdot \vec{S} + p^{0}P^{0} + \vec{p} \cdot \vec{P} \qquad (3.8.5)$$

Here the quark mass term in $\mathscr{L}_{QCD}$ is included in the scalar densities,

$$s^{0}(x) = (m_{u} + m_{d})/2 + \cdots, \qquad s_{3}(x) = (m_{u} - m_{d})/2 + \cdots. \qquad (3.8.6)$$

Earlier in Section 3.6 we imagined the mass term to transform as (3.6.5), when the full QCD Lagrangian becomes symmetric under G. In presence of external vector and axial-vector and scalar and pseudoscalar fields, this symmetry can now be promoted to a local one [12, 13]. To find the appropriate transformation rules, we decompose the quark field into its right- and left-handed components, when (3.8.1) becomes

$$\mathscr{L}_{ext} = i\bar{q}_R\gamma^\mu\{\partial_\mu - i(v_\mu + a_\mu)\}q_R + i\bar{q}_L\gamma^\mu\{\partial_\mu - i(v_\mu - a_\mu)\}q_L$$
$$-\bar{q}_R(s + ip)q_L - \bar{q}_L(s - ip)q_R. \tag{3.8.7}$$

Let the quark field transform locally as

$$q_R(x) \to q'_R(x) = V_R(x)q_R(x), \quad q_L(x) \to q'_L(x) = V_L(x)q_L(x), \tag{3.8.8}$$

where the group parameters in V_R and V_L are now coordinate dependent. If we also subject the external fields to the corresponding gauge transformations,

$$v_\mu + a_\mu \to v'_\mu + a'_\mu = V_R(v_\mu + a_\mu)V_R^\dagger + iV_R\partial_\mu V_R^\dagger$$
$$v_\mu - a_\mu \to v'_\mu - a'_\mu = V_L(v_\mu - a_\mu)V_L^\dagger + iV_L\partial_\mu V_L^\dagger$$
$$s + ip \to s' + ip' = V_R(s + ip)V_L^\dagger \tag{3.8.9}$$

the Lagrangian (3.8.7) remains invariant. Expanding V_R and V_L given by (3.1.4) for infinitesimal parameters θ_R and θ_L, (3.8.9) gives the infinitesimal transformations of the external fields as

$$\delta v_\mu = \partial_\mu\theta_V - i[v_\mu, \theta_V] - i[a_\mu, \theta_A]$$
$$\delta a_\mu = \partial_\mu\theta_A - i[v_\mu, \theta_A] - i[a_\mu, \theta_V]$$
$$\delta s = -i[s, \theta_V] - \{p, \theta_A\}$$
$$\delta p = -i[p, \theta_V] + \{s, \theta_A\}. \tag{3.8.10}$$

Let us analyze the transformation rules for the scalar and pseudoscalar external fields in the infinitesimal form. Equating the real and imaginary parts of the coefficients of $\mathbf{1}$ and $\vec{\tau}$ on both sides in the last two equations of (3.8.10), we get their transformation rules in four pieces,

$$\delta s^0 = -\vec{\theta}_A \cdot \vec{p}, \quad \delta\vec{p} = -\vec{\theta}_V \times \vec{p} + \vec{\theta}_A s^0$$
$$\delta p^0 = \vec{\theta}_A \cdot \vec{s}, \quad \delta\vec{s} = -\vec{\theta}_V \times \vec{s} - \vec{\theta}_A p^0 .$$

On comparing these with those for an $O(4)$ four-vector given by (2.5.5) and (2.5.6), we see that

$$h^+ \equiv (s^0, \vec{p}) \qquad \text{and} \qquad h^- \equiv (p^0, -\vec{s}) \tag{3.8.11}$$

transform as two independent $O(4)$ four-vectors. We see that the matrices s and p defined by (3.8.3) mix the two independent four-vectors. We can undo the mixing by replacing s and p with χ and $\tilde{\chi}$ defined as [12, 13]

$$\chi = 2B(s^0\mathbf{1} + i\vec{p}\cdot\vec{\tau}), \quad \tilde{\chi} = 2B(p^0\mathbf{1} - i\vec{s}\cdot\vec{\tau}) \tag{3.8.12}$$

which transform independently as

$$(\chi,\ \widetilde{\chi}) \to (\chi',\ \widetilde{\chi}') = V_R(\chi,\ \widetilde{\chi})V_L^\dagger\,. \tag{3.8.13}$$

The factor $2B$ in (3.8.12) is inserted for convenience. The transition from four-vectors $h^\pm$ to 2×2 matrices $\chi,\ \widetilde{\chi}$ illustrates the isomorphism of $O(4)$ and $SU(2)\times SU(2)$.

We now turn to the effective theory. The form of the transformation rule for the matrix field $U(x)$ remains the same,

$$U(x) \to U'(x) = V_R(x)U(x)V_L^\dagger(x) \tag{3.8.14}$$

with the group parameters being space–time dependent. The external fields in the underlying theory are readily carried over to the effective theory. The vector and axial-vector fields appear through the covariant derivative of U,

$$D_\mu U = \partial_\mu U - i(v_\mu + a_\mu)U + iU(v_\mu - a_\mu), \tag{3.8.15}$$

and their field strengths,

$$F_{R,L}^{\mu\nu} = \partial^\mu(v^\nu \pm a^\nu) - \partial^\nu(v^\mu \pm a^\mu) - i[v^\mu \pm a^\mu, v^\nu \pm a^\nu]. \tag{3.8.16}$$

It is simple to check that $D_\mu U$ transforms in the same way as U,

$$D_\mu U \to V_R(x)D_\mu U V_L^\dagger(x), \tag{3.8.17}$$

and the external field strengths as

$$F_{R,L}^{\mu\nu} \to F_{R,L}'^{\mu\nu} = V_{R,L}(x)F_{R,L}^{\mu\nu}V_{R,L}^\dagger(x)\,. \tag{3.8.18}$$

Gauge invariance implies that the external fields v_μ and a_μ can appear in the effective Lagrangian only through the covariant derivatives and field strengths.

Thus the most general effective Lagrangian has the form

$$\mathscr{L}_{\text{eff}} = \mathscr{L}_{\text{eff}}(U, D_\mu U, D_\mu D_\nu U, F_{R,L}^{\mu\nu}, \chi, \widetilde{\chi}, \cdots), \tag{3.8.19}$$

where each term is invariant under the above transformations of the symmetry group and also under parity

$$U \to U^\dagger, \quad \chi \to \chi^\dagger, \quad \widetilde{\chi} \to -\widetilde{\chi}^\dagger \quad F_{\mu\nu}^R \to F_{\mu\nu}^L, \tag{3.8.20}$$

with the arguments $(\boldsymbol{x}, t)$ changed to $(-\boldsymbol{x}, t)$. This extends (3.6.9) in presence of external fields. Earlier we ordered terms in the Lagrangian according to the number of derivatives and quark masses in it. This counting also extends in presence of these fields. We count the field $U(x)$ as a quantity of order p^0; v_μ, a_μ and $D_\mu U$ are of order p^1, and χ, $\widetilde{\chi}$ and $F_{\mu\nu}$ as order p^2, etc.

In presence of external fields, the earlier power counting formula (3.5.13) still holds, provided we count only the derivatives (d_i in the i-th vertex) involving external fields. If we also count the number of external fields e_i in the i-th vertex, we can introduce its order O_i as $O_i = d_i + e_i$. Noting $\sum_i e_i = n$ for an n-point function, we can eliminate d_i in (3.5.13) in terms of O_i to get

$$D = \sum_i V_i(O_i - 2) + 2L + 2 - n. \qquad (3.8.21)$$

The form of $\mathscr{L}_{\text{eff}}^{(2)}$ found earlier in Section 3.6 remains essentially unaltered by the introduction of external fields. Besides replacing ordinary derivatives by the covariant ones, we have to look for possible 'mass' terms. There are four invariants with definite parity that are linear in χ or $\widetilde{\chi}$,

$$\text{tr}\,(\chi U^\dagger \pm \chi^\dagger U), \qquad \text{tr}\,(\widetilde{\chi} U^\dagger \pm \widetilde{\chi}^\dagger U).$$

Requiring parity conservation we are left with $\text{tr}\,(\chi U^\dagger + \chi^\dagger U)$ and $\text{tr}\,(\widetilde{\chi} U^\dagger - \widetilde{\chi}^\dagger U)$, of which the latter is identically zero, as can be seen by inserting in it the expression (3.8.12) for $\widetilde{\chi}$. Also terms linear in $F_{\mu\nu}^R$ and $F_{\mu\nu}^L$ do not appear by Lorentz invariance. We thus get in presence of external fields

$$\mathscr{L}_{\text{eff}}^{(2)} = \frac{F^2}{4}\,\text{tr}\,(D_\mu U D^\mu U^\dagger + \chi U^\dagger + U\chi^\dagger). \qquad (3.8.22)$$

Clearly the form of the earlier equation of motion for U also remains the same.

$$* \quad * \quad *$$

We can extend the symmetry breaking piece in the Hamiltonian (3.6.4) to the case where external scalar and pseudoscalar fields are present. If the last four terms in (3.8.5) constitute $O(4)$ scalars with the four-vectors h^+ and h^- given by (3.8.11), then

$$\Phi^+ \equiv (S_0, -\vec{P}) \qquad \text{and} \qquad \Phi^- \equiv (P_0, \vec{S}) \qquad (3.8.23)$$

must also be two independent four-vectors, constructed out of quark fields. (It can also be checked directly by using the transformation rules for the quark fields. See Section 3.2.) Then (3.6.4) generalises to

$$H_1 = 2\int d^3x(h^+\Phi^+ - h^-\Phi^-).$$

3.9 Fourth Order Effective Lagrangian

As we saw in Section 3.5 all S-matrix elements (involving pions) can be calculated to leading (second) order in pion mass and momenta from tree graphs with vertices from $\mathscr{L}_{\text{eff}}^{(2)}$. To calculate quantum corrections to it, we have to include the next higher (fourth) order contributions from one-loop graphs, again with vertices from $\mathscr{L}_{\text{eff}}^{(2)}$. The momentum integrals over the loops would produce divergences. As the theory is non-renormalisable, we would require a new Lagrangian $\mathscr{L}_{\text{eff}}^{(4)}$ giving all vertices of order four to remove these divergences. In principle, the process can be continued to higher loops, encountering in each successive order more and more unknown parameters.

We now derive the most general Lagrangian $\mathscr{L}_{\text{eff}}^{(4)}$. Let us begin with terms containing only the first derivatives of U and no χ or $\widetilde{\chi}$. To find the number of independent terms in this case, we apply the same technique as we did for $\mathscr{L}_{\text{eff}}^{(2)}$. Ignoring Lorentz structure for the moment, they are of the form

$$f(U) \times \partial_\mu U \times \partial_\nu U \times \partial_\lambda U \times \partial_\sigma U$$

which we rewrite as before as

$$f'(U) \times \Delta_\mu \times \Delta_\nu \times \Delta_\lambda \times \Delta_\sigma$$

where $\Delta_\mu = U^\dagger \partial_\mu U$, as was introduced in Section 3.4. The number of independent scalars can be found by counting the singlets present in the decomposition of $D^{(1)} \times D^{(1)} \times D^{(1)} \times D^{(1)}$, which are three in number. These may be chosen as

$$\text{tr}\,(\Delta_\mu \Delta_\nu)\,\text{tr}\,(\Delta_\lambda \Delta_\sigma),\ \ \text{tr}\,(\Delta_\mu \Delta_\lambda)\,\text{tr}\,(\Delta_\nu \Delta_\sigma),\ \ \text{tr}\,(\Delta_\mu \Delta_\sigma)\,\text{tr}\,(\Delta_\nu \Delta_\lambda).$$

Requiring now that these terms be Lorentz scalars, this number reduces to two. Changing over to $\partial_\mu U$ notation, they are[8]

$$P_1 = \{\text{tr}\,(\partial_\mu U^\dagger \partial^\mu U)\}^2, \tag{3.9.1}$$

$$P_2 = \text{tr}\,(\partial_\mu U^\dagger \partial_\nu U)\,\text{tr}\,(\partial^\mu U^\dagger \partial^\nu U). \tag{3.9.2}$$

Before we proceed further we note two identities that follow from the classical equation of motion for $\mathscr{L}_{\text{eff}}^{(2)}$ relating some of the invariants. As derived in Problem 3.2 it reads

$$2\Box U + 2U\partial_\mu U^\dagger \partial^\mu U - \chi + U\chi^\dagger U = 0.$$

One of the identities may be obtained by multiplying this equation with $\Box U^\dagger$ from the left to get $\Box U^\dagger \Box U$ and other terms proportional to $\Box U^\dagger$, wherein we again use the adjoint of this equation to remove $\Box U^\dagger$. On taking trace we get,

$$\text{tr}\,(\Box U^\dagger \Box U) = \frac{1}{2}\{\text{tr}\,(\partial^\mu U^\dagger \partial_\mu U)\}^2 - \frac{1}{8}\{\text{tr}\,(\chi U^\dagger + \chi^\dagger U\}^2 + \text{tr}\,(\chi^\dagger \chi) \tag{3.9.3}$$

---

[8] Let us check that the invariants with a single trace can be expressed in terms of P_1 and P_2. Here we need the identity

$$\text{tr}\,(\Delta_\mu \Delta_\nu \Delta_\lambda \Delta_\sigma)$$
$$= \frac{1}{2}\{\text{tr}\,(\Delta_\mu \Delta_\nu)\,\text{tr}\,(\Delta_\lambda \Delta_\sigma) - \text{tr}\,(\Delta_\mu \Delta_\lambda)\,\text{tr}\,(\Delta_\nu \Delta_\sigma) + \text{tr}\,(\Delta_\mu \Delta_\sigma)\,\text{tr}\,(\Delta_\nu \Delta_\lambda)\}$$

which follows if we write $\Delta_\mu = \Delta_\mu^a \tau^a$ (as Δ_μ is traceless) and use anticommutation relation for Pauli matrices. It gives

$$\text{tr}\,(\partial_\mu U^\dagger \partial^\mu U \partial_\nu U^\dagger \partial^\nu U) = \frac{1}{2}P_1 \qquad \text{tr}\,(\partial_\mu U^\dagger \partial_\nu U \partial^\mu U^\dagger \partial^\nu U) = P_2 - \frac{1}{2}P_1\,.$$

where we use[9]

$$\mathrm{tr}\,(\chi U^\dagger \chi U^\dagger + U\chi^\dagger U\chi^\dagger) = \frac{1}{2}\{\mathrm{tr}\,(\chi U^\dagger + \chi^\dagger U)\}^2 - 2\,\mathrm{tr}\,(\chi^\dagger \chi). \tag{3.9.4}$$

The other identity is obtained by multiplying the classical equation with $\chi^\dagger$ from the left and taking the sum of the resulting equation and its adjoint,

$$\mathrm{tr}\,(\partial^\mu \chi^\dagger \partial_\mu U + \partial^\mu \chi \partial_\mu U^\dagger) - \frac{1}{2}\,\mathrm{tr}\,(\partial^\mu U^\dagger \partial_\mu U)\,\mathrm{tr}\,(\chi U^\dagger + \chi^\dagger U)$$

$$= \frac{1}{4}\{\mathrm{tr}\,(\chi U^\dagger + \chi^\dagger U)\}^2 - 2\,\mathrm{tr}\,(\chi^\dagger \chi). \tag{3.9.5}$$

Though classical, these identities may be used to eliminate two of the invariants, as any quantum correction to these will give rise to terms of order six and higher, which do not concern us here. It is interesting to note that the relations (3.9.3) and (3.9.5) also result from a suitable redefinition of the field variable U [16].

Returning to enumeration of the invariants, we now consider those containing $\Box U$. These are

$$\mathrm{tr}\,(U^\dagger \Box U \partial^\mu U^\dagger \partial_\mu U)$$

$$\mathrm{tr}\,(\Box U \Box U^\dagger).$$

(Invariants with $\partial_\mu \partial_\nu U$ can be reduced to these, up to total derivatives.) We remove these invariants by using the equation of motion or the identity (3.9.3) in terms of the others.

Let us next consider the set of invariants that are linear in χ,

$$\mathrm{tr}\,(\partial^\mu U^\dagger \partial_\mu U)\,\mathrm{tr}\,(\chi U^\dagger \pm U\chi^\dagger)$$

$$\mathrm{tr}\,(\partial^\mu \chi \partial_\mu U^\dagger \pm \partial^\mu \chi^\dagger \partial_\mu U).$$

(A trace like $\mathrm{tr}\,\{\partial^\mu U^\dagger \partial_\mu U(\chi^\dagger U \pm U^\dagger \chi)\}$ reduces to the product of two traces.) Retaining those that are even under parity, we get two more independent invariants,

$$P_3 = \mathrm{tr}\,(\partial^\mu U^\dagger \partial_\mu U)\,\mathrm{tr}\,(\chi U^\dagger + U\chi^\dagger) \tag{3.9.6}$$

$$P_4 = \mathrm{tr}\,(\partial^\mu \chi \partial_\mu U^\dagger + \partial^\mu \chi^\dagger \partial_\mu U). \tag{3.9.7}$$

A similar set linear in $\widetilde{\chi}$ does not contribute, as its elements are either of odd parity or identically zero. Next, independent invariants quadratic in χ or $\widetilde{\chi}$ are given by

$$P_5 = \{\mathrm{tr}\,(\chi^\dagger U + U^\dagger \chi)\}^2 \tag{3.9.8}$$

$$P_6 = \mathrm{tr}\,(\chi\chi^\dagger) \tag{3.9.9}$$

$$P_7 = \{\mathrm{tr}\,(\widetilde{\chi}^\dagger U + U^\dagger \widetilde{\chi})\}^2 \tag{3.9.10}$$

$$P_8 = \mathrm{tr}\,(\widetilde{\chi}\widetilde{\chi}^\dagger). \tag{3.9.11}$$

---

[9] Note that $\chi U^\dagger + U\chi^\dagger$, like $\partial_\mu U^\dagger \partial^\mu U$, is proportional to unit matrix.

Finally there are three terms involving $F_{\mu\nu}^{R,L}$,

$$P_9 = \text{tr}\,(U^\dagger F_{\mu\nu}^R U F^{L\mu\nu}) \tag{3.9.12}$$

$$P_{10} = \text{tr}\,(F_{\mu\nu}^R \partial^\mu U \partial^\nu U^\dagger + F_{\mu\nu}^L \partial^\mu U^\dagger \partial^\nu U) \tag{3.9.13}$$

$$P_{11} = \text{tr}\,(F_{\mu\nu}^R F^{R\mu\nu} + F_{\mu\nu}^L F^{L\mu\nu}). \tag{3.9.14}$$

We thus get eleven invariants, which are independent up to the linear relation (3.9.5), not used so far, which we rewrite in the above notation as

$$P_4 - \frac{1}{2}P_3 = \frac{1}{4}P_5 - 2P_6.$$

One can choose to eliminate P_3 and write the fourth order effective Lagrangian as

$$\begin{aligned}
\mathscr{L}_{\text{eff}}^{(4)} &= \frac{l_1}{4}P_1 + \frac{l_2}{4}P_2 + \frac{l_3}{16}P_5 + \frac{l_4}{4}P_4 + \frac{l_5}{2}P_9 \\
&\quad + \frac{l_6}{4}P_{10} + \frac{l_7}{16}P_7 + \frac{h_1}{2}P_6 + \frac{h_2}{4}P_{11} + \frac{h_3}{2}P_8
\end{aligned} \tag{3.9.15}$$

with coupling constants l_i and h_i in agreement with [12]. Equivalently[10] we can eliminate P_4 with (3.9.5) and rewrite (3.9.15) as [13, 23]

$$\begin{aligned}
\mathscr{L}_{\text{eff}}^{(4)} &= \frac{l_1}{4}\{\text{tr}\,(D^\mu U^\dagger D_\mu U)\}^2 + \frac{l_2}{4}\,\text{tr}\,(D^\mu U^\dagger D^\nu U)\,\text{tr}\,(D_\mu U^\dagger D_\nu U) \\
&\quad + \frac{l_3 + l_4}{16}\{\text{tr}\,(\chi U^\dagger + U\chi^\dagger)\}^2 + \frac{l_4}{8}\,\text{tr}\,(D^\mu U^\dagger D_\mu U)\,\text{tr}\,(\chi U^\dagger + U\chi^\dagger) \\
&\quad + \frac{l_5}{2}\,\text{tr}\,(U^\dagger F_{\mu\nu}^R U F^{L\mu\nu}) + \frac{l_6}{4}\,\text{tr}\,(F_{\mu\nu}^R D^\mu U D^\nu U^\dagger + F_{\mu\nu}^L D^\mu U^\dagger D^\nu U) \\
&\quad + \frac{l_7}{16}\{\text{tr}\,(\tilde{\chi}^\dagger U + U^\dagger \tilde{\chi})\}^2 + \frac{h_1 - l_4}{2}\,\text{tr}\,(\chi\chi^\dagger) \\
&\quad + \frac{h_2}{4}\,\text{tr}\,(F_{\mu\nu}^R F^{R\mu\nu} + F_{\mu\nu}^L F^{L\mu\nu}) + \frac{h_3}{2}\,\text{tr}\,(\tilde{\chi}\tilde{\chi}^\dagger)
\end{aligned} \tag{3.9.16}$$

where we also replace ordinary derivatives with covariant ones to include external vector and axial-vector fields. In the following we shall use this form of $\mathscr{L}_{\text{eff}}^{(4)}$. The coupling constants appear in renormalising the one-loop corrections to results at tree level. The low energy constants l_i can be determined from experiments. By contrast, the so-called high-energy constants h_i, not associated with particle fields, cannot be estimated from physical amplitudes at low energy.

---

[10] As far as calculations at one loop are concerned, the two versions of $\mathscr{L}_{\text{eff}}^{(4)}$ are strictly equivalent. But at two loops, the construction of $\mathscr{L}_{\text{eff}}^{(6)}$ will depend on how the terms of $\mathscr{L}_{\text{eff}}^{(4)}$ are written, because in getting one version from the other, we make use of the classical equation of motion (at the tree level), whose quantum corrections would contribute to $\mathscr{L}_{\text{eff}}^{(6)}$.

3.10 One-Loop Results (Pions Only)

We now use the effective theory to calculate correlation functions of currents and densities of the underlying theory, namely QCD. The link between the two theories is provided by the equality of their generating functionals [14]. In their original work, Gasser and Leutwyler evaluated the path integral with $\mathscr{L}_{\text{eff}}^{(2)}$ after expanding it around the classical solution in presence of all the external fields [12, 13]. Technically difficult as it is, this procedure yields simultaneously the complete one-loop result for all correlation functions of different currents and densities. Also they get the divergences by expanding the heat kernel at short distance.

Here we shall evaluate the correlation functions separately, one at a time, equating the generating functionals of QCD and effective theory in presence of a single external field. Taking functional derivatives with respect to the external field, we immediately get the correlation function of the corresponding quark current or density in the effective theory in the Heisenberg representation. We then follow the conventional method of evaluating Feynman graphs in the interaction representation.

For our work it will suffice to restrict the external fields to $s^0(x)$, $v^\mu(x)$ and $a^\mu(x)$ (in the covariant derivatives), so that $\chi = M^2 1$, $\tilde{\chi} = 0$, $F_{\mu\nu}^{R,L} = 0$. Then the effective Lagrangian simplifies to

$$\mathscr{L}_{\text{eff}}^{(2)} = \frac{F^2}{4} \operatorname{tr}\left\{ D_\mu U D^\mu U^\dagger + M^2(U + U^\dagger) \right\} \tag{3.10.1}$$

$$\mathscr{L}_{\text{eff}}^{(4)} = \frac{l_1}{4} \{\operatorname{tr}(D_\mu U^\dagger D^\mu U)\}^2 + \frac{l_2}{4} \operatorname{tr}(D_\mu U^\dagger D_\nu U)\operatorname{tr}(D^\mu U^\dagger D^\nu U)$$

$$+ \frac{l_4}{8} M^2 \operatorname{tr}(D_\mu U D^\mu U^\dagger)\operatorname{tr}(U + U^\dagger) + \frac{1}{16}(l_3 + l_4) M^4 \{\operatorname{tr}(U + U^\dagger)\}^2$$

$$+ (h_1 - l_4)M^4. \tag{3.10.2}$$

Here $l_1, \cdots, l_4$ are bare coupling constants. The (one-)loop integrals turn out to be divergent, which can be regulated by dimension d (see Appendix D). The divergences are proportional to λ defined by (D.13), which has a pole at $d = 4$. In the $\overline{MS}$ scheme, we have

$$l_i = l_i^r + \gamma_i \lambda, \quad i = 1, 2, 3, 4, \quad \gamma_1 = \frac{1}{3}, \quad \gamma_2 = \frac{2}{3}, \quad \gamma_3 = -\frac{1}{2}, \quad \gamma_4 = 2, \tag{3.10.3}$$

where l_i^r are the renormalised coupling constants. It is not necessary to know a priori these divergences; a specific loop calculation including the appropriate counterterm with the bare constant will show what divergence is required in it to cancel the loop divergence. The renormalised constants depend on the renormalisation scale μ appearing in (D.11). Following [12], we may replace them with the scale-independent ones, denoted by a bar

$$l_i^r = \frac{\gamma_i}{32\pi^2}(\bar{l}_i + \ln \frac{M^2}{\mu^2}), \quad i = 1, 2, 3, 4. \tag{3.10.4}$$

Note that $\bar{l}_i$ does not exist in the chiral limit ($\bar{l}_i \to -\ln M^2$ as $M \to 0$).

Reliable values for the low energy constants follow from matching the chiral perturbation theory result for $\pi\pi$ scattering amplitude with its dispersion theoretic phenomenological representation [24]. To one loop these are obtained as

$$\bar{l}_1 = -1.9 \pm 0.2, \quad \bar{l}_2 = 5.25 \pm 0.04, \quad \bar{l}_3 = 2.9 \pm 2.4, \quad \bar{l}_4 = 4.3 \pm 0.9. \quad (3.10.5)$$

(These constants are also obtained to two loops, but the one-loop result will suffice for us.)

Axial-Vector Correlator

We work out the correction to the pion pole term in the two-point function of axial-vector current

$$T_{\mu\nu}^{ab}(k) = i \int d^4x \, e^{ik\cdot x} \, \langle 0| \, T \, A_\mu^a(x) \, A_\nu^b(0) \, |0\rangle \equiv \delta^{ab}\{k_\mu k_\nu T(k^2) + R_{\mu\nu}\} \quad (3.10.6)$$

where $R_{\mu\nu}$ denotes the remaining amplitude. Retaining only the axial-vector as the external field, we write the QCD and effective Lagrangians respectively as

$$\mathscr{L} = \mathscr{L}_{QCD} + \vec{a}^\mu \cdot \vec{A}_\mu, \quad \mathscr{L}_{\text{eff}} = \mathscr{L}_0 + \mathscr{L}_{\text{int}}^{(2)} + \mathscr{L}_{\text{int}}^{(4)} + \vec{a}^\mu \cdot \vec{\mathscr{A}}_\mu + \cdots$$

where dots denote the second order terms in $\vec{a}^\mu$. The pieces in $\mathscr{L}_{\text{eff}}$ can be read off from (3.10.1) and (3.10.2). The first two pieces are already given by (3.7.15). The remaining ones are

$$\mathscr{L}_{\text{int}}^{(4)} = \frac{M^2}{F^2}\{l_4 \partial_\mu \vec{\phi} \cdot \partial^\mu \vec{\phi} - (l_3 + l_4)M^2 \vec{\phi} \cdot \vec{\phi}\}$$

$$\vec{\mathscr{A}}_\mu = -F\partial_\mu \vec{\phi} + \frac{2}{3F}(\vec{\phi} \cdot \vec{\phi}\,\partial_\mu \vec{\phi} - \vec{\phi} \cdot \partial_\mu \vec{\phi}\,\vec{\phi}) - 2l_4 \frac{M^2}{F}\partial_\mu \vec{\phi}. \quad (3.10.7)$$

We now equate the generating functionals of the QCD and effective theory[11]

$$\langle 0|T \exp\left(i \int d^4x \, \vec{a}^\mu(x) \cdot \vec{A}_\mu(x)\right) |0\rangle = \langle 0|T \exp\left(i \int d^4x \, \vec{a}^\mu(x) \cdot \vec{\mathscr{A}}_\mu(x)\right) |0\rangle$$

$$(3.10.8)$$

and take the second functional derivative with respect to $\vec{a}^\mu(x)$ on both sides; the right-hand side of the resulting equation allows us to find the two-point function (3.10.6) from the effective theory. Converting the fields from the Heisenberg to interaction representation, we can generate the conventional perturbation expansion to evaluate contributions in terms of Feynman graphs.

---

[11] It should be noted that we are actually using three different ground states in writing (3.10.8) and (3.10.9), namely the QCD ground state and two ground states of the effective theory, the interacting and free ones. However, we denote all these states with the same symbol.

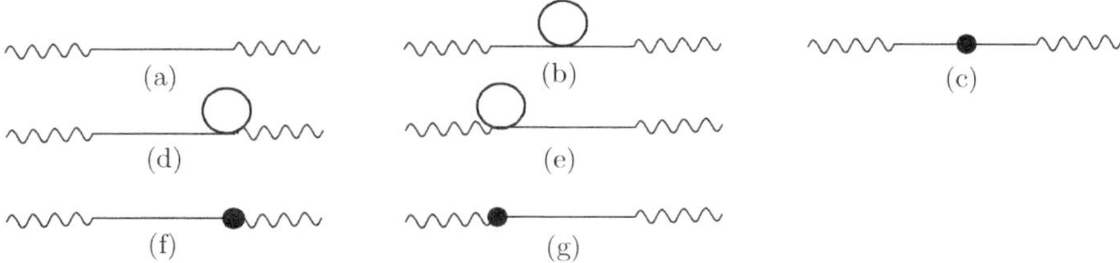

Figure 3.2 Pion pole graphs for the two-point function of axial-vector current, giving free propagation (a) and one-loop corrections (b,d,e) along with renormalisation counter-graphs (c,f,g) with vertices from $\mathscr{L}_{\text{int}}^{(4)}$ denoted by solid circles. Currents and pions are denoted respectively by wavy and straight lines.

The graphs for the two-point function containing the pion pole are shown in Figure 3.2 up to one loop. They depend only on linear couplings to the external field. For the graphs (a), (b) and (c), we have

$$T_{\mu\nu}^{ab}(k)_{(a+b+c)} = F^2 i \int d^4x\, e^{ik\cdot x}\, \langle 0|\, T\, \partial_\mu \phi^a(x)\, \partial_\nu \phi^b(0)\, e^{i\int d^4y \mathscr{L}_{\text{int}}(y)}|0\rangle \quad (3.10.9)$$

Expanding the exponential

$$\exp\left(i\int d^4y \mathscr{L}_{\text{int}}(y)\right) = 1 + i\int d^4y \mathscr{L}_{\text{int}}^{(2)}(y) + i\int d^4y \mathscr{L}_{\text{int}}^{(4)}(y) + \cdots,$$

the first three terms give respectively

$$T(k^2)_{(a)} = F^2 \Delta_F(k^2) \tag{3.10.10}$$

$$T(k^2)_{(b)} = -\frac{F^2}{6}\eta\{4(k^2 - M^2) + 3M^2\}\Delta_F^2(k)G \tag{3.10.11}$$

$$T(k^2)_{(c)} = 2F^2\eta\{l_4(k^2 - M^2) - l_3 M^2\}\Delta_F^2(k) \tag{3.10.12}$$

where $\eta \equiv M^2/F^2$ is an expansion parameter of the effective theory and $G(M)$ is defined by (D.11) of Appendix D. The vertex correction graphs may be evaluated in a similar way.

The self-energy and vertex graphs yield, respectively,

$$T(k^2)_{(b+c)} = F^2\, \eta\, (M^2\sigma\Delta_F^2 + \sigma'\Delta_F) \tag{3.10.13}$$

$$T(k^2)_{(d+e+f+g)} = F^2\, \eta\, \gamma\Delta_F(k) \tag{3.10.14}$$

where

$$\sigma = -\frac{1}{2}(4l_3 + G), \qquad \sigma' = -\frac{2}{3}(3l_4 - G), \qquad \gamma = \frac{4}{3}(3l_4 - 2G). \tag{3.10.15}$$

In (3.10.13) the term linear in Δ_F arises from cancelling $(k^2 - M^2)$ with one propagator. To order η, we can now put this term in the vertex and collect the contribution of all graphs as

$$T(k^2) = F^2\{1 + \eta(\gamma + \sigma')\}\Delta_F(k)\{1 + M^2\eta\sigma\Delta_F + O(\eta^2)\}$$
$$= \frac{-F^2\{1 + \eta(\gamma + \sigma')\}}{k^2 - M^2(1 - \eta\sigma)}$$
$$\equiv \frac{-F_\pi^2}{k^2 - M_\pi^2} \tag{3.10.16}$$

where M_π and F_π are the one-loop corrected mass and decay constant of pion,

$$M_\pi^2 = M^2\{1 + \frac{1}{2}\eta(4l_3 + G)\} \tag{3.10.17}$$

$$F_\pi = F\{1 + \eta(l_4 - G)\}. \tag{3.10.18}$$

Inserting the evaluation of G from (D.12), we get

$$M_\pi^2 = M^2\left\{1 + \frac{M^2}{F^2}\left(2l_3 + \lambda + \frac{1}{32\pi^2}\ln\frac{M^2}{\mu^2}\right)\right\} \tag{3.10.19}$$

$$F_\pi = F\left\{1 + \frac{M^2}{F^2}\left(l_4 - 2\lambda - \frac{1}{16\pi^2}\ln\frac{M^2}{\mu^2}\right)\right\}. \tag{3.10.20}$$

As seen from (3.10.3) the bare couplings l_3, l_4 absorb the (infinite) λ, giving

$$M_\pi^2 = M^2\left\{1 + \frac{M^2}{F^2}\left(2l_3^r + \frac{1}{32\pi^2}\ln\frac{M^2}{\mu^2}\right)\right\} \tag{3.10.21}$$

$$F_\pi = F\left\{1 + \frac{M^2}{F^2}\left(l_4^r - \frac{1}{16\pi^2}\ln\frac{M^2}{\mu^2}\right)\right\}. \tag{3.10.22}$$

Thus, for sufficiently small quark mass and hence M, the logarithmic terms dominate the corrections, whatever the values of l_i^r, giving $M_\pi < M$, $F_\pi > F$. In terms of the scale independent couplings defined by (3.10.4), these become

$$M_\pi^2 = M^2\left(1 - \frac{M^2}{32\pi^2 F^2}\bar{l}_3\right), \qquad F_\pi = F\left(1 + \frac{M^2}{16\pi^2 F^2}\bar{l}_4\right). \tag{3.10.23}$$

With the values (3.10.5) for the low energy constants, it turns out that the one-loop corrections to M and F are a few percent only.

Quark Condensate

In Section 3.7 the quark condensate was obtained in the tree approximation. We shall now get it to one loop. The external scalar field $s^0(x)$ in the extended QCD Lagrangian (3.8.1) may be varied in the neighbourhood of the quark mass given by (3.8.6). As $(m_u + m_d)/2 = M^2/(2B)$ by (3.7.3), we may equivalently extend the pion mass M^2 to an external field $M^2(x)$ and vary it around M^2,

$$s^0(x) \equiv \frac{1}{2B}M^2(x) = \frac{1}{2B}\{M^2 + \delta M^2(x)\} \tag{3.10.24}$$

when we get

$$\mathscr{L}_{\text{ext}} = \mathscr{L}_{QCD} - \delta M^2(x)\frac{1}{2B}\bar{q}q. \tag{3.10.25}$$

We also vary $M^2(x)$ in the effective Lagrangian (3.10.1) and (3.10.2) to get

$$\mathscr{L}_{\text{eff}} = C + \mathscr{L}_0 + \delta M^2(x)\mathscr{N}(x) \tag{3.10.26}$$

where $C = M^2 F^2$ is an irrelevant constant. As there are no external pion lines and we work to one loop, it suffices to have $\mathscr{N}(x)$ up to quadratic terms in ϕ,

$$\mathscr{N}(x) = F^2\left(1 - \frac{1}{2F^2}\vec{\phi}\cdot\vec{\phi}\right) + 2M^2(l_3 + h_1). \tag{3.10.27}$$

We now equate the resulting generating functionals of QCD and effective theory,

$$\langle 0|T\,\exp\left(i\int d^4x(-)\delta M^2\frac{1}{2B}\bar{q}q\right)|0\rangle = \langle 0|T\,\exp\left(i\int d^4x(\delta M^2\mathscr{N})\right)|0\rangle. \tag{3.10.28}$$

Differentiating both sides functionally with respect to δM^2, we get

$$-\frac{1}{2B}\langle 0|\bar{q}q|0\rangle = \langle 0|\mathscr{N}|0\rangle$$

$$= F^2\left(1 - \frac{1}{2F^2}\langle 0|\vec{\phi}\cdot\vec{\phi}|0\rangle + 2(l_3 + h_1)\frac{M^2}{F^2}\right) \tag{3.10.29}$$

where the three terms correspond respectively to graphs (a), (b) and (c) of Figure 3.3. On putting the regularised expression (D.12) of Appendix D for the pion propagator at the origin, it gives

$$\langle 0|\bar{u}u|0\rangle = \langle 0|\bar{d}d|0\rangle = -F^2 B\left\{1 - \frac{3M^2}{2F^2}\left(2\lambda + \frac{1}{16\pi^2}\ln\frac{M^2}{\mu^2} - \frac{4}{3}(l_3 + h_1)\right)\right\} \tag{3.10.30}$$

We have already carried out the renormalisation of l_3. So now we can isolate the divergent piece from h_1 uniquely to define h_1^r as before to get

$$h_1 = h_1^r + 2\lambda. \tag{3.10.31}$$

Introducing the scale independent coupling $\bar{h}_1$,

$$h_1^r = \frac{1}{16\pi^2}\left(\bar{h}_1 + \ln\frac{M^2}{\mu^2}\right) \tag{3.10.32}$$

we get

$$\langle 0|\bar{u}u|0\rangle = \langle 0|\bar{d}d|0\rangle = -F^2 B\left\{1 + \frac{M^2}{32\pi^2 F^2}(4\bar{h}_1 - \bar{l}_3)\right\}. \tag{3.10.33}$$

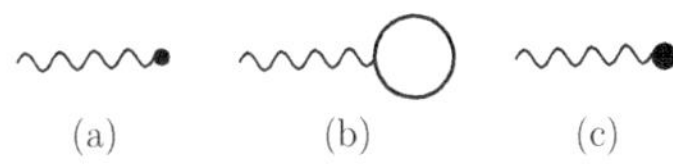

(a)   (b)   (c)

Figure 3.3 Graphs for the one-point function of the quark condensate to one loop. The solid circle denotes vertex from $\mathscr{L}_{\text{eff}}^{(4)}$.

As mentioned earlier, $\bar{h}_1$ cannot be estimated from experimental data at low energy, so that the correction in (3.10.33) becomes convention dependent. However, in the chiral limit, such corrections vanish and so does the convention dependence.

3.11 Non-Goldstone Fields

So far we have been concerned with Goldstone fields (pions) and have derived the effective Lagrangian involving these fields alone. We now include non-Goldstone fields interacting with pion fields. The general transformation rule of these heavy fields was obtained in (2.7.14); it transforms with the isospin subgroup $H \equiv SU(2)_V$ with group parameters depending on the Goldstone fields. In the following we deal with only nucleon ($J^P = \frac{1}{2}^+$) and vector ($J^{PC} = 1^{--}$) and axial-vector ($J^{PC} = 1^{++}$) meson fields, though the same procedure applies to other fields, like the scalar and pseudoscalar mesons.

Under the action of the isospin subgroup, an isosinglet field $S_\mu(x)$ remains, of course, unchanged,

$$S_\mu \to S_\mu .$$

The nucleon isodoublet $\psi(x)$ transforms as

$$\psi \to \psi' = h\psi \tag{3.11.1}$$

where from now on, $h(x)$ will denote the two-dimensional representation of $SU(2)_V$. The isovector field may be put in the form of a traceless 2×2 matrix,

$$R_\mu(x) = \frac{1}{\sqrt{2}} \sum_a R_\mu^a \tau^a, \tag{3.11.2}$$

when it transforms as

$$R_\mu \to R'_\mu = hR_\mu h^\dagger . \tag{3.11.3}$$

Because the parameters of h are x-dependent, we need an appropriate connection Γ_μ to define the covariant derivative of heavy fields. In the absence of external fields, it is

$$\Gamma_\mu = \frac{1}{2}[u^\dagger, \partial_\mu u] \tag{3.11.4}$$

where u is defined and its transformation rule derived in Section 3.3. (The subgroup elements V are now denoted by h.) The derivative for the nucleon field becomes

$$D_\mu \psi = \partial_\mu \psi + \Gamma_\mu \psi \tag{3.11.5}$$

while for the isovector fields it is

$$\nabla_\mu R_\nu = \partial_\mu R_\nu + [\Gamma_\mu, R_\nu], \tag{3.11.6}$$

which transform as the fields themselves,

$$D_\mu \psi \to (D_\mu \psi)' = h D_\mu \psi \qquad (3.11.7)$$

$$\nabla_\mu R_\nu \to (\nabla_\mu R_\nu)' = h \nabla_\mu R_\nu h^\dagger . \qquad (3.11.8)$$

To include external vector and axial-vector fields introduced in Section 3.8, the above connection must be extended to

$$\Gamma_\mu = \frac{1}{2} \left(u^\dagger [\partial_\mu - i(v_\mu + a_\mu)] u + u[\partial_\mu - i(v_\mu - a_\mu)] u^\dagger \right) \qquad (3.11.9)$$

In constructing invariants with the heavy fields, the mesonic and external variables can be so modified as to transform under the subgroup only [31]. Thus instead of $D_\mu U$ one can construct

$$u_\mu \equiv i u^\dagger D_\mu U u^\dagger = u_\mu^\dagger \qquad (3.11.10)$$

transforming as

$$u_\mu \to u_\mu{}' = h u_\mu h^\dagger.$$

In the same way we replace $F_{R,L}^{\mu\nu}$ and χ respectively by

$$f_\pm^{\mu\nu} \equiv u F_L^{\mu\nu} u^\dagger \pm u^\dagger F_R^{\mu\nu} u, \qquad \text{and} \qquad \chi_\pm \equiv u^\dagger \chi u^\dagger \pm u \chi^\dagger u \qquad (3.11.11)$$

transforming as

$$(f_\pm^{\mu\nu}, \chi_\pm) \to (f_\pm^{\mu\nu}{}', \chi_\pm') = h(f_\pm^{\mu\nu}, \chi_\pm) h^\dagger.$$

Once we construct terms for the Lagrangian with the above variables that are invariant under the subgroup, they will be automatically invariant under the full group [11]. We now describe below this construction for the nucleon and the vector and axial-vector mesons.

Nucleon

Unlike the pion case, where only even powers arise in the derivative expansion, the nucleon Lagrangian will give rise to odd as well as even powers,

$$\mathscr{L}_N = \mathscr{L}_N^{(1)} + \mathscr{L}_N^{(2)} + \cdots$$

To order $O(p)$ we have

$$\mathscr{L}_N^{(1)} = \overline{\psi}(i\not{D} - m_N)\psi + \frac{1}{2} g_A \overline{\psi} \not{u} \gamma_5 \psi. \qquad (3.11.12)$$

To order $O(p^2)$, we have a term $\sim \text{tr } \chi_+ \overline{\psi}\psi$ where $\chi_+ = u^\dagger \chi u^\dagger + u \chi^\dagger u$ and terms involving two u_μs. Noting that $u_\mu(x)$ is traceless and so can be expanded over the Pauli matrices, we can have three independent terms quadratic in u_μ. Following Refs. [25, 26] we write

$$\mathscr{L}_N^{(2)} = c_1 \text{ tr } \chi_+ \overline{\psi}\psi - \frac{c_2}{4m_N^2} \text{ tr } (u_\mu u_\nu)(\overline{\psi} D^\mu D^\nu \psi + \text{ h.c.})$$

$$+ \frac{c_3}{2} \text{ tr } (u_\mu u^\mu) \overline{\psi}\psi - \frac{c_4}{4} \overline{\psi} \gamma^\mu \gamma^\nu [u_\mu, u_\nu] \psi \qquad (3.11.13)$$

where c_i are low energy constants in the πN system. They are obtained by comparing the expansion around the subthreshold, $\nu \equiv (s-u)/(4m_N) = 0$, $t = 0$ of amplitudes from chiral perturbation theory and dispersion relations. (Here s, t and u are the Mandelstam variables.) Originally determined in [26], the values were improved in [27]. The latest determination is [28]

$$c_1 = -0.74 \pm 0.02, \quad c_2 = 1.81 \pm 0.03, \quad c_3 = -3.61 \pm 0.05, \quad c_4 = 2.17 \pm 0.03$$
$$(3.11.14)$$

all in units of GeV^{-1}. With $\mathscr{L}_N^{(1)}$ we shall calculate the correlation function of nucleon current in Section 3.12 and use both $\mathscr{L}_N^{(1)}$ and $\mathscr{L}_N^{(2)}$ in Section 6.5 to calculate the axial-vector two-point function in nuclear matter.

We can include nucleon in the power counting described in Section 3.5 with pions alone. Let there be I_π and I_N internal pion and nucleon lines respectively in a graph. Each of the nucleon propagators has the form

$$-\frac{\not{p} + \not{q} + m_N}{(p+q)^2 - m_N^2}$$

where p is an on-shell nucleon four-momentum and q a soft momentum it absorbs by interacting with pions. Then to lowest order it is $\sim (\not{p} + m)/2p \cdot q \sim O(q^{-1})$. Thus the index (3.5.4) for the graph changes to

$$\nu = -2I_\pi - I_N + \sum V_i d_i + 4L. \tag{3.11.15}$$

The topological relation (3.5.5) now becomes

$$L = I_\pi + I_N - \sum V_i + 1. \tag{3.11.16}$$

Also counting the nucleon fields in the graph, we get a further topological relation

$$2I_N + E_N = \sum V_i n_i \tag{3.11.17}$$

where E_N is the number of external nucleon lines and n_i are the number of nucleon fields in the interaction vertex of type i. Eliminating the number of internal lines, we get

$$\nu = \sum V_i \left(d_i + 2m_i + \frac{n_i}{2} - 2 \right) + 2L - \frac{E_N}{2} + 2 \tag{3.11.18}$$

where we also include pion mass factors in the interaction terms as in (3.7.9). Though appearing general, this power counting rule cannot be applied to processes involving two or more nucleons. As pointed out by Weinberg [29] the contribution of graphs for such processes actually grows with increasing number of loops, in gross violation to the above counting rule. Phenomenologically this is only to be expected, as there are shallow bound states in such systems. However, we shall not meet this problem, as our applications concern only single nucleon amplitudes.

Vector and Axial-Vector Mesons

There is a problem in applying chiral perturbation theory to vector mesons in general. In processes such as $\rho \to \pi + \pi$, there is no vector meson in the final state and so the pions are not soft. Strictly speaking, the theory should be applied where the vector mesons are conserved and so the Goldstone boson momenta are low enough [30]. However, if we are interested in amplitudes at finite temperature, the chiral Lagrangian approach for spin-one mesons finds a partial justification: since we consider temperatures $\lesssim 100$ MeV, the distribution function for pions present in the amplitude effectively cuts off their momenta around this value.

We review the construction of the Lagrangian involving vector and axial-vector mesons [31, 32]. We already obtained in Section 1.4 the kinetic term for an isosinglet vector/axial-vector field $S_\mu(x)$, which can be extended to the isotriplet $R_\mu(x)$ defined by (3.11.2). Also defining

$$S_{\mu\nu} = \partial_\mu S_\nu - \partial_\nu S_\mu, \qquad R_{\mu\nu} = \nabla_\mu R_\nu - \nabla_\nu R_\mu \tag{3.11.19}$$

the kinetic terms take the form

$$\mathscr{L}_{kin}(S, R) = \left(-\frac{1}{4} S_{\mu\nu} S^{\mu\nu} + \frac{1}{2} m_S^2 S_\mu S^\mu \right)$$
$$+ \left(-\frac{1}{4} \operatorname{tr}(R_{\mu\nu} R^{\mu\nu}) + \frac{1}{2} m_R^2 \operatorname{tr}(R_\mu R^\mu) \right). \tag{3.11.20}$$

In writing the interaction terms we spell out the fields we are interested in. As vector fields, we take $\rho(770)$ and $\omega(787)$, while for the axial-vector fields we take $a_1(1280)$ and $f_1(1290)$, where in each set we have an isotriplet and an isosinglet. Though they belong to different $SU(2)$ multiplets, we take the triplet and singlet masses to be same, considering that they belong to the same $SU(3)$ multiplets.

As parity and charge conjugation (and hence time reversal) are symmetries of strong interaction, we require the terms to be invariant under these operations [31]. Considering derivatives and external fields as $O(p)$, the leading interaction terms linear in the spin-one fields are $O(p^3)$. Those with vector fields give

$$\mathscr{L}_{int}(V) = \frac{1}{2\sqrt{2}m_V} \left(F_\rho \operatorname{tr}(\rho_{\mu\nu} f_+^{\mu\nu}) + iG_\rho \operatorname{tr}(\rho_{\mu\nu}[u^\mu, u^\nu]) \right.$$
$$\left. + iH_\rho \operatorname{tr}(\rho_\mu[u_\nu, f_-^{\mu\nu}]) + H_\omega \epsilon_{\mu\nu\lambda\sigma} \omega^\mu \operatorname{tr}(u^\nu f_+^{\lambda\sigma}) \right), \tag{3.11.21}$$

while those with axial vector fields give

$$\mathscr{L}_{int}(A) = \frac{1}{2\sqrt{2}m_A} \left(F_{a_1} \operatorname{tr}(a_{1\,\mu\nu} f_-^{\mu\nu}) + iH_{a_1} \operatorname{tr}(a_{1\,\mu}[u_\nu, f_+^{\mu\nu}]) \right.$$
$$\left. + H_{f_1} \epsilon_{\mu\nu\lambda\sigma} f_1^\mu \operatorname{tr}(u^\nu f_-^{\lambda\sigma}) \right). \tag{3.11.22}$$

Terms containing three u_μs are also possible, but will not contribute to vertices of the graphs to be considered here. Finally we write down the quadratic couplings of the triplets with the singlets and between themselves, which are $O(p^2)$

$$\mathscr{L}_{int}(V,A) = \frac{1}{\sqrt{2}}\epsilon_{\mu\nu\lambda\sigma}\left\{g_1\omega^\mu\,\mathrm{tr}\,(\rho^{\nu\lambda}u^\sigma) + g_2 f_1^\mu\,\mathrm{tr}\,(a_1^{\nu\lambda}u^\sigma)\right\}$$

$$+\frac{i}{2}g_3\,\mathrm{tr}\,(\rho^{\mu\nu}[a_{1\,\mu},u_\nu]). \tag{3.11.23}$$

(The term $\mathrm{tr}\,(a_1^{\mu\nu}[\rho_\mu,u_\nu])$ is equivalent to the third term above to leading order in pion momentum.) We find the values of the above coupling constants in Appendix F.

In Sections 3.12 and 3.13 below we shall calculate the one-loop graphs of nucleon and meson correlation functions with vertices from Lagrangians (3.11.12) and (3.11.21–23) respectively. We shall be interested only in the structure of the amplitudes and not evaluate them. As we shall see in Sections 6.3 and 6.4 these vacuum amplitudes can be immediately extended to finite temperature contributing to the thermal correlation functions.

3.12 Nucleon Current Correlation Function

Here we consider the two-point function of the so-called 'nucleon current', to be denoted by $\eta(x)$, which is composed of three quark fields, having the quantum numbers of the nucleon [33]. Two different quark structures can be built to represent $\eta(x)$, but these details are not of interest here. It suffices to assume that $\eta(x)$ transforms like a Dirac field. Then we can construct it in the effective field theory in terms of pion and nucleon fields.

Following the same method of external fields as for the mesonic currents (Section 3.8), we introduce external fermionic field $f(x)$ coupling to the nucleon current, extending the QCD Lagrangian to

$$\mathscr{L}_{QCD} \longrightarrow \mathscr{L}_{QCD} + \bar{f}\eta + \bar{\eta}f. \tag{3.12.1}$$

To find the transformation rule for the external field, we split the nucleon current and the external field into their right- and left-handed components,

$$\eta = \eta_R + \eta_L, \quad f = f_R + f_L.$$

Under the symmetry group $G \equiv SU(2)_R \times SU(2)_L$ the nucleon current is assumed to transform as

$$\eta_R \longrightarrow V_R\eta_R, \quad \eta_L \longrightarrow V_L\eta_L.$$

Then the transformation rule for $f(x)$ follows from requiring the invariance of $\bar{f}\eta = \bar{f}_R\eta_L + \bar{f}_L\eta_R$. Thus $f(x)$ must transform oppositely to $\eta(x)$,

$$f_R \longrightarrow V_L f_R, \quad f_L \longrightarrow V_R f_L.$$

It is now possible to construct variables involving the external field that transform only under the isospin subgroup $H \equiv SU(2)_V$. These are $u f_R$ and $u^\dagger f_L$ transforming as

$$u f_R \longrightarrow h u f_R, \quad u^\dagger f_L \longrightarrow h u^\dagger f_L.$$

Then the terms in the Lagrangian involving the external field that are linear in the nucleon field $\psi(x)$ have the form

$$\mathscr{L}_f = \lambda \overline{u f_R} \psi + \lambda' \overline{u^\dagger f_L} \psi + h.c. \tag{3.12.2}$$

where λ and λ' are constants. Expressing $f_{R,L}$ in terms of f, it becomes

$$\mathscr{L}_f = \lambda \bar{f} \frac{1}{2}(1 - \gamma^5) u^\dagger \psi + \lambda' \bar{f} \frac{1}{2}(1 + \gamma^5) u \psi + h.c. \tag{3.12.3}$$

Requiring parity invariance, under which

$$u \to u^\dagger, \qquad \psi \to \gamma^0 \psi, \qquad f \to \gamma^0 f,$$

we get $\lambda = \lambda'$. Then $\mathscr{L}_f$ simplifies to

$$\mathscr{L}_f = \frac{\lambda}{2} \bar{f} \{u + u^\dagger + (u - u^\dagger)\gamma^5\} \psi + h.c. \tag{3.12.4}$$

We could now equate the generating functionals of QCD and effective theory to derive the two-point function of nucleon current in the effective theory. Equivalently we can directly read off $\eta(x)$ by comparing (3.12.1) and (3.12.4) to get

$$\eta = \frac{\lambda}{2} \{u + u^\dagger + (u - u^\dagger)\gamma^5\} \psi$$
$$= \lambda \left(1 + \frac{i \vec{\phi} \cdot \vec{\tau}}{2F} \gamma^5 - \frac{\vec{\phi} \cdot \vec{\phi}}{8F^2} + \cdots \right) \psi. \tag{3.12.5}$$

We now find the vacuum two-point function

$$W(p) = i \int d^4x\, e^{ip \cdot x} \langle 0|T\eta(x)\bar{\eta}(0)|0\rangle \tag{3.12.6}$$

using the explicit form of $\eta(x)$ given by (3.12.5) and perturbing the amplitude with the pion–nucleon interaction obtained from (3.11.12),

$$\mathscr{L}_{int} = -\frac{g_A}{2F} \bar{\psi} \gamma_\mu \gamma^5 \vec{\tau} \cdot \partial^\mu \vec{\phi} \psi.$$

Up to one loop the contributing graphs are shown in Figure 3.4. The free amplitude of graph (a) is

$$W(p)_{(a)} = \lambda^2 S_F(p) \tag{3.12.7}$$

where $S_F(p)$ is the free nucleon propagator (1.3.15). The self-energy correction to the free propagator from graph (b) is

$$W(p)_{(b)} = \lambda^2 S_F(p)\Sigma(p)S_F(p) \tag{3.12.8}$$

where the self-energy function is

$$\Sigma(p) = -\frac{3i}{4} \left(\frac{g_A}{F} \right)^2 \int \frac{d^4k}{(2\pi)^4} \not{k}\gamma^5 S_F(p - k)\not{k}\gamma^5 \Delta_F(k). \tag{3.12.9}$$

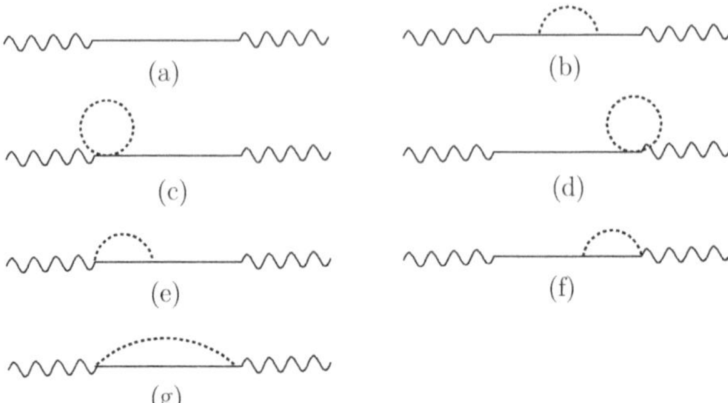

Figure 3.4 Graphs up to one loop contributing to nucleon current correlation function. Wavy lines denote nucleon current. Solid lines are nucleons; dotted lines are pions.

The two graphs (c) and (d) contribute equally to the momentum-independent vertex correction

$$W(p)_{(c+d)} = -\frac{3}{4}\left(\frac{\lambda}{F}\right)^2 M^2 G(M) S_F(p) \tag{3.12.10}$$

where $G(M)$ is defined in (D.10). Then the contribution of graphs (a) to (d) may be put in the form

$$W(p)_{(a+b+c+d)} = \lambda^2 \left(1 - \frac{3M^2 G}{4F^2}\right)\left(S_F(p) + S_F(p)\Sigma(p)S_F(p) + \cdots\right)$$

$$= \lambda^2 \left(1 - \frac{3M^2 G}{4F^2}\right)\frac{-1}{\not{p} - m + \Sigma + i\epsilon}. \tag{3.12.11}$$

We next consider the momentum-dependent vertex correction from graphs (e) and (f)

$$W(p)_{(e+f)} = i\lambda^2 \frac{3g_A}{4F^2}\{S_F(p)\Gamma^{(1)}(p) - \Gamma^{(2)}S_F(p)\} \tag{3.12.12}$$

with

$$\Gamma^{(1)}(p) = \int \frac{d^4 k}{(2\pi)^4} \not{k}\gamma^5 S_F(p-k)\gamma^5 \Delta_F(k), \tag{3.12.13}$$

and $\Gamma^{(2)}$ obtained by interchanging $\not{k}\gamma^5$ and γ^5 in $\Gamma^{(1)}$. Finally graph (g) is

$$W(p)_{(g)} = i\frac{3}{4}\left(\frac{\lambda}{F}\right)^2 \int \frac{d^4 k}{(2\pi)^4} \Delta_F(k)\gamma^5 S_F(p-k)\gamma^5. \tag{3.12.14}$$

In the above formulae G, Σ and Γ are divergent. We need renormalisation counterterms (3.11.13) to cancel these divergences. But here we shall not carry out this procedure to find finite corrections to m_N and λ. Later in Section 6.3 we shall find the temperature dependence of the physical mass and coupling to leading order.

3.13 Meson Current Correlation Functions

In Section 3.10 we found the axial-vector current correlation function in the neighbourhood of pion pole to determine corrections to pion pole parameters. We shall now find the vector current correlation function in a similar way correcting pole parameters of the vector meson ρ. (For the axial-vector meson a_1, the axial-vector current correlation function can be analysed in the same way, but at momenta higher than in Section 3.10, so as to be in the neighbourhood of a_1 meson.) As with nucleon in the previous section, this section will prepare us to find the temperature dependence of pole parameters in Section 6.4 and of the spectral function in Chapter 8.

To get the vertices in the one-loop graphs contributing to the correlation functions (Figure 3.5), we expand out the pieces (3.11.21–23) of the effective Lagrangian in the particle and external fields. We retain terms up to four fields, including linear and quadratic terms in the pion field and only linear terms in the external fields. Also in the four field terms, we avoid derivatives on pion fields. With these restrictions we get

$$\mathscr{L}(V) = \frac{F_\rho}{m_V} \left\{ \partial^\mu \vec{v}^\nu \cdot (\partial_\mu \vec{\rho}_\nu - \partial_\nu \vec{\rho}_\mu) \left(1 - \frac{1}{3F^2} \vec{\phi} \cdot \vec{\phi} \right) + \frac{1}{F} \partial^\mu \vec{a}^\nu \cdot (\partial_\mu \vec{\rho}_\nu - \partial_\nu \vec{\rho}_\mu) \times \vec{\phi} \right\}$$
$$- \frac{2G_\rho}{m_V F^2} \partial_\mu \vec{\rho}_\nu \cdot \partial^\mu \vec{\phi} \times \partial^\nu \vec{\phi} - \frac{\sqrt{2}H_\omega}{m_V F} \epsilon_{\mu\nu\lambda\sigma} \omega^\mu \partial^\nu \vec{\phi} \cdot \partial^\lambda \vec{v}^\sigma \,, \tag{3.13.1}$$

$$\mathscr{L}(A) = -\frac{F_{a_1}}{m_A} \left\{ \partial^\mu \vec{a}^\nu \cdot (\partial_\mu \vec{a}_{1\nu} - \partial_\nu \vec{a}_{1\mu}) \left(1 - \frac{1}{3F^2} \vec{\phi} \cdot \vec{\phi} \right) \right.$$
$$\left. + \frac{1}{F} \partial^\mu \vec{v}^\nu \cdot (\partial_\mu \vec{a}_{1\nu} - \partial_\nu \vec{a}_{1\mu}) \times \vec{\phi} \right\} + \frac{\sqrt{2}H_{f_1}}{m_A F} \epsilon_{\mu\nu\lambda\sigma} f_1^\mu \partial^\nu \vec{\phi} \cdot \partial^\lambda \vec{a}^\sigma \,, \tag{3.13.2}$$

and

$$\mathscr{L}(V, A) = -2\epsilon_{\mu\nu\lambda\sigma} \left(\frac{g_1}{F} \omega^\mu \partial^\nu \vec{\rho}^\lambda \cdot \partial^\sigma \vec{\phi} + \frac{g_2}{F} f_1^\mu \partial^\nu \vec{a}_1^\lambda \cdot \partial^\sigma \vec{\phi} \right)$$
$$+ \frac{g_3}{F} \partial^\mu \vec{\rho}^\nu \cdot (\vec{a}_{1\mu} \times \partial_\nu \vec{\phi} - \vec{a}_{1\nu} \times \partial_\mu \vec{\phi}). \tag{3.13.3}$$

Terms with H_ρ and H_{a_1} present in (3.11.21) and (3.11.22) are omitted here, as their leading terms are already present in those proportional to F_ρ and F_{a_1}.

Finally the $\rho\rho\pi\pi$ vertex appearing in the seagull graph of Figure 3.5(c) can be obtained from the kinetic term (3.11.20). Noting

$$\nabla_\mu \rho_\nu = \partial_\mu \rho_\nu + \frac{1}{2}[u^\dagger \partial_\mu u + u \partial_\mu u^\dagger, \rho_\nu],$$

we get

$$\mathscr{L}_{\text{kin}} \to \frac{-1}{8F^2} \langle (\partial_\mu \rho_\nu - \partial_\nu \rho_\mu)[[\phi, \partial^\mu \phi], \rho^\nu] \rangle.$$

It reproduces the current algebra result for the leading behaviour of pion scattering off a heavy target, which is ρ in the present case [34]. However, for the seagull graph we need to contract the two pion fields at the same point, getting

$$\langle 0|\phi\partial^\mu\phi|0\rangle = \int \frac{d^4k}{(2\pi)^4}k^\mu\Delta_F(k)$$

which is zero by symmetry.

We find the ρ meson pole in the two-point vector current correlation function

$$M_{\mu\nu}^{ab}(q) = i\int d^4x\, e^{iq\cdot x}\langle 0|TV_\mu^a(x)V_\nu^b(0)|0\rangle. \tag{3.13.4}$$

Let us first obtain the constraint from current conservation [35][12]. Consider

$$q^\mu M_{\mu\nu}^{ab}(q) = \int d^4x\,(\partial^\mu e^{iq\cdot x})\langle 0|TV_\mu^a(x)V_\nu^b(0)|0\rangle. \tag{3.13.5}$$

Integrating by parts and ignoring the surface term, it gives

$$q^\mu M_{\mu\nu}^{ab}(q) = -\int d^4x\, e^{iq\cdot x}\langle 0|\partial^\mu\{TV_\mu^a(x)V_\nu^b(0)\}|0\rangle. \tag{3.13.6}$$

Because of current conservation, non-vanishing terms arise only from theta functions in the time ordering,

$$\partial^\mu\{TV_\mu^a(x)V_\nu^b(0)\} = \delta(x^0)[V_0^a(x),V_\nu^b(0)].$$

The equal time commutator works out to give

$$[V_0^a(x),V_\nu^b(0)]_{x^0=0} = i\epsilon^{abc}V_\nu^c(x)\delta^3(\boldsymbol{x}).$$

Then (3.13.6) gives

$$q^\mu M_{\mu\nu}^{ab}(q) = -i\epsilon^{abc}\langle 0|V_\nu^c(0)|0\rangle \tag{3.13.7}$$

As there is no four-vector in vacuum, the vacuum expectation value of the vector current must be zero. Also isospin symmetry, which we are assuming here, leads to the same result. By either reasoning, we get

$$q^\mu M_{\mu\nu}^{ab}(q) = 0, \tag{3.13.8}$$

so that the tensor $M(q)$ is transverse to the four-momentum q. Along with Lorentz covariance, it puts the tensor in the form

$$M_{\mu\nu}^{ab}(q) = \delta^{ab}\left(-g_{\mu\nu} + \frac{q_\mu q_\nu}{q^2}\right)M(q^2). \tag{3.13.9}$$

As we shall see below, all loop graphs satisfy (3.13.8) automatically.

We now equate, as in the case of the axial-vector current in Section 3.10, the generating functionals of QCD and the effective theory for the vector current,

$$\langle 0|T\exp\left(i\int d^4x\, v_{\mu a}V_a^\mu\right)|0\rangle = \langle 0|T\exp\left(i\int d^4x\, v_{\mu a}\mathcal{V}_a^\mu\right)|0\rangle. \tag{3.13.10}$$

Here $\mathcal{V}^\mu$ is the hadronic vector current, which can be read off as the coefficient of v_μ in the effective Lagrangian above. The two-point function results from the

---

[12] For an alternative derivation without using equal time commutator, see Appendix B.

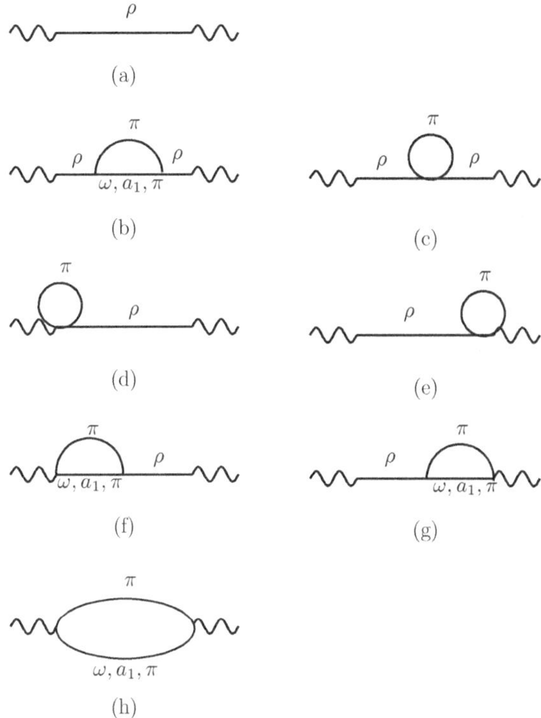

Figure 3.5 Graphs for the vector current two-point function up to one loop, which may contribute to ρ meson pole. Currents and particles are denoted by wavy and ordinary lines respectively.

second functional derivative with respect to v_μ. The terms in the Lagrangian without v_μ give interactions among particles, in which we do the perturbation expansion. In this way we can work out graphs of Figure 3.5.

The ρ propagation originating from coupling of the external field to the ρ field is

$$\langle 0|TV_\mu^a(x)V_\nu^b(y)|0\rangle$$

$$= \left(\frac{F_\rho}{m_V}\right)^2 \partial_x^\rho \partial_y^\sigma \langle 0|T(\partial_\rho \rho_\mu^a(x) - \partial_\mu \rho_\rho^a(x))(\partial_\sigma \rho_\nu^b(y) - \partial_\nu \rho_\sigma^b(y))|0\rangle. \quad (3.13.11)$$

To zeroth order in interaction the ρ field is free, whose contraction gives the vector meson propagator $\Delta_{\mu\nu}^F$ given by (1.4.19). Taking Fourier transform with the derivatives in (3.13.11) we get for graph (a),

$$M_{\mu\nu}^{ab}(q)_{(a)} = \delta^{ab}\left(\frac{F_\rho q^2}{m_V}\right)^2 \widetilde{\Delta}_{\mu\nu}(q), \quad (3.13.12)$$

where

$$\widetilde{\Delta}_{\mu\nu}(q) = \left(-g_{\mu\nu} + \frac{q_\mu q_\nu}{q^2}\right)\frac{-1}{q^2 - m_V^2 + i\epsilon}. \qquad (3.13.13)$$

An advantage of working with the two-point function of vector current, rather than of the ρ meson field itself, is that we get $\widetilde{\Delta}_{\mu\nu}(q)$ instead of $\Delta_{\mu\nu}^F(q)$: the former is transverse to q_μ ($q^\mu \widetilde{\Delta}_{\mu\nu} = 0$), while the latter is not ($q^\mu \Delta_{\mu\nu}^F = -q_\nu/m_V^2$).

The one-loop graphs of Figure 3.5 can be grouped into self-energies (b,c), vertices (d,e,f,g) and intermediate states (h). The self-energy graphs (b) may be evaluated by expanding the matrix element (3.13.11) to second order in the interaction and carrying out the required contractions. Each of the graphs gives a contribution of the form

$$M_{\mu\nu}(q)_{(\text{each b})} = \left(\frac{F_\rho q^2}{m_V}\right)^2 \widetilde{\Delta}_{\mu\rho}(q)\Pi^{\rho\sigma}(q)\widetilde{\Delta}_{\sigma\nu}(q) \qquad (3.13.14)$$

where the self-energy tensor Π is a loop intergral,

$$\Pi_{\mu\nu}(q) = i \int \frac{d^4k}{(2\pi)^4} K_{\mu\nu}(q,k)\Delta_F(k,M)\Delta_F(q-k,m). \qquad (3.13.15)$$

Here M and m are respectively the masses of pion and the other particle in the loop, which may be the pion itself. Introducing the following conserved tensors

$$A_{\alpha\beta}(q) = -g_{\alpha\beta} + q_\alpha q_\beta/q^2,$$
$$B_{\alpha\beta}(q,k) = q^2 k_\alpha k_\beta - q\cdot k(q_\alpha k_\beta + k_\alpha q_\beta) + (q\cdot k)^2 g_{\alpha\beta},$$
$$C_{\alpha\beta}(q,k) = q^4 k_\alpha k_\beta - q^2(q\cdot k)(q_\alpha k_\beta + k_\alpha q_\beta) + (q\cdot k)^2 q_\alpha q_\beta \qquad (3.13.16)$$

the tensor K in (3.13.15) may be written separately for the three different loops in Figure 3.5 (b),

$$-\left(\frac{2G_\rho}{m_V F^2}\right)^2 C_{\mu\nu}, \qquad (\pi\pi \text{ loop})$$

$$\left(\frac{2g_1}{F}\right)^2 (B_{\mu\nu} + M^2 q^2 A_{\mu\nu}), \qquad (\pi\omega \text{ loop})$$

$$2\left(\frac{g_3}{F}\right)^2 (B_{\mu\nu} - C_{\mu\nu}/m_{a_1}^2). \qquad (\pi a_1 \text{ loop}) \qquad (3.13.17)$$

As we saw earlier graph (c) gives zero. The momentum independent vertex in graphs (d) and (e) can be evaluated simply by contracting the two pion fields in the first term of the interaction Lagrangian (3.13.1) to get

$$M_{\mu\nu}(q)_{(d+e)} = -\frac{F_\rho q^2}{m_V}\frac{M^2 G}{F^2}\widetilde{\Delta}_{\mu\nu}(q) \qquad (3.13.18)$$

where G is given by (D.10). Including the self-energy and constant vertex graphs, the two-point function becomes

$$M_{\mu\nu}(q)_{(a+\cdots+e)} = \left(\frac{F_\rho q^2}{m_V}\right)^2 \left(1 - \frac{GM^2}{F^2}\right)\widetilde{\Delta}'_{\mu\nu} \qquad (3.13.19)$$

where $\widetilde{\Delta}'_{\mu\nu}$ is the one-loop corrected propagator

$$\widetilde{\Delta}'_{\mu\nu}(q) = \widetilde{\Delta}_{\mu\nu}(q) + \widetilde{\Delta}_{\mu\rho}\Pi^{\rho\sigma}_{\text{tot}}\widetilde{\Delta}_{\sigma\nu} + \cdots \qquad (3.13.20)$$

with Π_{tot} denoting the sum of self-energy contributions. The other momentum-dependent vertex corrections from (f) and (g) may be written as

$$M_{\mu\nu}(q)_{(f+g)} = \frac{F_\rho q^2}{m_V^2}\Gamma_{\mu\rho}\widetilde{\Delta}^\rho_\nu \qquad (3.13.21)$$

where the tensor Γ is a loop integral like the tensor Π. The intermediate state graphs (h) do not have the vector meson pole.

Having considered all the graphs of Figure 3.5, we can simplify the tensor structures of Π and Γ. As the self-energy multiplies the (transverse) tilde propagators and q_μ is the only available four-vector, it must also be transverse

$$\Pi_{\mu\nu}(q) = \left(-g_{\mu\nu} + \frac{q_\mu q_\nu}{q^2}\right)\Pi(q^2) \qquad (3.13.22)$$

when we can solve (3.13.20) for the one-loop corrected propagator as

$$\widetilde{\Delta}'_{\mu\nu}(q) = \frac{-1}{q^2 - m^2 + \Pi_{\text{tot}}}\left(-g_{\mu\nu} + \frac{q_\mu q_\nu}{q^2}\right). \qquad (3.13.23)$$

Similarly the momentum dependent vertex has the form

$$\Gamma_{\mu\nu} = \left(-g_{\mu\nu} + \frac{q_\mu q_\nu}{q^2}\right)\Gamma(q^2). \qquad (3.13.24)$$

As in the previous section, the loop elements G, Π and Γ are divergent, and their divergences can be removed with counterterms. But as we are interested in finding the thermal effects to leading order, we do not need to go through this step.

Problems

Problem 3.1: Assuming that Q_a^5 acting on vacuum creates a pion of zero momentum, prove the result (3.2.12) directly from the commutation relation (3.2.6).

Solution: We take the matrix element of (3.2.6) between vacuum,

$$\langle 0|Q_a^5\Phi_b^\dagger|0\rangle - \langle 0|\Phi_a^+ Q_b^5|0\rangle = i\delta_{ab}\langle 0|\bar{q}q|0\rangle. \qquad (\text{P3.1})$$

As given in the problem

$$Q_a^5|0\rangle = C|\pi_a(\boldsymbol{k}=0)\rangle \qquad (\text{P3.2})$$

where C is a normalisation constant. To find it we extend this definition to non-zero momentum by writing [36]

$$\int d^3x\, e^{i\boldsymbol{k}\cdot\boldsymbol{x}} A_a^0(\boldsymbol{x})|0\rangle = C|\pi_a(\boldsymbol{k})\rangle \tag{P3.3}$$

which goes over to (P3.2) as $\boldsymbol{k} \to 0$. Then we compute

$$\begin{aligned}
\langle\pi_b(\boldsymbol{k}')|\pi_a(\boldsymbol{k})\rangle &= \frac{1}{C}\int d^3x\, e^{i\boldsymbol{k}\cdot\boldsymbol{x}}\langle\pi_b(\boldsymbol{k}')|A_a^0(\boldsymbol{x})|0\rangle \\
&= \frac{1}{C}(2\pi)^3\delta(\boldsymbol{k}'-\boldsymbol{k})\langle\pi_b(\boldsymbol{k})|A_a^0(0)|0\rangle \\
&= -i\frac{F}{2C}(2\pi)^3 2\omega\delta(\boldsymbol{k}'-\boldsymbol{k})
\end{aligned} \tag{P3.4}$$

where the matrix element in the second line is obtained from (3.2.9). Comparing it with the normalisation (1.1.1), we get $C = -iF/2$. Now insert (P3.2) in (P3.1) to get

$$F\langle\pi_a(\boldsymbol{k}=0)|\Phi_b^\dagger|0\rangle = \delta_{ab}\langle 0|\bar{q}q|0\rangle. \tag{P3.5}$$

Finally using (3.2.10) it gives the result

$$-FG = \langle 0|\bar{q}q|0\rangle \tag{P3.6}$$

which we derived earlier from the spectral function.

Problem 3.2: Find the classical equation of motion for U from the second order effective Lagrangian.

Solution: The general form of the effective Lagrangian to order p^2 is given by

$$\mathscr{L}_{\text{eff}}^{(2)} = \frac{F^2}{4}\,\text{tr}\,(\partial_\mu U\partial^\mu U^\dagger + \chi U^\dagger + U\chi^\dagger). \tag{P3.7}$$

If the action were stationary under independent variation of the matrix fields U and $U^\dagger$, we would get

$$0 = \int d^4x\,\text{tr}\,\{\partial_\mu U\partial^\mu(\delta U^\dagger) + \partial_\mu(\delta U)\partial^\mu U^\dagger + \chi\delta U^\dagger + \delta U\chi^\dagger\}. \tag{P3.8}$$

But the variations δU and $\delta U^\dagger$ are not independent, the constraint $UU^\dagger = 1$ relating them as $\delta U = -U\delta U^\dagger U$. Integrating the first two terms partially and ignoring the surface terms, we get the Euler–Lagrange equation,

$$\Box U - U\Box U^\dagger U - \chi + U\chi^\dagger U = 0. \tag{P3.9}$$

Carrying out a differentiation in the second term, it reduces to

$$2\Box U + 2U\partial_\mu U^\dagger\partial^\mu U - \chi + U\chi^\dagger U = 0. \tag{P3.10}$$

References

[1] H. Fritzsch and M. Gell-Mann, Proc. of the XVI Int. Conf. on High Energy Physics, Chicago, vol. 2, p. 135 (J.D. Jackson and A. Roberts, eds.) See also H. Fritzsch, M. Gell-Mann and H. Leutwyler, Phys. Lett. **47 B**, 365 (1973).

[2] Y. Nambu, Phys. Rev. Lett., **4**, 380 (1960).

[3] Y. Nambu and D. Lurie, Phys. Rev. **125**, 1429 (1961).

[4] M. Gell-Mann, Physics, **1**, 63 (1964).

[5] S.L. Adler, Phys. Rev. Lett. **14**, 1051 (1965); Phys. Rev. **140**, B736 (1965).

[6] W.I. Weisberger, Phys. Rev. Lett. **14**, 1047 (1965); Phys. Rev. **143**, 1302 (1965).

[7] S. Weinberg, Phys. Rev. Lett. **18**, 188 (1967).

[8] S. Weinberg, Phys. Rev. **166**, 1568 (1968).

[9] H. Leutwyler, Physics Latters, **B48**, 431 (1974).

[10] S. Weinberg, Physica, **96A**, 327 (1979).

[11] S. Weinberg, *The Quantum Theory of Fields*, vol. 2, Cambridge University Press (1995).

[12] J. Gasser and H. Leutwyler, Ann. Phys. (NY) **158**, 142 (1984).

[13] J. Gasser and H. Leutwyler, Nucl. Phys. **B250**, 465 (1985).

[14] H. Leutwyler, Ann. Phys. **235**, 165 (1994).

[15] C. Itzykson and J.-B. Zuber, *Quantum Field Theory*, McGraw Hill (1980).

[16] H. Leutwyler, *Quark Masses and Chiral Symmetry*, lectures given at the XXII international school of subnuclear Physics, Erice, Aug. 5–15, 1984.

[17] H. Leutwyler, *Chiral Effective Lagrangians*, lectures given at the Workshop on topics in QCD, Waischenfeld, Germany (1991).

[18] H. Leutwyler, *Principles of chiral perturbation theory*, lectures given at the workshop, 'Hadrons 1994', Gramado, RS, Brasil (1994).

[19] F.J. Dyson, Phys. rev. **75**, 1736 (1949).

[20] W. Zimmermann, Brandeis Lectures, vol. 1 (1970).

[21] M. Gell-Mann, R.J. Oakes and B. Renner, Phys. Rev. **175**, 2195 (1968).

[22] FLAG Working Group, Eur. Phys. J., **C 74**, 2890 (2014).

[23] P. Gerber and H. Leutwyler, Nucl. Phys. **B 321**, 387 (1989).

[24] G. Colangelo, J. Gasser and H. Leutwyler, Nucl. Phys. **B 603**, 125 (2001).

[25] J. Gasser, M.E. Sainio and A. Svarc, Nucl. Phys. **B 307**, 779 (1988).

[26] T. Becher and H. Leutwyler, JHEP, 0106, 017 (2001).

[27] U.G. Meissner, J.A. Oller and A. Wirzba, Ann. Phys. **297**, 27 (2002).

[28] J.R. de Elvira, proceedings of Chiral Dynamics (2015).

[29] S. Weinberg, Nucl. Phys. **B 363**, 3 (1991).

[30] E. Jenkins and A.V. Manohar, Phys. Rev. Lett. **75**, 2272 (1995).

[31] G. Ecker, J. Gasser, A. Pich and E. de Rafael, Nucl. Phys. **B 321**, 311 (1989).

[32] G. Ecker, J. Gasser, H. Leutwyler, A. Pich and E. de Rafael, Phys. Lett. **B 223**, 425 (1989).

[33] B.L. Ioffe, Nucl. Phys. **B 188**, 317 (1981).

[34] S. Weinberg, Phys. Rev. Lett. **17**, 616 (1966).

[35] S. Weinberg, *The Quantum Theory of Fields*, vol. 1, Cambridge University Press (1995).

[36] A. Zee, *Quantum Field Theory in a Nutshell*, Princeton University Press (2003).

4

Thermal Propagators

Methods of thermal field theory can broadly be put into two categories, according to their use of imaginary time or real time in the ensemble averages. The method of imaginary time was first proposed by Matsubara [1]. The real time method was introduced later, originally by Schwinger [2] and Keldysh [3] and followed by others [4], who applied it to study non-equilibrium processes. It was also applied to equilibrium situations by Mills [5]. The real time method was reformulated in the so-called thermofield dynamics by Umezawa and collaborators [6], who suggested a doubling of field variables in the statistical context. Niemi and Semenoff [7] clarified this apparent doubling in terms of the time contour needed in the real time formulation. A survey of early literature is available in [8]. Physical applications including later works are presented also in books [9, 10].

For systems in thermal equilibrium, one can develop perturbative expansion in both methods in parallel to the conventional method in vacuum. The difference shows up in the form of the propagators. Because of the finite interval of imaginary time, the frequencies are discrete, giving the propagator as a sum over these frequencies. On the other hand, the real time method has two infinite intervals of time, giving rise to a 2×2 matrix for the propagator. The majority of calculations reported in the literature are done in the imaginary time method, avoiding matrix propagators in the real time method.

For systems out of thermal equilibrium, the real time method suggests itself as the appropriate framework to calculate physical quantities. However, for quantities such as the response functions (we address this problem in the last chapter) the end results relate to thermal equilibrium and so can be evaluated by either method. But, to follow the time dependence of quantities during non-equilibrium phases, such as in the early universe [11–13] – a topic too distant to discuss in this book – the real time method is the only option.

In this book we adopt the real time method to calculate thermal quantities in equilibrium. This chapter is devoted to deriving the matrix form of propagators arising in this method. Early works with fields of low spin show that free thermal propagators are given essentially by the corresponding vacuum propagators. More precisely, the thermal propagator, when diagonalised, has just the vacuum propagator as elements. One can generalise this result to fields of arbitrary spin on the basis of spectral representations of thermal two-point functions [14]. In view of this generalisation, the derivation of free thermal propagators for individual fields may appear superfluous.

However, it will be instructive to see the thermal elements by deriving first the low spin propagators in more elementary ways. So we derive the propagators for the scalar and Dirac fields in two ways directly from their definitions. In one way we solve their differential equations without using the fields directly [7]. In the other way we evaluate the thermal expectation values of fields requiring no boundary condition explicitly [6]. We then derive the spectral representation for general operators, which for free fields relates the thermal and vacuum propagators as stated above.

4.1 Time Path

The interval in which the time coordinate varies in different formulations of thermal field theory, can already be seen in quantum mechanics. Considering the vacuum theory, we are generally interested in transition amplitudes. For example, if a system is prepared at time t to be in an eigenstate of, say, the position operator with eigenvalue q, we ask for the transition amplitude for the system to be found at q' at a later time t'. Using base kets in the Heisenberg representation[1], it is

$$\langle q', t' | q, t \rangle = \langle q', t | e^{-iH(t'-t)} | q, t \rangle,$$

where H is the Hamiltonian of the system. At temperature β^{-1}, the analogous quantity of interest is the partition function

$$\mathrm{Tr}\, e^{-\beta H} = \int_{-\infty}^{+\infty} dq \langle q, t | e^{-\beta H} | q, t \rangle.$$

We compare the time evolution operator and the Boltzmann weight in the two cases. The operator $e^{-iH(t'-t)}$ evolves the system from time t to t' with both the times on the real axis. In the same way, if we write the operator $e^{-\beta H} = e^{-iH(\tau-i\beta-\tau)}$, where τ may take complex values, it may be thought to evolve the system from 'time' τ to $\tau - i\beta$, both taking values on a contour C starting from, say, τ_0 and ending at $\tau_0 - i\beta$ in the complex time plane (Figure 4.1a).

As in the case of vacuum field theory, the basic ingredient of the thermal theory is the propagator. To define it on C, the latter must satisfy certain conditions, which we now find. As the Lorentz structure is not relevant here, we shall work with the scalar field $\phi(x)$ with $x = (\tau, \boldsymbol{x})$. Denoting the (canonical) ensemble average of an operator $\mathcal{O}$ by

$$\langle \mathcal{O} \rangle = \mathrm{Tr}\,(\rho_0 \mathcal{O}), \qquad \rho_0 = e^{-\beta H}/Z, \qquad Z = \mathrm{Tr}\, e^{-\beta H} \tag{4.1.1}$$

---

[1] Recall that a base ket in the Heisenberg representation evolves in time in the opposite direction to a state ket in the Schrödinger representation.

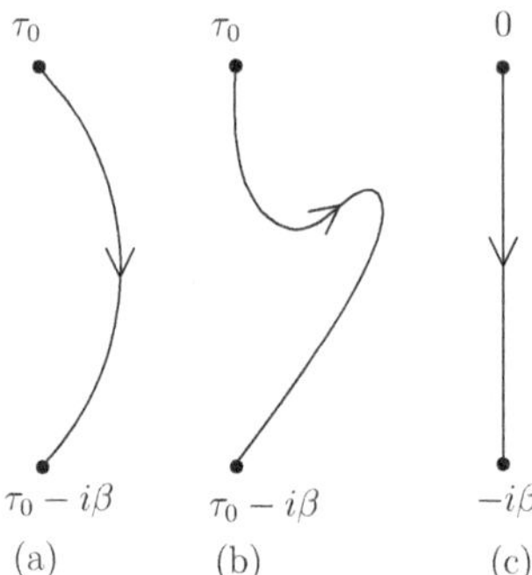

Figure 4.1 Possible contours in time plane. Analyticity requirement allows (a) and (c), but forbids (b).

the thermal propagator for the scalar field is defined as

$$
\begin{aligned}
D(x, x') &\equiv i\langle T_c \phi(x)\phi(x')\rangle \\
&= i\theta_c(\tau - \tau')\langle \phi(x)\phi(x')\rangle + i\theta_c(\tau' - \tau)\langle \phi(x')\phi(x)\rangle \\
&\equiv \theta_c(\tau - \tau')D_+(x, x') + \theta_c(\tau' - \tau)D_-(x, x')
\end{aligned}
\tag{4.1.2}
$$

where τ and τ' are any two points on the contour. Here and in the following T_c, $\theta_c(\tau - \tau')$ and $\delta_c(\tau - \tau')$ are generalisations to the contour of respectively the time ordering symbol, theta- and delta-functions on the real line[2].

We now obtain the regions in which D_+ and D_- are defined in the complex time plane [5]. Consider first $D_+(x, x')$. To extract the τ dependence of fields in it, take a complete set of eigenstates $|m\rangle$ of H with eigenvalues E_m and use it twice; in addition to inserting it between the fields, as we did for the vacuum expectation value (2.3.9), also evaluate the thermal trace in the same basis

$$
D_+(x, x') = \frac{i}{Z}\sum_{m,n} e^{-\beta E_m}\langle m|\phi(\tau, \boldsymbol{x})|n\rangle\langle n|\phi(\tau', \boldsymbol{x}')|m\rangle.
\tag{4.1.3}
$$

Restricting to translation in τ only, (2.3.10) gives for $\phi(x)$

$$
\phi(\tau, \boldsymbol{x}) = e^{iH\tau}\phi(0, \boldsymbol{x})e^{-iH\tau}.
\tag{4.1.4}
$$

Using it in (4.1.3) we get

$$
D_+(x, x') = \frac{i}{Z}\sum_{m,n} e^{iE_m(\tau-\tau'+i\beta)}\, e^{-iE_n(\tau-\tau')}\langle m|\phi(0, \boldsymbol{x})|n\rangle\langle n|\phi(0, \boldsymbol{x}')|m\rangle.
$$

$$
\tag{4.1.5}
$$

---

[2] As the notation indicates, theta (and delta) functions are understood only with respect to a contour: $\theta_c(\tau - \tau') = 1$ or 0, according to whether τ occurs later or earlier than τ' on the contour. To write it with a real argument, the variable τ may be put in one-to-one correspondence with a real parameter λ, say, increasing as the contour is traversed from τ_0 to $\tau_0 - i\beta$. (One simple parameterization is to take λ as the real (or imaginary) part of τ itself.) Then the theta (or delta) are functions of the difference of such parameters.

The spectrum of H, which is unbounded above, determines the convergence of the sums and hence the domain of definition in the τ plane. The sums over m and n converge respectively for $0 < \mathrm{Im}(\tau - \tau' + i\beta)$ and $\mathrm{Im}(\tau - \tau') < 0$. Including the boundary in both the inequalities, we get this domain as

$$- \beta \leq \mathrm{Im}(\tau - \tau') \leq 0. \tag{4.1.6}$$

It restricts the time contour C. The theta function multiplying $D_+(x, x')$ requires τ to appear later than τ' on C. So the condition (4.1.6) is fulfilled for C, if it is drawn from τ_0 to $\tau_0 - i\beta$ in such a way that nowhere is it directed upward in the complex τ plane as in Figure 4.1 (a) and (c). The second term in (4.1.2) has the domain and theta function the same as the first term with τ and τ' interchanged and so gives the same result. A similar analysis holds for an n-point function [5].

With this restriction, a variety of convenient choices of the time contour is possible. One choice, where the contour runs over the imaginary axis from 0 to $-i\beta$ (Figure 4.1c) gives rise to the imaginary time method (see Appendix G). However, to compute Green's functions with real time arguments directly, it is necessary to draw one segment of the contour running over the whole of the real axis. Here two choices are generally made[3]. One covers the real time axis from $-\bar{t}$ to $+\bar{t}$ twice (where $\bar{t}$ is large, eventually to tend to ∞), once in the positive direction and then in the negative direction, always with infinitesimal downward slope and finally straight vertically down to $-\bar{t} - i\beta$ [3]. The other traces the real axis from $-\bar{t}$ to $+\bar{t}$ as before, but continues parallel to the imaginary axis to $\bar{t} - i\beta/2$, then parallel to the real axis to $-\bar{t} - i\beta/2$ and finally again vertically to $-\bar{t} - i\beta$ [6, 7]. The latter choice, shown in Figure 4.2, produces a symmetrical propagator matrix in the absence of chemical potential and will be used throughout this book.

∗   ∗   ∗

We have a comment on the Lorentz invariance or covariance of two-point functions. As we have already seen for vacuum propagators in Chapter 1, it can

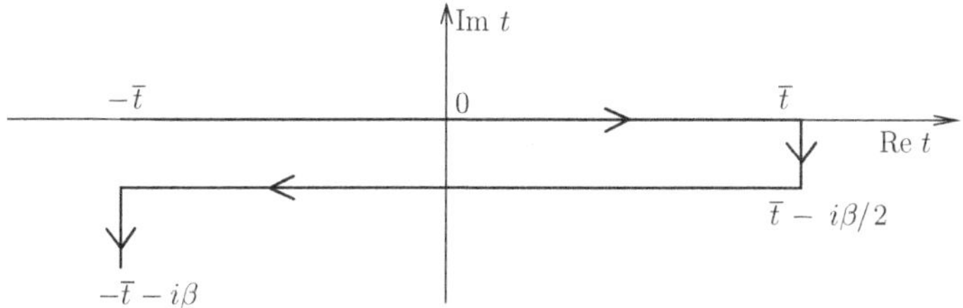

Figure 4.2 The contour C in time plane for real time formalism.

---

[3] In principle, one can choose the return horizontal path at an arbitrary value of $\mathrm{Im}\,\tau = -\sigma \, (0 \leq \sigma \leq \beta)$, so that it traces this path from $\bar{t} - i\sigma$ to $-\bar{t} - i\sigma$; then the vertical paths run from $\bar{t}$ to $\bar{t} - i\sigma$ and $-\bar{t} - i\sigma$ to $-\bar{t} - i\beta$ [10]. The two particular choices made above correspond to $\sigma = 0$ and $\sigma = \beta/2$.

be disturbed by time ordering of fields, if they belong to spin ≥ 1. This non-covariance is, however, only by local terms and so can be cancelled by adding a local, non-covariant term in the interaction. In the thermal context, we have apparently another non-covariance due to the Boltzmann weight in the form $e^{-\beta H}$. It can again be remedied by introducing the four-velocity u^μ ($u^\mu u_\mu = 1$) of the heat bath, when we can replace the Hamiltonian H by $u^\mu P_\mu$, where P_μ is the energy-momentum operator. Kinematically we then have a four-vector u^μ in addition to k^μ, the Fourier variable conjugate to x^μ. So we have another Lorentz scalar $\omega = u^\mu k_\mu$ besides k^2. Let us choose ω and $\bar{k} = \sqrt{\omega^2 - k^2}$ as the two independent variables. Then in the rest frame of the heat bath, $u^\mu = (1, \vec{0})$, which we shall actually use in computations, they reduce to time and space components of k^μ.

4.2 Scalar Propagator

We find the scalar field propagator in two different ways, namely by solving its differential equation and by evaluating the thermal expectation value of field product.

Differential Equation Method

As the simplest case, we begin with a hermitian (real) scalar field $\phi(x)$. Its vacuum propagator satisfies the differential equation (P1.4) of Problem 1.2. As the differential operator has no effect on the state, the thermal propagator (4.1.2) also satisfies the same equation

$$(\Box_c + m^2)D(x, x') = \delta_c^4(x - x'). \qquad (4.2.1)$$

To get the thermal boundary condition on the solution, we transform any one of $D_\pm$, say D_- as

$$\begin{aligned}
D_-(\boldsymbol{x}, \boldsymbol{x}'; \tau, \tau') &\equiv iZ^{-1} \operatorname{Tr} e^{-\beta H} \phi(\boldsymbol{x}', \tau')\phi(\boldsymbol{x}, \tau) \\
&= iZ^{-1} \operatorname{Tr} e^{-\beta H} \{e^{\beta H} \phi(\boldsymbol{x}, \tau)e^{-\beta H}\}\phi(\boldsymbol{x}', \tau'), \qquad (4.2.2)
\end{aligned}$$

on using the cyclicity of the trace and then inserting $e^{-\beta H}e^{\beta H} = 1$ within it. By (4.1.4) the quantity within the curly bracket gives the field with τ shifted to $\tau - i\beta$, when (4.2.2) gives the required condition

$$D_-(\boldsymbol{x}, \boldsymbol{x}'; \tau, \tau') = D_+(\boldsymbol{x}, \boldsymbol{x}'; \tau - i\beta, \tau') \qquad (4.2.3)$$

known as Kubo [15], Martin and Schwinger [16] (KMS) condition. It is a periodicity condition on the propagator in the time variable τ (see G.2 of Appendix G) and points to the essentially complex nature of τ in the thermal context. As the boundary condition does not involve the space coordinates, it is convenient to take the spatial Fourier transform of the propagator

$$D(x, x') = \int \frac{d^3k}{(2\pi)^3} e^{ik\cdot(x-x')} D(k; \tau, \tau'). \tag{4.2.4}$$

Then (4.2.1) gives the differential equation for the transform

$$\left(\frac{d^2}{d\tau^2} + \omega^2\right) D(k; \tau, \tau') = \delta_c(\tau - \tau'), \qquad \omega^2 = k^2 + m^2 \tag{4.2.5}$$

with the boundary condition following from (4.2.3)

$$D_-(k; \tau, \tau') = D_+(k; \tau - i\beta, \tau') \tag{4.2.6}$$

We now find $D(k; \tau, \tau')$ as the Green's function of the differential equation (4.2.5) subject to (4.2.6), in a way we did for the vacuum case in Problem 1.3. For $\tau \neq \tau'$, D satisfies the homogeneous equation, having solutions $e^{\pm i\omega\tau}$. So each of $D_\pm$ is expressible as linear combination of these solutions with constant (τ' dependent) coefficients

$$D_+(k; \tau, \tau') = A_1(\tau')e^{-i\omega\tau} + A_2(\tau')e^{i\omega\tau}, \quad \tau > \tau' \tag{4.2.7}$$
$$D_-(k; \tau.\tau') = B_1(\tau')e^{-i\omega\tau} + B_2(\tau')e^{i\omega\tau}, \quad \tau < \tau'. \tag{4.2.8}$$

Further, $D(k; \tau, \tau')$ must satisfy two conditions at $\tau = \tau'$, namely that it is continuous and its first derivative has unit discontinuity at this point. Using these conditions, let us choose to eliminate B_1 and B_2 in terms of A_1 and A_2, when D_- becomes

$$D_-(k; \tau, \tau') = D_+(k; \tau, \tau') - \frac{i}{2\omega}\left(e^{-i\omega(\tau-\tau')} - e^{i\omega(\tau-\tau')}\right). \tag{4.2.9}$$

Next we insert (4.2.9) for D_- and (4.2.7) for D_+ in the KMS relation (4.2.6) and equate separately the coefficients of $\exp(\pm i\omega\tau)$ on both sides. The resulting two equations determine A_1 and A_2. We thus get

$$D(k; \tau - \tau') = \frac{i}{2\omega}\left[\{\theta_c(\tau - \tau') + n\}e^{-i\omega(\tau-\tau')} + \{\theta_c(\tau' - \tau) + n\}e^{i\omega(\tau-\tau')}\right] \tag{4.2.10}$$

where n turns out to be the single particle thermal distribution function

$$n(\omega) = \frac{1}{e^{\beta\omega} - 1}. \tag{4.2.11}$$

As expected $D(k; \tau - \tau')$ goes over to the vacuum result, $\Delta_F(\omega, t - t')$ given by (1.2.11), if we take τ and τ' on the real axis and set n to zero.

So far we have left the time contour unspecified. As already stated, we shall choose the one of Figure 4.2. Accordingly the propagator breaks up into pieces with points on different segments of the contour. Let us now take $\bar{t} \to \infty$ for the contour to cover the entire real axis. Then if τ_h and τ_v are points respectively on the horizontal and vertical segments, the pieces, $D(k, \tau_h - \tau_v) \to 0$ by Riemann–Lebesgue lemma [17]. We are interested in the pieces of the propagator with points on the real axis and those that arise in perturbative calculations. Then the pieces with both points on vertical segments drop out. The contour thus reduces essentially to two parallel lines, one being the real axis and the other shifted from it by $-i\beta/2$, points on which will be denoted respectively by subscripts 1 and 2, so that $\tau_1 = t$, $\tau_2 = t - i\beta/2$. The propagator then consists of four pieces, which may be put in the form of a 2×2 matrix. For clarity, we rewrite (4.2.10) as the ij-th element of this matrix

$$D(\boldsymbol{k}; \tau_i, \tau_j') = \frac{i}{2\omega}\left[\{\theta_c(\tau_i - \tau_j') + n\}e^{-i\omega(\tau_i - \tau_j')} + \{\theta_c(\tau_j' - \tau_i) + n\}e^{i\omega(\tau_i - \tau_j')}\right]$$

$$(4.2.12)$$

The contour-ordered theta functions may now be converted to the usual time-ordered ones. If τ, τ' are both on line 1 (the real axis), the τ and t orderings coincide, $\theta_c(\tau_1 - \tau_1') = \theta(t - t')$. If they are on two different lines, the τ orderings are fixed, $\theta_c(\tau_1 - \tau_2') = 0$, $\theta_c(\tau_2 - \tau_1') = 1$. Finally, if they are both on line 2, the two orderings are opposite, $\theta_c(\tau_2 - \tau_2') = \theta(t' - t)$.

Back to real time, we can take the temporal Fourier transform of the components as usual,

$$D_{ij}(\boldsymbol{k}, k_0) = \int_{-\infty}^{\infty} dt\, e^{ik_0(t-t')} D(\boldsymbol{k}, \tau_i, \tau_j') \ . \qquad (4.2.13)$$

Thus

$$D_{11} = \frac{i}{2\omega}\left\{\int_0^{\infty} dt\, e^{i(k_0 - \omega + i\eta)t} + \int_{-\infty}^0 dt e^{i(k_0 + \omega - i\eta)t}\right\}$$

$$+ \frac{in}{2\omega}\int_{-\infty}^{\infty} dt\{e^{i(k_0 - \omega)t} + e^{i(k_0 + \omega)t}\}$$

$$= \Delta_F(k) + 2\pi in(\omega)\delta(k^2 - m^2) \qquad (4.2.14)$$

where the integrals in the first line are defined by the $i\eta$ terms ($\eta \to 0$) in the exponential and $\Delta_F(k)$ is the vacuum propagator (1.2.14). Also

$$D_{12} = \frac{i}{2\omega}\left\{ne^{\beta\omega/2}\int_{-\infty}^{+\infty} dt\, e^{i(k_0 - \omega)t} + (1 + n)e^{-\beta\omega/2}\int_{-\infty}^{+\infty} dt\, e^{i(k_0 + \omega)t}\right\}$$

$$= 2\pi i\sqrt{n(1 + n)}\delta(k^2 - m^2). \qquad (4.2.15)$$

The other two elements can be expressed in terms of them

$$D_{21} = D_{12}, \qquad\qquad D_{22} = -D_{11}^*. \qquad (4.2.16)$$

Collecting the elements, we get the Fourier transform of the matrix propagator

$$\begin{pmatrix} D(\boldsymbol{x}, \boldsymbol{x}'; t, t') & D(\boldsymbol{x}, \boldsymbol{x}'; t, t' - i\beta/2) \\ D(\boldsymbol{x}, \boldsymbol{x}'; t - i\beta/2, t') & D(\boldsymbol{x}, \boldsymbol{x}'; t', t) \end{pmatrix} = \int \frac{d^4k}{(2\pi)^4} e^{-ik\cdot(x-x')} \boldsymbol{D}(\boldsymbol{k}, k_0)$$

where

$$\boldsymbol{D}(\boldsymbol{k}, k_0) = \begin{pmatrix} \Delta_F + 2\pi in\delta(k^2 - m^2) & 2\pi i\sqrt{n(1 + n)}\delta(k^2 - m^2) \\ 2\pi i\sqrt{n(1 + n)}\delta(k^2 - m^2) & -\Delta_F^* + 2\pi in\delta(k^2 - m^2) \end{pmatrix}.$$

$$(4.2.17)$$

Replacing $2\pi i\delta(k^2 - m^2)$ by $\Delta_F(k) - \Delta_F^*(k)$, it becomes

$$\boldsymbol{D}(\boldsymbol{k}, k_0) = \begin{pmatrix} (1+n)\Delta_F - n\Delta_F^* & \sqrt{n(1+n)}(\Delta_F - \Delta_F^*) \\ \sqrt{n(1+n)}(\Delta_F - \Delta_F^*) & n\Delta_F - (1+n)\Delta_F^* \end{pmatrix} \quad (4.2.18)$$

which can be put in the diagonal form

$$\boldsymbol{D}(\boldsymbol{k}, k_0) = \boldsymbol{U}(\omega) \begin{pmatrix} \Delta_F(k) & 0 \\ 0 & -\Delta_F^*(k) \end{pmatrix} \boldsymbol{U}(\omega) \quad (4.2.19)$$

by

$$\boldsymbol{U}(\omega) = \begin{pmatrix} \sqrt{1+n} & \sqrt{n} \\ \sqrt{n} & \sqrt{1+n} \end{pmatrix}. \quad (4.2.20)$$

Notice that the diagonalising matrix $\boldsymbol{U}$ is not orthogonal, $\boldsymbol{U}\boldsymbol{U}^T \neq 1$.

We see the incompleteness in representing the real time thermal propagator by the 11-component only as in [18]. The time contour C lies inherently in the complex plane, beginning from a point at $-\bar{t}$ on the real axis and ending at $-\bar{t} - i\beta$ within the domain of definition of the propagator. We have just seen that if this contour has to traverse the real axis, it must consist of (at least) two lines giving rise to the 2×2 matrix for the propagator.

$$* \quad * \quad *$$

The above derivation for the hermitian scalar field can be readily extended to a non-hermitian field. From the conserved current j_μ given by (P1.16) of Problem 1.4, we get the associated charge

$$Q = i \int d^3x \phi^\dagger(x) \overleftrightarrow{\partial_0} \phi(x). \quad (4.2.21)$$

The statistical averaging (4.1.1) of an operator $\mathcal{O}$ can then be extended to the grand canonical ensemble

$$\langle \mathcal{O} \rangle = \mathrm{Tr}\left[e^{-\beta(H-\mu Q)}\mathcal{O}\right]/Z, \qquad Z = \mathrm{Tr}\, e^{-\beta(H-\mu Q)} \quad (4.2.22)$$

where μ is called the chemical potential. Retaining the earlier notation, the thermal propagator is

$$D(x, x') = i\langle T_c \phi(x)\phi^\dagger(x')\rangle. \quad (4.2.23)$$

It can be evaluated as before to give (Problem 4.1)

$$\boldsymbol{D}(\boldsymbol{k}, k_0) = \boldsymbol{U}' \begin{pmatrix} \Delta_F & 0 \\ 0 & -\Delta_F^* \end{pmatrix} \boldsymbol{U}' \quad (4.2.24)$$

with $\boldsymbol{U}'$ differing from $\boldsymbol{U}$ by factors involving the chemical potential

$$\boldsymbol{U}' = \begin{pmatrix} N_2 & N_1 e^{\beta\mu/2} \\ N_1 e^{-\beta\mu/2} & N_2 \end{pmatrix} \quad (4.2.25)$$

where

$$N_1(k_0) = \theta(k_0)\sqrt{n_+} + \theta(-k_0)\sqrt{n_-}$$
$$N_2(k_0) = \theta(k_0)\sqrt{1 + n_+} + \theta(-k_0)\sqrt{1 + n_-}. \qquad (4.2.26)$$

Here $n_\pm$ denotes the distribution functions for particles and antiparticles

$$n_\pm(\omega) = \frac{1}{e^{\beta(\omega \mp \mu)} - 1}. \qquad (4.2.27)$$

Operator Method

In the above derivation we first find the propagator with respect to the entire time contour, which then breaks up effectively to four pieces forming the 2×2 matrix. Once we realise that it is only the two horizontal pieces of the contour that are relevant for the construction, we can derive the matrix propagator directly from its definition (4.2.23) in terms of the field operator, requiring no explicit use of the thermal boundary condition. Restricting to these segments of the contour, the ij-th elements

$$i\langle T_c\phi(\boldsymbol{x}, \tau_i)\phi^\dagger(\boldsymbol{x}', \tau_j')\rangle$$
$$= i\theta_c(\tau_i - \tau_j')\langle\phi(\boldsymbol{x}, \tau_i)\phi^\dagger(\boldsymbol{x}', \tau_j')\rangle + i\theta_c(\tau_j' - \tau_i)\langle\phi^\dagger(\boldsymbol{x}', \tau_j')\phi(\boldsymbol{x}, \tau_i)\rangle \quad (4.2.28)$$

build the matrix

$$i\begin{pmatrix} \langle T\phi(\boldsymbol{x}, t)\phi^\dagger(\boldsymbol{x}', t')\rangle & \langle\phi^\dagger(\boldsymbol{x}', t' - i\beta/2)\phi(\boldsymbol{x}, t)\rangle \\ \langle\phi(\boldsymbol{x}, t - i\beta/2)\phi^\dagger(\boldsymbol{x}', t')\rangle & \langle\widetilde{T}\phi(\boldsymbol{x}, t - i\beta/2)\phi^\dagger(\boldsymbol{x}', t' - i\beta/2)\rangle \end{pmatrix} \qquad (4.2.29)$$

where $\widetilde{T}$ stands for anti-time ordering. We now find the rationale in introducing a doubling of field variables, $\phi_1(x) \equiv \phi(\boldsymbol{x}, t)$ and $\phi_2(x) \equiv \phi(\boldsymbol{x}, t - i\beta/2)$ in earlier works on statistical mechanics [6].

To work out the ensemble averages in (4.2.29), we put the system in a finite box of volume V following Appendix E. In analogy to the definition (1.1.2) in the continuum, we define the creation operator $A_k^\dagger$ to generate a one-particle box state from the vacuum

$$|\boldsymbol{k}\rangle^{box} = A_k^\dagger|0\rangle$$

and similarly the creation operator $B_k^\dagger$ for a one-antiparticle box state. Then the correspondence (E.4) between the continuum and box states gives the correspondence between the operators

$$a^\dagger(\boldsymbol{k}), \, b^\dagger(\boldsymbol{k}) \to \sqrt{2\omega V}(A_k^\dagger, \, B_k^\dagger) \qquad (4.2.30)$$

and of course between their adjoints, with the non-vanishing commutation relations

$$[A_k, A_{k'}^\dagger] = \delta_{k,k'}, \qquad\qquad [B_k, B_{k'}^\dagger] = \delta_{k,k'}.$$

Using (4.2.30) and (E.2) the continuum expansion (1.2.3) extended to the non-hermitian field becomes

$$\phi(x) \rightarrow \sum \frac{1}{\sqrt{2\omega V}} (A_k e^{-ik\cdot x} + B_k^\dagger e^{ik\cdot x}). \tag{4.2.31}$$

On inserting the expansion (4.2.31) in the components of the propagator, we meet thermal traces quadratic in creation and annihilation operators. These are (Problem 4.2)

$$\langle A_k^\dagger A_{k'} \rangle = n_+(\omega)\delta_{kk'}, \qquad \langle A_k A_{k'}^\dagger \rangle = \{1 + n_+(\omega)\}\delta_{kk'},$$
$$\langle B_k^\dagger B_{k'} \rangle = n_-(\omega)\delta_{kk'}, \qquad \langle B_k B_{k'}^\dagger \rangle = \{1 + n_-(\omega)\}\delta_{kk'}. \tag{4.2.32}$$

Using these traces, we get, for example the 11-component as

$$i\langle T\phi(\boldsymbol{x},t)\phi^\dagger(\boldsymbol{x}',t')\rangle = \sum_k \frac{i}{2\omega V} \Big[\{\theta(t-t') + n_+\}e^{-ik\cdot(x-x')}$$
$$+ \{\theta(t'-t) + n_-\}e^{ik\cdot(x-x')}\Big]. \tag{4.2.33}$$

Going over to infinite volume limit and changing the sign of $\boldsymbol{k}$ in the second term, we get

$$i\langle T\phi(\boldsymbol{x},t)\phi^\dagger(\boldsymbol{x}',t')\rangle$$
$$= \int \frac{d^3k}{(2\pi)^3} e^{ik\cdot(x-x')} \left\{ \Delta_F(\omega, t-t') + \frac{i}{2\omega}\left(n_+ e^{-i\omega(t-t')} + n_- e^{i\omega(t-t')} \right) \right\} \tag{4.2.34}$$

where Δ_F, given by (1.2.11), reproduces the vacuum part. For another component, say 12, we get

$$i\langle \phi^\dagger(\boldsymbol{x}',t'-i\beta/2)\phi(\boldsymbol{x},t)\rangle$$
$$= \int \frac{d^3k}{(2\pi)^3} e^{ik\cdot(x-x')}\frac{i}{2\omega} \left\{ n_+ e^{\beta\omega/2}e^{-i\omega(t-t')} + (1+n_-)e^{-\beta\omega/2}e^{i\omega(t-t')} \right\}$$
$$\tag{4.2.35}$$

Taking temporal Fourier transform, these expressions lead to (P4.5) of Problem 4.1, reproducing the propagator (4.2.24).

4.3 Dirac Propagator

The thermal propagator for the Dirac field may also be obtained by the same methods we applied to the scalar propagator. Corresponding to the conserved (vector) current V_μ given by (P1.20) of Problem 1.5, we have the charge

$$Q = \int d^3x \psi^\dagger(x)\psi(x). \tag{4.3.1}$$

The Dirac thermal propagator averaged over the grand canonical ensemble reads

$$S(x,x') \equiv i\,\mathrm{Tr}\ e^{-\beta(H-\mu Q)}T_c\psi(x)\overline{\psi}(x')/Z. \tag{4.3.2}$$

Differential Equation Method

Like the scalar field case, $S(x, x')$ satisfies the same differential equation as for the vacuum propagator

$$(i\partial\!\!\!/_c - m)S(x, x') = -\delta_c^4(x - x'). \tag{4.3.3}$$

To derive the thermal boundary condition we write the propagator as

$$S(x, x') = \theta_c(\tau - \tau')S_+(x, x') + \theta_c(\tau' - \tau)S_-(x, x') \tag{4.3.4}$$

with

$$S_+(x, x') = i\langle\psi(x)\overline{\psi}(x')\rangle, \qquad S_-(x, x') = -i\langle\overline{\psi}(x')\psi(x)\rangle.$$

Noting $[Q, \psi(x)] = -\psi(x)$, we can use (P4.2) of Problem (4.1) to get

$$S_-(\boldsymbol{x}, \boldsymbol{x}'; \tau, \tau') = -e^{\beta\mu}S_+(\boldsymbol{x}, \boldsymbol{x}'; \tau - i\beta, \tau') \tag{4.3.5}$$

The derivation of the Dirac propagator follows the scalar field case, if we write

$$S(x, x') = (i\partial\!\!\!/_c + m)D(x, x'). \tag{4.3.6}$$

Applying $(i\partial\!\!\!/_c - m)$ on (4.3.6) and noting (4.3.3), we find that $D(x, x')$ is indeed the Green's function satisfying (4.2.1) for the scalar propagator. Again taking spatial Fourier transform

$$S(x, x') = \int \frac{d^3p}{(2\pi)^3} e^{ip\cdot(\boldsymbol{x}-\boldsymbol{x}')} S(\boldsymbol{p}, \tau, \tau'), \tag{4.3.7}$$

the relation (4.3.6) gives

$$S(\boldsymbol{p}, \tau, \tau') = (i\gamma^0\partial_0 - \vec{\gamma}\cdot\boldsymbol{p} + m)D(\boldsymbol{p}, \tau, \tau') . \tag{4.3.8}$$

As a Green's function D has the same solution as in the scalar field case, with D_+ and D_- given by (4.2.7) and (4.2.9). Inserting this solution in (4.3.8) and noting that the time derivative acting on the theta-functions cancels out, we read off $S_\pm$ as

$$S_+(\boldsymbol{p}, \tau, \tau') = A_1(\tau')c_+e^{-i\omega\tau} + A_2(\tau')c_-e^{i\omega\tau}$$

$$S_-(\boldsymbol{p}; \tau, \tau') = S_+(\boldsymbol{p}, \tau, \tau') - \frac{i}{2\omega}\left\{c_+e^{-i\omega(\tau-\tau')} - c_-e^{i\omega(\tau-\tau')}\right\} \tag{4.3.9}$$

where $c_\pm = \pm\gamma^0\omega - \gamma\cdot\boldsymbol{p} + m$. The KMS relation (4.3.5) now determines A_1 and A_2, getting the propagator as

$$S(\boldsymbol{p}; \tau - \tau')$$

$$= \frac{i}{2\omega}\left[c_+e^{-i\omega(\tau-\tau')}\{\theta_c(\tau - \tau') - \tilde{n}_+\} + c_-e^{i\omega(\tau-\tau')}\{\theta_c(\tau' - \tau) - \tilde{n}_-\}\right] \tag{4.3.10}$$

where $\tilde{n}_\pm$ are the distribution functions for particles and antiparticles,

$$\tilde{n}_\pm(\omega) = \frac{1}{e^{\beta(\omega\mp\mu)} + 1}, \qquad \omega = \sqrt{\boldsymbol{p}^2 + m^2}. \tag{4.3.11}$$

As in the scalar case, we break up (4.3.10) into four components with real times and evaluate their temporal Fourier transforms. For example,

$$
\begin{aligned}
S_{11} &= \frac{i}{2\omega}\left[\int_0^\infty dt\{c_+ e^{i(p_0-\omega+i\eta)t} + c_- e^{-i(p_0+\omega-i\eta)t}\}\right.\\
&\quad \left.- \int_{-\infty}^\infty dt\{c_+\tilde{n}_+ e^{i(p_0-\omega)t} - c_-\tilde{n}_- e^{i(p_0+\omega)t}\}\right]\\
&= (\not{p}+m)\{\Delta_F(p) - 2\pi i\tilde{N}_1^2\delta(p^2-m^2)\}
\end{aligned}
\tag{4.3.12}
$$

$$
\begin{aligned}
S_{12} &= \frac{i}{2\omega}\int_{-\infty}^\infty dt\, e^{ip_0 t}\{-e^{-i\omega t}e^{\beta\omega/2}c_+\tilde{n}_+ + e^{i\omega t}e^{-\beta\omega/2}c_-(1-\tilde{n}_-)\}\\
&= -2\pi i(\not{p}+m)e^{\beta\mu/2}\tilde{N}_1\tilde{N}_2\delta(p^2-m^2)
\end{aligned}
\tag{4.3.13}
$$

where

$$
\begin{aligned}
\tilde{N}_1(p_0) &= \theta(p_0)\sqrt{\tilde{n}_+} + \theta(-p_0)\sqrt{\tilde{n}_-}\\
\tilde{N}_2(p_0) &= \theta(p_0)\sqrt{1-\tilde{n}_+} - \theta(-p_0)\sqrt{1-\tilde{n}_-}.
\end{aligned}
\tag{4.3.14}
$$

We thus get the matrix propagator

$$
\begin{pmatrix} S(\boldsymbol{x},\boldsymbol{x}';t,t') & S(\boldsymbol{x},\boldsymbol{x}';t,t'-i\beta/2)\\ S(\boldsymbol{x},\boldsymbol{x}';t-i\beta/2,t') & S(\boldsymbol{x},\boldsymbol{x}';t',t) \end{pmatrix} = \int \frac{d^4p}{(2\pi)^4} e^{-ip\cdot(x-x')}\boldsymbol{S}(\boldsymbol{p},p_0)
$$

where

$$
\boldsymbol{S}(\boldsymbol{p},p_0) = (\not{p}+m)\begin{pmatrix} s_{11} & s_{12}\\ s_{21} & s_{22} \end{pmatrix}
\tag{4.3.15}
$$

with the matrix elements

$$
\begin{aligned}
s_{11} &= -s_{22}^* = \Delta_F(p,m) - 2\pi i\tilde{N}_1^2\delta(p^2-m^2)\\
s_{12} &= -2\pi i e^{\beta\mu/2}\tilde{N}_1\tilde{N}_2\delta(p^2-m^2)\\
s_{21} &= 2\pi i e^{-\beta\mu/2}\tilde{N}_1\tilde{N}_2\delta(p^2-m^2).
\end{aligned}
\tag{4.3.16}
$$

The matrix can be diagonalised as

$$
\boldsymbol{S}(\boldsymbol{p},p_0) = (\not{p}+m)\boldsymbol{V}\begin{pmatrix} \Delta_F & 0\\ 0 & -\Delta_F^* \end{pmatrix}\boldsymbol{V}
\tag{4.3.17}
$$

where

$$
\boldsymbol{V} = \begin{pmatrix} \tilde{N}_2 & -\tilde{N}_1 e^{\beta\mu/2}\\ \tilde{N}_1 e^{-\beta\mu/2} & \tilde{N}_2 \end{pmatrix}.
\tag{4.3.18}
$$

Operator Method

Like the matrix (4.2.29) for the scalar propagator, the definition of the Dirac propagator (4.3.4) may be split to form the matrix

$$
i \begin{pmatrix} \langle T\psi(\boldsymbol{x},t)\overline{\psi}(\boldsymbol{x}',t')\rangle & -\langle \overline{\psi}(\boldsymbol{x}',t'-i\beta/2)\psi(\boldsymbol{x},t)\rangle \\ \langle \psi(\boldsymbol{x},t-i\beta/2)\overline{\psi}(\boldsymbol{x}',t')\rangle & \langle \widetilde{T}\psi(\boldsymbol{x},t-i\beta/2)\overline{\psi}(\boldsymbol{x}',t'-i\beta/2)\rangle \end{pmatrix} . \tag{4.3.19}
$$

In the finite volume the expansion of the field operator reads

$$
\psi(x) = \sum \frac{1}{\sqrt{2\omega V}} \sum_\sigma \left(C_{p,\sigma} u_{p,\sigma} e^{-ip\cdot x} + D_{p,\sigma}^\dagger v_{p,\sigma} e^{ip\cdot x} \right), \tag{4.3.20}
$$

where the creation and annihilation operators are normalised according to

$$
\{C_{p,\sigma}, C_{p,\sigma}^\dagger\} = \{D_{p,\sigma}, D_{p,\sigma}^\dagger\} = \delta_{p,p'}\delta_{\sigma,\sigma'}.
$$

Their thermal traces are again defined by (P4.6) of Problem 4.2, but with $m_i = 0, 1$, as we now deal with fermions. We thus get

$$
\langle C_{p,\sigma}^\dagger C_{p',\sigma'}\rangle = \widetilde{n}_+ \delta_{p,p'}\delta_{\sigma.\sigma'}, \quad \langle C_{p,\sigma} C_{p',\sigma'}^\dagger\rangle = (1-\widetilde{n}_+)\delta_{p,p'}\delta_{\sigma.\sigma'}
$$

$$
\langle D_{p,\sigma}^\dagger D_{p',\sigma'}\rangle = \widetilde{n}_- \delta_{p,p'}\delta_{\sigma.\sigma'}, \quad \langle D_{p,\sigma} D_{p',\sigma'}^\dagger\rangle = (1-\widetilde{n}_-)\delta_{p,p'}\delta_{\sigma.\sigma'}. \tag{4.3.21}
$$

As an example, let us find the 11 element of the propagator. Using (4.3.21) and the polarisation sums (1.3.11), we get it as

$$
\int \frac{d^3p}{(2\pi)^3}\frac{i}{2\omega} \left\{ \theta(t-t')(\not{p}+m)e^{-ip\cdot(x-x')} - \theta(t'-t)(\not{p}-m)e^{ip\cdot(x-x')} \right.
$$

$$
\left. -\widetilde{n}_+(\not{p}+m)e^{-ip\cdot(x-x')} + \widetilde{n}_-(\not{p}-m)e^{ip\cdot(x-x')} \right\} \tag{4.3.22}
$$

where the first two terms will be recognised as the vacuum propagator (1.3.14). We can express $(\not{p}\pm m)$ in terms of $c_\pm$ appearing in (4.3.9) and take the temporal Fourier transform to find this element as (4.3.12) in momentum space.

4.4 Time-Ordered Spectral Representation

From (4.2.24) and (4.3.17) we see that the free thermal propagators for scalar and Dirac fields turn out to be simply related to their vacuum counterparts. Though the explicit construction of propagators may be difficult for fields of high spin, these relations continue to hold for any spin. To prove this result, we first derive the spectral representations for thermal two-point functions of general operators. Applied to free fields, these representations yield the relations between thermal and vacuum propagators. Besides proving this result, the spectral representations are interesting in their own right and will find extensive applications in later chapters.

These spectral forms are similar to the Källén–Lehmann representation for vacuum expectation values, which we already obtained for a particular case in

Section 2.3. However, because of our choice of a reference frame (in which the medium is at rest), the Lorentz covariance in the thermal context is not explicit any more. So, unlike the vacuum case, the vanishing of the commutator outside the light cone can no longer be used to relate the two spectral functions appearing in the ensemble average. But as we shall see below, the two functions still become related by the KMS condition.

In Appendix A, we denote general (multicomponent) boson and fermion fields respectively by $\Phi_l(x)$ and $\Psi_l(x)$. In this section they will represent (local) general operators, which may be elementary fields representing particles or composite ones built out of these fields. But we shall also call the two-point functions propagators. Below we treat boson and fermion cases separately, setting the chemical potential to zero in the boson case for simplicity.

Boson Two-Point Function

We begin with the thermal expectation value of the operator product $\Phi_l(x)\Phi_{l'}^\dagger(x')$. As we did with the time coordinate in Section 4.1, we now extract the full space–time dependence of the operators to get

$$\langle\Phi_l(x)\Phi_{l'}^\dagger(x')\rangle = \frac{1}{Z}\sum_{m,n} e^{-\beta E_m}\,e^{i(k_m - k_n)\cdot(x-x')}\langle m|\Phi_l(0)|n\rangle\langle n|\Phi_{l'}^\dagger(0)|m\rangle. \quad (4.4.1)$$

As in Section 4.2, we let τ and τ' be on the contour C in the complex time plane and take the spatial Fourier transform

$$\int d^3x\, e^{-i\boldsymbol{k}\cdot(\boldsymbol{x}-\boldsymbol{x}')}\langle\Phi_l(x)\Phi_{l'}^\dagger(x')\rangle$$

$$= \frac{(2\pi)^3}{Z}\sum_{m,n} e^{-\beta E_m}\,e^{i(E_m - E_n)(\tau-\tau')}\delta^3(\boldsymbol{k}_m - \boldsymbol{k}_n + \boldsymbol{k})\langle m|\Phi_l(0)|n\rangle\langle n|\Phi_{l'}^\dagger(0)|m\rangle.$$

$$(4.4.2)$$

We now insert unity in the form $\int_{-\infty}^{\infty} dk_0'\delta(E_m - E_n + k_0') = 1$. (We reserve k_0 for later use in the temporal Fourier transform.) Then (4.4.2) may be written as

$$\int d^3x\, e^{-i\boldsymbol{k}\cdot(\boldsymbol{x}-\boldsymbol{x}')}\langle\Phi_l(x)\Phi_{l'}^\dagger(x')\rangle = \int\frac{dk_0'}{2\pi}e^{-ik_0'(\tau-\tau')}M_{ll'}^+(k_0',\boldsymbol{k}) \quad (4.4.3)$$

where $M_{ll'}^+$ is given by $[k_\mu' = (k_0',\boldsymbol{k})]$

$$M_{ll'}^+(k') = \frac{(2\pi)^4}{Z}\sum_{m,n} e^{-\beta E_m}\,\delta^4(k_m - k_n + k')\langle m|\Phi_l(0)|n\rangle\langle n|\Phi_{l'}^\dagger(0)|m\rangle. \quad (4.4.4)$$

In just the same way, we can work out the Fourier transform of $\langle\Phi_{l'}^\dagger(x')\Phi_l(x)\rangle$

$$\int d^3x\, e^{-i\boldsymbol{k}\cdot(\boldsymbol{x}-\boldsymbol{x}')}\langle\Phi_{l'}^\dagger(x')\Phi_l(x)\rangle = \int\frac{dk_0'}{2\pi}e^{-ik_0'(\tau-\tau')}M_{ll'}^-(k_0',\boldsymbol{k}) \quad (4.4.5)$$

with $M_{ll'}^-$ given by

$$M_{ll'}^-(k') = \frac{(2\pi)^4}{Z} \sum_{m,n} e^{-\beta E_m} \delta^4(k_n - k_m + k') \langle m|\Phi_{l'}^\dagger(0)|n\rangle\langle n|\Phi_l(0)|m\rangle. \quad (4.4.6)$$

The two functions $M_{ll'}^\pm$ defined by (4.4.4) and (4.4.6) are not independent, being related to each other by

$$M_{ll'}^+(k) = e^{\beta k_0} M_{ll'}^-(k), \quad (4.4.7)$$

as can be seen by interchanging the dummy indices m, n in one of $M_{ll'}^\pm(k)$ and using the energy conserving δ-function. This relation is actually the momentum space version of the KMS condition in coordinate space. Also, their definitions are free from any reference to the time contour, allowing us to rewrite them as four-dimensional Fourier transforms of the operator products we started with

$$M_{ll'}^+(k_0, \boldsymbol{k}) = \int d^4y\, e^{ik\cdot(y-y')}\langle\Phi_l(y)\Phi_{l'}^\dagger(y')\rangle, \quad (4.4.8)$$

$$M_{ll'}^-(k_0, \boldsymbol{k}) = \int d^4y\, e^{ik\cdot(y-y')}\langle\Phi_{l'}^\dagger(y')\Phi_l(y)\rangle. \quad (4.4.9)$$

Here the time components of y and y' are on the real axis in the τ-plane. So the physical functions $M_{ll'}^\pm$ (and hence the spectral function $\rho_{ll'}$ below) are defined, as usual, by the 11-component of the respective matrices, the other components appearing in their perturbative evaluation.

So far we have not committed to boson or fermion fields. We now introduce the *boson* spectral function as the Fourier transform of the *commutator*[4]

$$\rho_{ll'}(k) \equiv \int d^4y\, e^{ik\cdot(y-y')}\langle[\Phi_l(y), \Phi_{l'}^\dagger(y')]\rangle \quad (4.4.10)$$

$$= M_{ll'}^+(k) - M_{ll'}^-(k). \quad (4.4.11)$$

From (4.4.7) and (4.4.11) we can relate $M_{ll'}^\pm(k)$ to the spectral function

$$M_{ll'}^+(k) = \{1 + f(k_0)\}\rho_{ll'}(k), \quad M_{ll'}^-(k) = f(k_0)\rho_{ll'}(k) \quad (4.4.12)$$

where

$$f(k_0) = \frac{1}{e^{\beta k_0} - 1}. \quad (4.4.13)$$

We shall presently interpret this function.

We are now in a position to build the spectral representation of the complete (interacting) time-ordered two-point function

$$D_{ll'}'(x, x') = i\langle T_c \Phi_l(x)\Phi_{l'}^\dagger(x')\rangle$$

$$= \theta_c(\tau - \tau')i\langle\Phi_l(x)\Phi_{l'}^\dagger(x')\rangle + \theta_c(\tau' - \tau)i\langle\Phi_{l'}^\dagger(x')\Phi_l(x)\rangle. \quad (4.4.14)$$

---

[4] Complete spectral functions, in contrast to complete two-point functions, do not need any prime to distinguish them from free ones, as the latter are denoted by subscript or superscript zero.

Using (4.4.3), (4.4.5) and (4.4.12), its spatial Fourier transform is

$$D'_{ll'}(\tau, \tau'; \boldsymbol{k}) = i \int_{-\infty}^{\infty} \frac{dk'_0}{2\pi} \rho_{ll'}(k'_0, \boldsymbol{k}) e^{-ik'_0(\tau - \tau')} \{\theta_c(\tau - \tau') + f(k'_0)\}. \quad (4.4.15)$$

Following the procedure of Section 4.2, we introduce components with real times. The usual temporal Fourier transform of the components then gives the matrix spectral representation

$$\boldsymbol{D}'_{ll'}(k_0, \boldsymbol{k}) = \int_{-\infty}^{\infty} \frac{dk'_0}{2\pi} \rho_{ll'}(k'_0, \boldsymbol{k}) \boldsymbol{\Lambda}(k'_0, k_0) \quad (4.4.16)$$

with the elements of the matrix $\boldsymbol{\Lambda}$ given by

$$\Lambda_{11} = -\Lambda_{22}^* = \frac{1}{k'_0 - k_0 - i\eta} + 2\pi i f(k_0)\delta(k'_0 - k_0)$$

$$\Lambda_{12} = \Lambda_{21} = 2\pi i e^{\beta k_0/2} f(k_0)\delta(k'_0 - k_0). \quad (4.4.17)$$

Below we shall diagonalise this matrix, but the original components are also needed in calculations. We shall often use the 11-component, which we rewrite in the form[5]

$$D'_{ll'}(k_0, \boldsymbol{k})_{11} = \int_{-\infty}^{\infty} \frac{dk'_0}{2\pi} \rho_{ll'}(k'_0, \boldsymbol{k}) \left(\frac{1 + f(k'_0)}{k'_0 - k_0 - i\eta} - \frac{f(k'_0)}{k'_0 - k_0 + i\eta} \right) \quad (4.4.18)$$

on using $2\pi i \delta(k'_0 - k_0) = (k'_0 - k_0 - i\eta)^{-1} - (k'_0 - k_0 + i\eta)^{-1}$ to remove the delta function in Λ_{11}.

The function $f(k_0)$ given by (4.4.13) apparently has the form of a single particle distribution, but it cannot be interpreted as such, as k_0 also runs over negative values. However, we can express it in terms of $f(|k_0|)$

$$\begin{aligned}
f(k_0) &= f(k_0)\{\theta(k_0) + \theta(-k_0)\} \\
&= f(|k_0|)\theta(k_0) + f(-|k_0|)\theta(-k_0) \\
&= f(|k_0|)\epsilon(k_0) - \theta(-k_0), \quad (4.4.19)
\end{aligned}$$

where we convert $f(-|k_0|)$ to $f(|k_0|)$ by using (4.4.13) itself. To avoid new symbols we put $f(|k_0|) \equiv n(|k_0|)$. As we have seen in Section 4.2 and shall verify again presently, $|k_0|$ in $n(|k_0|)$ can be replaced by ω, when the propagator is free. With this understanding, we shall also omit the argument of n to write (4.4.19) simply as

$$f(k_0) = n\epsilon(k_0) - \theta(-k_0). \quad (4.4.20)$$

We use it to rewrite (4.4.17) in terms of n with a different pole prescription

$$\Lambda_{11} = -\Lambda_{22}^* = \frac{1}{k'_0 - k_0 - i\eta\epsilon(k_0)} + 2\pi i n\epsilon(k_0)\delta(k'_0 - k_0)$$

$$\Lambda_{12} = \Lambda_{21} = 2\pi i \sqrt{n(1 + n)}\epsilon(k_0)\delta(k'_0 - k_0). \quad (4.4.21)$$

---

[5] A second form may be obtained after diagonalisation as in Section 4.2. Which one is more convenient to use depends on what we want to calculate.

It is interesting to note that in the process of converting $f(k_0)$ to $n(\omega)$, the pole prescription is also changed to one for the time-ordered propagator.

Recognising the similarity of $\boldsymbol{\Lambda}$ with the free scalar propagator $\boldsymbol{D}$ as given by (4.2.17), we can diagonalise it by the same $\boldsymbol{U}$

$$\boldsymbol{\Lambda} = \boldsymbol{U}(|k_0|)) \begin{pmatrix} \dfrac{1}{k_0' - k_0 - i\eta\epsilon(k_0)} & 0 \\[2ex] 0 & \dfrac{-1}{k_0' - k_0 + i\eta\epsilon(k_0)} \end{pmatrix} \boldsymbol{U}(|k_0|), \qquad (4.4.22)$$

leading in turn to the diagonal form of $\boldsymbol{D}_{ll'}'$ of (4.4.16)

$$\boldsymbol{D}_{ll'}'(k_0, \boldsymbol{k}) = \boldsymbol{U}(|k_0|)) \begin{pmatrix} \overline{D}_{ll'} & 0 \\ 0 & -\overline{D}_{ll'}^* \end{pmatrix} \boldsymbol{U}(|k_0|) \qquad (4.4.23)$$

with the diagonal element

$$\overline{D}_{ll'}(k_0, \boldsymbol{k}) = \int_{-\infty}^{\infty} \frac{dk_0'}{2\pi} \frac{\rho_{ll'}(k_0', \boldsymbol{k})}{k_0' - k_0 - i\eta\epsilon(k_0)} . \qquad (4.4.24)$$

Thus (4.4.23) and (4.4.24) with (4.4.10) give the complete spectral representation for the thermal two-point function of a general boson operator $\Phi_l(x)$.

It is now simple to recover the result for the general free thermal propagator. Let $\Phi_l(x)$ represent a free field describing an elementary particle. If the field is *free*, its commutator (4.4.10) is a c-number and so the thermal spectral function coincides with the vacuum spectral function. Hence by (4.4.24) and (1.6.13) we get

$$\overline{D}_{ll'}(k_0, \boldsymbol{k}) = \int_{-\infty}^{\infty} \frac{dk_0'}{2\pi} \frac{\rho_{ll'}^{(0)}(k_0', \boldsymbol{k})}{k_0' - k_0 - i\eta\epsilon(k_0)} = \Delta_{ll'}^{F}(k) , \qquad (4.4.25)$$

If we now multiply out the three matrices in (4.4.23), we find that $n(|k_0|)$ is always multiplied by the mass-shell δ-function, so that it may be replaced by $n(\omega)$

$$\boldsymbol{D}_{ll'}(k_0, \boldsymbol{k}) = \boldsymbol{U}(\omega) \begin{pmatrix} \Delta_{ll'}^{F}(k) & 0 \\ 0 & -\Delta_{ll'}^{F*}(k) \end{pmatrix} \boldsymbol{U}(\omega) \qquad (4.4.26)$$

showing that the diagonalised element of the free thermal propagator for a general Bose field is indeed the same as the vacuum propagator.

Finally let us note a general property of these matrices. To get the diagonal element $\overline{D}_{ll'}$, it suffices to know any one, say the 11-component of the matrix $\boldsymbol{D}_{ll'}$. Omitting the field indices ll', (4.4.23) gives

$$\mathrm{Re}\overline{D} = \mathrm{Re}D_{11}'$$

$$\mathrm{Im}\overline{D} = \frac{1}{1 + 2n}\mathrm{Im}D_{11}' = \epsilon(p_0)\tanh(\beta p_0)\mathrm{Im}D_{11}' \qquad (4.4.27)$$

where we multiply the last equation on the right by $\epsilon(p_0)$ to convert $\tanh(\beta|p_0|)$ to $\tanh(\beta p_0)$.

Fermion Two-Point Function

The derivation of fermion representation follows the same line of reasoning as the previous boson case, except that we now define the spectral function with the *anticommutator* of operators. Nevertheless we go through the steps once again without repeating the arguments. We now include a chemical potential in the ensemble average.

The three-dimensional Fourier transform of $\langle \Psi_l(x)\overline{\Psi}_{l'}(x')\rangle$ is given by

$$\int d^3x\, e^{-i\boldsymbol{p}\cdot(\boldsymbol{x}-\boldsymbol{x}')}\langle \Psi_l(x)\overline{\Psi}_{l'}(x')\rangle = \int \frac{dp_0'}{2\pi} e^{-ip_0'(\tau-\tau')}\widetilde{M}_{ll'}^+(p_0',\boldsymbol{p}) \qquad (4.4.28)$$

where

$$\widetilde{M}_{ll'}^+(p') = \frac{(2\pi)^4}{Z}\sum_{m,n} e^{-\beta(E_m-\mu N_m)}\delta^{(4)}(p'+p_m-p_n)\langle m|\Psi_l(0)|n\rangle\langle n|\overline{\Psi}_{l'}(0)|m\rangle.$$

$$(4.4.29)$$

The integer N_m is the eigenvalue of the charge operator Q in the state $|m\rangle$. Thus the matrix element $\langle m|\Psi_l(0)|n\rangle$ is non-zero only for $N_m = N_n - 1$, assuming $\Psi_l(x)$ has fermion number unity. Similarly,

$$\int d^3x\, e^{-i\boldsymbol{p}\cdot(\boldsymbol{x}-\boldsymbol{x}')}\langle \overline{\Psi}_{l'}(x')\Psi_l(x)\rangle = \int \frac{dp_0'}{2\pi} e^{-ip_0'(\tau-\tau')}\widetilde{M}_{ll'}^-(p_0',\boldsymbol{p}) \qquad (4.4.30)$$

where

$$\widetilde{M}_{ll'}^-(p') = \frac{(2\pi)^4}{Z}\sum_{m,n} e^{-\beta(E_m-\mu N_m)}\delta^{(4)}(p'+p_n-p_m)\langle m|\overline{\Psi}_{l'}(0)|n\rangle\langle n|\Psi_l(0)|m\rangle$$

$$(4.4.31)$$

The two functions defined by (4.4.29) and (4.4.31) are related as

$$\widetilde{M}_{ll'}^+(p) = e^{\beta(p_0-\mu)}\widetilde{M}_{ll'}^-(p) \qquad (4.4.32)$$

which is the momentum representation of the KMS condition, usually written in coordinate space. We can rewrite the double sums as four-dimensional Fourier transforms of the two operator products. The *fermion* spectral function is now given by the Fourier transform of the *anticommutator*

$$\sigma_{ll'}(p) \equiv \int d^4y\, e^{ik\cdot(y-y')}\langle\{\Psi_l(y),\overline{\Psi}_{l'}(y')\}\rangle \qquad (4.4.33)$$

$$= \widetilde{M}_{ll'}^+(p) + \widetilde{M}_{ll'}^-(p). \qquad (4.4.34)$$

We solve (4.4.32) and (4.4.34) for $\widetilde{M}_{ll'}^\pm$ to get

$$\widetilde{M}_{ll'}^+(p) = \{1 - \tilde{f}(p_0)\}\sigma_{ll'}(p), \qquad \widetilde{M}_{ll'}^-(p) = \tilde{f}(p_0)\sigma_{ll'}(p) \qquad (4.4.35)$$

where

$$\tilde{f}(p_0) = \frac{1}{e^{\beta(p_0-\mu)}+1}. \qquad (4.4.36)$$

The complete two-point function for the Fermi operator is

$$S'_{ll'}(x - x') = i\langle T_c \Psi_l(x)\overline{\Psi}_{l'}(x')\rangle$$
$$= \theta_c(\tau - \tau')i\langle \Psi_l(x)\overline{\Psi}_{l'}(x')\rangle - \theta_c(\tau' - \tau)i\langle \overline{\Psi}_{l'}(x')\Psi_l(x)\rangle \quad (4.4.37)$$

whose three-dimensional Fourier transform may be written by noting (4.4.28), (4.4.30) and (4.4.35)

$$S'_{ll'}(\tau - \tau', \boldsymbol{p}) = i \int_{-\infty}^{+\infty} \frac{dp'_0}{2\pi} \sigma_{ll'}(p'_0, \boldsymbol{p})e^{-ip'_0(\tau-\tau')}\{\theta_c(\tau - \tau') - \widetilde{f}(p'_0)\}. \quad (4.4.38)$$

Breaking it into components, we get the spectral representation for the complete fermion thermal two-point function

$$\boldsymbol{S}'_{ll'}(p_0, \vec{p}) = \int_{-\infty}^{\infty} \frac{dp'_0}{2\pi} \sigma_{ll'}(p'_0, \boldsymbol{p})\boldsymbol{\Omega}(p'_0, p_0) \quad (4.4.39)$$

where the matrix $\boldsymbol{\Omega}$ has the elements,

$$\Omega_{11} = -\Omega_{22}^* = \frac{1}{p'_0 - p_0 - i\eta} - 2\pi i\widetilde{f}(p'_0)\delta(p'_0 - p_0) \quad (4.4.40)$$

$$\Omega_{12} = -e^{\beta\mu}\Omega_{21} = -2\pi ie^{\beta p'_0/2}\widetilde{f}(p_0)\delta(p'_0 - p_0). \quad (4.4.41)$$

The 11-component may be written in a convenient form for loop calculations

$$S'_{ll'}(p_0, \vec{p})_{11} = \int_{-\infty}^{\infty} \frac{dp'_0}{2\pi} \sigma_{ll'}(p'_0, \boldsymbol{p}) \left(\frac{1 - \widetilde{f}(p'_0)}{p'_0 - p_0 - i\eta} + \frac{\widetilde{f}(p'_0)}{p'_0 - p_0 + i\eta} \right). \quad (4.4.42)$$

The distribution-like function $\widetilde{f}$ may be expressed in terms of the true distribution functions $\widetilde{n}_{\pm}$,

$$\widetilde{f}(p_0) = \widetilde{n}_+\theta(p_0) - \widetilde{n}_-\theta(-p_0) + \theta(-p_0) = \widetilde{N}_1^2\epsilon(p_0) + \theta(-p_0) \quad (4.4.43)$$

where $\widetilde{N}_1$ appearing here and $\widetilde{N}_2$ below are defined in (4.3.14). The matrix elements of $\boldsymbol{\Omega}$ may be written in terms of $\widetilde{N}_{1,2}$ by a change in the pole prescription

$$\Omega_{11} = -\Omega_{22}^* = \frac{1}{p'_0 - p_0 - i\eta\epsilon(p_0)} - 2\pi i\widetilde{N}_1^2\epsilon(p_0)\delta(p'_0 - p_0)$$

$$\Omega_{12} = -e^{\beta\mu}\Omega_{21} = -2\pi ie^{\beta\mu/2}\widetilde{N}_1\widetilde{N}_2\epsilon(p_0)\delta(p'_0 - p_0). \quad (4.4.44)$$

The matrix $\boldsymbol{\Omega}$ can be diagonalised by the matrix $\boldsymbol{V}$ given by (4.3.18),

$$\boldsymbol{\Omega} = \boldsymbol{V}(|p_0|) \begin{pmatrix} \dfrac{1}{p'_0 - p_0 - i\eta\epsilon(p_0)} & 0 \\ 0 & \dfrac{-1}{p'_0 - p_0 + i\eta\epsilon(p_0)} \end{pmatrix} \boldsymbol{V}(|p_0|). \quad (4.4.45)$$

It follows from (4.4.39) that $\boldsymbol{S}'$ is also diagonalised by $\boldsymbol{V}$,

$$\boldsymbol{S}'_{ll'}(p_0, \boldsymbol{p}) = \boldsymbol{V}(|p_0|) \begin{pmatrix} \overline{S}_{ll'} & 0 \\ 0 & -\overline{S}_{ll'}^* \end{pmatrix} \boldsymbol{V}(|p_0|) \quad (4.4.46)$$

where $\overline{S}'$ has the spectral representation

$$\overline{S}_{ll'}(\boldsymbol{p}, p_0) = \int \frac{dp'_0}{2\pi} \frac{\sigma_{ll'}(p'_0, \boldsymbol{p})}{p'_0 - p_0 - i\eta\epsilon(p_0)}. \tag{4.4.47}$$

Here the $*$ on $\overline{S}_{ll'}$ is understood to change the sign of $i\eta\epsilon$ only.

We can now get the free thermal propagator by letting $\Psi_l(x)$ denote an elementary Fermi field. By the same reasoning as in the Bose case, the free thermal spectral function is the same as the free vacuum spectral function, getting

$$\overline{S}_{ll'}(\boldsymbol{p}, p_0) = \int \frac{dp'_0}{2\pi} \frac{\sigma^{(0)}_{ll'}(p'_0, \boldsymbol{p})}{p'_0 - p_0 - i\epsilon(p_0)} = S^F_{ll'}(p) \tag{4.4.48}$$

where $S^F_{ll'}$ stands for the free fermion propagator. As before, we multiply out the matrices in (4.4.46) to establish that the argument $|p_0|$ of $\widetilde{n}_\pm$ becomes ω. Thus it reduces for a free general field to

$$\boldsymbol{S}_{ll'}(p_0, \boldsymbol{p}) = \boldsymbol{V}(\omega) \begin{pmatrix} S^F_{ll'} & 0 \\ 0 & -S^{F*}_{ll'} \end{pmatrix} \boldsymbol{V}(\omega), \tag{4.4.49}$$

again showing that the diagonalised thermal propagator for a general free Fermi field is given by the vacuum propagator.

As in the boson case, we can relate the diagonalised element to the 11-component. Again omitting the field indices, we get from (4.4.46)

$$\mathrm{Re}\overline{S} = \mathrm{Re}S'_{11}$$
$$\mathrm{Im}\overline{S} = \left[\frac{1}{1 - 2\widetilde{n}_+}\theta(p_0) + \frac{1}{1 - 2\widetilde{n}_-}\theta(-p_0) \right] \mathrm{Im}S'_{11}$$
$$= \epsilon(p_0)\coth[\beta(p_0 - \mu)/2]\mathrm{Im}S'_{11}. \tag{4.4.50}$$

*   *   *

Having derived the spectral representations for boson and fermion two-point functions, we make the following observations on their structures:

1) By defining the spectral function $\rho(\sigma)$ with the commutator (anti-commutator) for operators of integer (half-integer) spin, we get the Bose–Einstein (Fermi–Dirac) statistical distribution function, in agreement with the spin-statistics theorem of Pauli and others.

2) The thermal matrices $\boldsymbol{U}$ and $\boldsymbol{V}$ are free from the Lorentz and Dirac indices (ll'), these being confined entirely to the spectral function. As a result, $\boldsymbol{U}$ and $\boldsymbol{V}$ are universal, appearing respectively for all boson and fermion two-point functions.

3) Written in terms of f and $\widetilde{f}$, the two matrices $\boldsymbol{\Lambda}$ and $\boldsymbol{\Omega}$ are symmetrical under the replacement $f \leftrightarrow -\widetilde{f}$. Using this symmetry we can unify writing the boson and fermion two-point functions, which we shall denote by the common symbol $R_{ll'}(p_0, \boldsymbol{p})$, but retain $\rho_{ll'}$ to denote any of the spectral functions. Then

the 11-component of the propagator, which will be often used later, may be written as

$$R_{ll'}(p_0, \boldsymbol{p})_{11} = \int_{-\infty}^{\infty} \frac{dp_0'}{2\pi} \rho_{ll'}(p_0', \boldsymbol{p}) \left(\frac{1 + g(p_0')}{p_0' - p_0 - i\eta} - \frac{g(p_0')}{p_0' - p_0 + i\eta} \right) \quad (4.4.51)$$

where $g(p_0)$ represents $f(p_0)$ or $-\tilde{f}(p_0)$ according as it refers to boson or fermion operators.

4.5 Retarded Spectral Representation

As we shall see in Chapter 9, the (linear) response of a system to a (weak) external perturbation is expressed naturally in terms of retarded (two-point) correlation functions. However, the conventional perturbative framework cannot calculate such functions directly, as it applies only to time-ordered ones. The procedure followed here is to calculate the associated time-ordered correlation function and relate it to the retarded one [19]. Had we worked in the imaginary time formulation, it would require an analytic continuation in the process. One of the advantages of the real time formulation is that no such continuation is necessary: we can relate the two two-point functions simply by changing a prescription in the k_0 plane. We saw this relation for the free field in Section 1.2. We now prove it for the interacting case. Let us first derive the thermal spectral representation of the complete, retarded two-point function.

Consider bosonic fields. The *retarded* thermal propagator is

$$D_{ll'}^R(x.x') = i\theta_c(\tau - \tau')\langle[\Phi_l(\boldsymbol{x}, \tau), \Phi_{l'}(\boldsymbol{x}', \tau')]\rangle, \quad (4.5.1)$$

where again τ, τ' are on the contour C (Figure 4.2). Its representation can be found in the same way as for the time-ordered one in the previous section. Using (4.4.3), (4.4.5) and (4.4.11) we find its spatial Fourier transform as

$$D_{ll'}'^R(\tau - \tau', \boldsymbol{k}) = i\theta_c(\tau - \tau') \int \frac{dk_0'}{2\pi} e^{-ik_0'(\tau - \tau')} \rho_{ll'}(k_0', \boldsymbol{k}). \quad (4.5.2)$$

For the components of the matrix propagator, the complex times get related to real times as we saw in Section 4.2, when we can take the temporal Fourier transform. Thus for the 11-component, $\theta_c(\tau - \tau') = \theta(t - t')$ and we get

$$D_{ll'}'^R(k_0, \boldsymbol{k})_{11} = \int \frac{dk_0'}{2\pi} \frac{\rho_{ll'}(k_0', \boldsymbol{k})}{k_0' - k_0 - i\eta}. \quad (4.5.3)$$

The 12-component is zero, as $\theta_c(\tau_1 - \tau_2') = 0$. For the 21-component, $\theta_c(\tau_2 - \tau_1') = 1$ and we get

$$D_{ll'}'^R(k_0, \boldsymbol{k})_{21} = i\rho_{ll'}(k_0, \boldsymbol{k})e^{-\beta k_0/2}. \quad (4.5.4)$$

Here the factor $\exp(-\beta k_0/2)$ can be expressed through the distribution function after splitting k_0 into positive and negative values, as we did in (4.4.19). In this way we find the retarded matrix propagator as

$$\boldsymbol{D}'^R_{ll'}(k_0, \boldsymbol{k}) = \begin{pmatrix} D^R_{ll'}(k)_{11} & 0 \\ i\rho_{ll'}(k)\left\{\sqrt{\frac{n}{n+1}}\theta(k_0) + \sqrt{\frac{n+1}{n}}\theta(-k_0)\right\} & -D^{R*}_{ll'}(k)_{11} \end{pmatrix}.$$

$$(4.5.5)$$

We are now in a position to find the required relation. As stated already, the linear response is given by the 11-element of the retarded propagator, $D'^R_{ll'}(k_0, \boldsymbol{k})_{11}$, given by (4.5.3), which cannot be calculated perturbatively. However we can find the element of the *diagonalised* time-ordered two-point function, $\overline{D}'_{ll'}(k_0, \boldsymbol{k})$ given by (4.4.24). The two quantities agree, except for the difference in how to approach the real line in the k_0 plane (causing a difference in sign of their imaginary parts for negative values of k_0). So we can relate $D'^R_{ll'}$ with $\overline{D}'_{ll'}$ as

$$D'^R_{ll'}(k_0 + i\eta, \boldsymbol{k})_{11} = \overline{D}'_{ll'}(k_0 + i\eta\epsilon(k_0) \to k_0 + i\eta, \boldsymbol{k}). \qquad (4.5.6)$$

The point to note here is that no diagonalisation is involved for the retarded propagator; it is only the diagonal element of the time-ordered propagator which enters the above relation. We shall use this continuation in Section 9.4.

Problems

Problem 4.1: Find the thermal propagator for a non-hermitian scalar field by the differential equation method.

Solution: The propagator satisfies the same differential equation (4.2.1) as for the hermitian case. To get the thermal boundary condition we proceed as before

$$\begin{aligned} D_-(\boldsymbol{x}, \boldsymbol{x}'; \tau, \tau') &\equiv iZ^{-1} \operatorname{Tr} e^{-\beta(H-\mu Q)} \phi^\dagger(\boldsymbol{x}', \tau')\phi(\boldsymbol{x}, \tau) \\ &= iZ^{-1} \operatorname{Tr} e^{-\beta(H-\mu Q)} e^{\beta H}\left\{e^{-\beta\mu Q}\phi(\boldsymbol{x}, \tau)e^{\beta\mu Q}\right\}e^{-\beta H}\phi^\dagger(\boldsymbol{x}', \tau'). \end{aligned}$$

$$(P4.1)$$

The curly bracket can be worked out by noting the expansion formula (C.9) and the commutator $[Q, \phi(x)] = -\phi(x)$

$$\begin{aligned} e^{-\beta\mu Q}\phi(x)e^{\beta\mu Q} &= \phi(x) - \beta\mu[Q, \phi(x)] + \frac{(\beta\mu)^2}{2}[Q,[Q,\phi]] - \cdots \\ &= e^{\beta\mu}\phi(x). \end{aligned} \qquad (P4.2)$$

The operators $e^{\pm\beta H}$ multiplying $\phi(\boldsymbol{x}, \tau)$ on both sides translate the time variable as in (4.2.2) and we get the boundary condition

$$D_-(\boldsymbol{x}, \boldsymbol{x}'; \tau, \tau') = e^{\beta\mu}D_+(\boldsymbol{x}, \boldsymbol{x}'; \tau - i\beta, \tau'), \qquad (P4.3)$$

which is the same as (4.2.6) except for the μ-dependent multiplicative factor.

Proceeding in the same way as before to solve the differential equation, we get

$$D(\boldsymbol{k};\tau,\tau') = \frac{i}{2\omega}\left[\{\theta_c(\tau-\tau')+n_+(\omega)\}e^{-i\omega(\tau-\tau')} + \{\theta_c(\tau'-\tau)+n_-(\omega)\}e^{i\omega(\tau-\tau')}\right]$$

$$\text{(P4.4)}$$

replacing (4.2.10) for the hermitian case, where $n_\pm$ are the distribution functions for particles and antiparticles defined by (4.2.27).

In getting the temporal Fourier transform of the propagator components from (4), we shall now have different coefficients of $\delta(k_0-\omega)$ and $\delta(k_0+\omega)$. We multiply these terms respectively by $\theta(k_0)$ and $\theta(-k_0)$, when both delta functions can be replaced by $2\omega\delta(k^2-m^2)$. Finally, we get for the propagator components

$$D_{11} = -D_{22}^* = \Delta_F(k) + 2\pi i N_1^2 \delta(k^2-m^2)$$
$$D_{12} = 2\pi i e^{\beta\mu/2} N_1 N_2 \delta(k^2-m^2)$$
$$D_{21} = 2\pi i e^{-\beta\mu/2} N_1 N_2 \delta(k^2-m^2)$$

$$\text{(P4.5)}$$

where N_1 and N_2 are defined by (4.2.26). Diagonalising this matrix we get (4.2.24).

Problem 4.2: Evaluate thermal traces of quadratic forms in creation and annihilation operators.

Solution: We consider a system described by a non-hermitian field $\phi(x)$, confined to a box of volume V. Then as indicated in Section 4.2 under 'Operator method', the free Hamiltonian H and charge Q become sums over discrete momenta

$$H = \int d^3x \frac{1}{2}\left[\partial_0\phi^\dagger\partial_0\phi + \vec{\nabla}\phi^\dagger\cdot\vec{\nabla}\phi + m^2\phi^\dagger\phi\right] \to \sum_k \omega_k(A_k^\dagger A_k + B_k^\dagger B_k)$$

$$Q = i\int d^3x\phi^\dagger \overleftrightarrow{\partial_0}\phi \to \sum_k(A_k^\dagger A_k - B_k^\dagger B_k)$$

up to irrelevant additive constants.

The thermal traces are easily calculated in the occupation number basis, in which matrix elements are evaluated in the states $|m_1, m_2, \cdots\rangle$, which are direct products of states $|m_k\rangle$. To begin with, m_k stands for two numbers, of particles and antiparticles in momentum state k. On this basis the ensemble average (4.2.22) of an operator $\mathcal{O}$ becomes

$$\langle\mathcal{O}\rangle = \frac{\sum_{m_1,m_2,\cdots}\langle m_1, m_2, \cdots|e^{-\beta(H-\mu Q)}\mathcal{O}|m_1, m_2, \cdots\rangle}{\sum_{m_1,m_2,\cdots}\langle m_1, m_2, \cdots|e^{-\beta(H-\mu Q)}|m_1, m_2, \cdots\rangle}. \tag{P4.6}$$

Let us take $\mathcal{O} = A_k^\dagger A_{k'}$ with operators for particles. Its matrix elements survive only if $k = k'$, when $\mathcal{O}$ becomes the number operator for particles in the state with momentum k. Then the numerator (and also the denominator) factorises; each factor belongs to a particular momentum and is a sum over its occupation number. All such factors cancel out between the numerator and denominator, except the one with momentum k. We thus get

$$\langle A_k^\dagger A_{k'} \rangle = \frac{\sum_{m_k=1,2\cdots} \langle m_k | e^{-\beta(\omega-\mu)A_k^\dagger A_k} A_k^\dagger A_k | m_k \rangle}{\sum_{m_k=1,2\cdots} \langle m_k | e^{-\beta(\omega-\mu)A_k^\dagger A_k} | m_k \rangle} \delta_{k,k'}$$

$$= \frac{x + 2x^2 + 3x^3 + \cdots}{1 + x + x^2 + \cdots} \delta_{k,k'}, \qquad x = e^{-\beta(\omega-\mu)}$$

$$= \frac{x}{1-x} \delta_{k,k'}.$$

Similarly we can find $\langle B_k^\dagger B_{k'} \rangle$. The results are given in (4.2.32).

References

[1] T. Matsubara, Prog. Theor. Phys. **14**, 351 (1955).

[2] J. Schwinger, Jour. Math. Phys. **2**, 407 (1961).

[3] L.V. Keldysh, Sov. Phys. JETP 20, 1018 (1965).

[4] R.A. Craig, J. Math. Phys. **9**, 605 (1968).

[5] R. Mills, *Propagators for Many-Particle Systems*, Gordon and Breach, New York (1969).

[6] Y. Takahashi and H. Umezawa, Collective Phenomena, **2**, 55 (1975); G.W. Semenoff and H. Umezawa, Nucl. Phys. **B 220**, 196 (1983).

[7] A.J. Niemi and G.W. Semenoff, Ann. Phys. **152**, 105 (1984).

[8] N.P. Landsman and Ch.G. van Weert, Phys. Rep. **145**, 141 (1987).

[9] J.I. Kapusta, *Finite Temperature Field Theory*, first edition, Cambridge University Press (1989). See also C. Gale and J.I. Kapusta, ibid., second edition, Cambridge University Press (2006).

[10] M. Le Bellac, *Thermal Field Theory*, Cambridge University Press (1996).

[11] A.H. Guth and S.-Y. Pi, Phys. Rev. **D32**, 1899 (1985).

[12] G. Semenoff and N. Weiss, Phys. Rev. **D31**, 689 (1985).

[13] H. Leutwyler and S. Mallik, Ann. Phys. **205**, 1 (1991).

[14] S. Mallik and S. Sarkar, Eur. Phys. J. **C 61**, 489 (2009).

[15] R. Kubo, J. Phys. Soc. Japan, 12, 570 (1957).

[16] P.C. Martin and J. Schwinger, Phys. Rev. 115, 1342 (1959).

[17] E.C. Titchmarsh, *Introduction to the Theory of Fourier Integrals*, Oxford University Press (1937).

[18] L. Dolan and R. Jackiw, Phys. Rev. **D9**, 3320 (1974).

[19] A.L. Fetter and J.D. Walecka, *Quantum Theory of Many-Particle Systems*, Dover Publications Inc. (2003).

5

Thermal Perturbation Theory

Having derived the thermal propagator matrices for different fields in the last chapter, we now describe how to adapt the conventional (vacuum) perturbative framework to calculate one-point and two-point functions[1] with these matrices. The point here is that although the real line in the time contour (Figure 4.2) gives the physical amplitude, it is not possible to stay with it alone in perturbative calculations; the shifted parallel line also enters through the interaction vertex. Thus, rather than restricting to the original physical amplitude, it is convenient to consider the associated matrix for two-point (and column vector for one-point) functions, involving both segments of the contour. After including perturbative corrections, the result can be diagonalised, giving the corrected amplitiude.

Once the matrix structure is accommodated, the advantages of the real time method relative to the imaginary time one become apparent. The frequency sums in the latter method can be tedious, particularly for graphs with more than one loop. On the other hand, dealing with momentum integrals, the real time method is closer to the vacuum theory as to computation and renormalisation. Further, computations are done in Minkowski space, so no question of analytic continuation to real time is involved.

5.1 Matrix Structure

As expected, the perturbative evaluation of thermal amplitudes follows essentially the same steps as for vacuum amplitudes. In particular, the topological structure and combinatorial factors of Feynman graphs remain the same. However, we need to take into account the time contour C, leading to an effective doubling of field variables, as already mentioned in Section 4.2. Thus the thermal perturbation theory replaces the time integral of the interaction Lagrangian (density) of the vacuum theory by [1]

---

[1] We shall be concerned mostly with these functions, leaving higher-point functions to be considered in vacuum, possibly including thermal corrections to masses and couplings.

$$\int_{-\infty}^{+\infty} dt\,\mathscr{L}_{int}(\phi) \longrightarrow \int_C d\tau\,\mathscr{L}_{int}(\phi) = \int_{-\infty}^{+\infty} dt\{\mathscr{L}_{int}(\phi_1) - \mathscr{L}_{int}(\phi_2)\} \quad (5.1.1)$$

where we leave out the vertical segments of the contour, following earlier discussion. The two types of fields $\phi_1(x)$ and $\phi_2(x)$ differ only in their time variables lying on two different segments of C. The minus sign before $\mathscr{L}_{int}(\phi_2)$ arises from the reverse direction of the returning time path.

Given a Feynman amplitude in vacuum, it is simple to write the corresponding thermal matrix amplitude. We illustrate below this procedure in detail by writing the amplitudes for some representative graphs in both vacuum and medium. To focus on thermal indices, we avoid Lorentz and isospin indices by taking a single component scalar field with $\mathscr{L}_{int}(\phi) \sim \phi^4(x)$. Anticipating realistic cases to be considered later, we take a scalar density $S(x)$ and a 'current' $J(x)$ of the form

$$S(x) \sim a\phi^2(x) + b\phi^4(x), \qquad J(x) \sim c\phi(x) + d\phi^3(x)$$

where a, b, c and d are constants, which will be omitted in the following. We consider graphs for the one- and two-point functions in coordinate space and write their amplitudes omiting all numerical constants, showing only the Wick contractions.

One-Point Functions

We first consider the one-point function in *vacuum*,

$$E^{(0)}(x) = \langle 0|TS(x)|0\rangle \quad (5.1.2)$$

and evaluate graphs of Figure 5.1. Graph (a) gives

$$E^{(0)}(x)_{(a)} \sim \langle 0|\phi^2(x)|0\rangle \sim M^2 G(M) \quad (5.1.3)$$

which is the vacuum propagator evaluated at the origin, already encountered in Section 3.10,

$$M^2 G(M) = -i \int \frac{d^4k}{(2\pi)^4} \Delta_F(k). \quad (5.1.4)$$

Similarly, graph (b) gives

$$E^{(0)}(x)_{(b)} \sim \langle 0|\phi^4(x)|0\rangle \sim M^4 G^2(M). \quad (5.1.5)$$

Graph (c) has an interaction vertex, giving

$$E^{(0)}(x)_{(c)} \sim \langle 0|T\phi^2(x)\phi^4(u)|0\rangle \sim M^2 G(M)\Delta_F(x-u)\Delta_F(u-x). \quad (5.1.6)$$

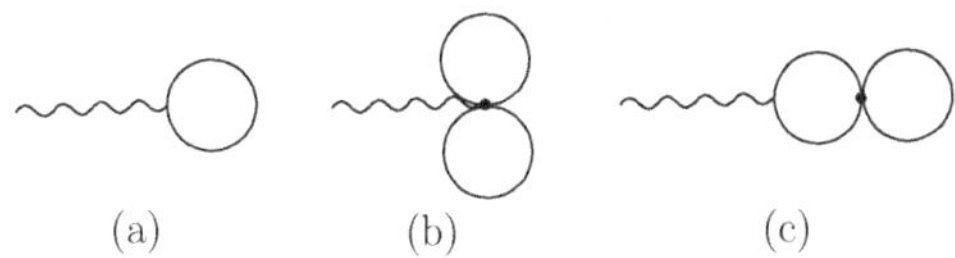

Figure 5.1 Some illustrative graphs for the one-point function.

We now evaluate the same graphs for the *thermal* one-point function, which is now a two-component object

$$E_i(x) = \langle TS_i(x) \rangle, \qquad i = 1, 2 \tag{5.1.7}$$

with type 1 and type 2 fields. Note that contractions of type 1 and of type 2 fields at the origin give rise to the same quantity,

$$-i \int \frac{d^4k}{(2\pi)^4} D_{11}(k) = -i \int \frac{d^4k}{(2\pi)^4} D_{22}(k) \equiv M^2\overline{G}(M) \tag{5.1.8}$$

where

$$M^2\overline{G}(M) = M^2 G(M) + N(\beta, M), \tag{5.1.9}$$

with

$$N(\beta, M) = \int \frac{d^4k}{(2\pi)^3} \delta(k^2 - M^2) n(\omega), \tag{5.1.10}$$

evaluated in Appendix D. Thus the first two graphs give

$$E_i(x)_{(a)} \sim M^2\overline{G}(M) \tag{5.1.11}$$

$$E_i(x)_{(b)} \sim \langle T\phi_i^4(x) \rangle \sim \{M^2\overline{G}\}^2 \tag{5.1.12}$$

which are independent of the index i. With an interaction vertex, graph (c) gives

$$\begin{aligned}
E_i(x)_{(c)} &\sim \langle T\phi_i^2(x)\{\phi_1^4(u) - \phi_2^4(u)\} \rangle \\
&\sim M^2\overline{G}(M)\{D_{i1}(x-u)D_{1i}(u-x) - D_{i2}(x-u)D_{2i}(u-x)\} \\
&\sim M^2\overline{G}(M)\{\boldsymbol{D}(x-u)\boldsymbol{\tau}\boldsymbol{D}(u-x)\}_{ii}, \qquad \text{no sum over } i \tag{5.1.13}
\end{aligned}$$

where the minus sign between the terms in the second line is taken care of by the matrix $\boldsymbol{\tau}$,

$$\boldsymbol{\tau} = \begin{pmatrix} 1 & 0 \\ 0 & -1 \end{pmatrix}.$$

The result (5.1.13) is also independent of the index i, as can be seen by going over to momentum space and noting the mass derivative formula (5.1.36) below.

Two-Point Functions

We first evaluate the graphs of Figure 5.2 for the *vacuum* two-point function

$$F^{(0)}(x, y) = \langle 0|TJ(x)J(y)|0 \rangle. \tag{5.1.14}$$

Graph (a) gives just the propagator,

$$F^{(0)}(x, y)_{(a)} \sim \langle 0|T\phi(x)\phi(y)|0 \rangle \sim \Delta_F(x - y). \tag{5.1.15}$$

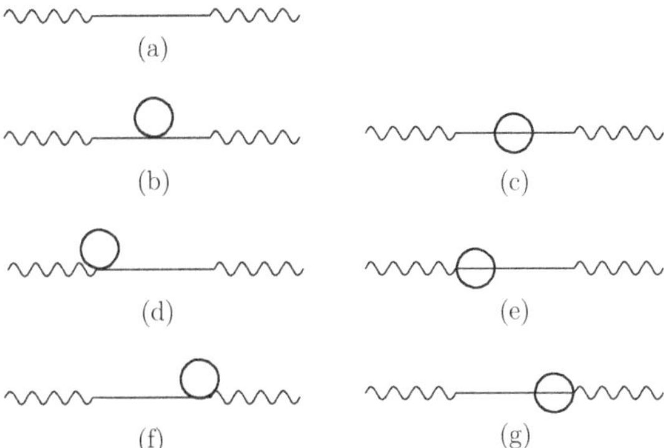

Figure 5.2 Some illustrative graphs for the two-point function.

Graph (b) contributes a self-energy to first order in perturbation

$$F^{(0)}(x,y)_{(b)} \sim \langle 0|T\phi(x)\phi^4(u)\phi(y)|0\rangle \sim M^2 G(M)\Delta_F(x-u)\Delta_F(u-y). \quad (5.1.16)$$

The self-energy arising in second order is given by graph (c),

$$F^{(0)}(x,y)_{(c)} \sim \langle 0|T\phi(x)\phi^4(u)\phi^4(v)\phi(y)|0\rangle \sim \Delta_F(x-u)\Sigma(u-v)\Delta_F(v-y), \quad (5.1.17)$$

where

$$\Sigma(u-v) = \Delta_F^3(u-v).$$

Graph (d) contributes a correction to the vertex

$$F^{(0)}(x,y)_{(d)} \sim \langle 0|T\phi^3(x)\phi(y)|0\rangle \sim M^2 G(M)\Delta_F(x-y). \quad (5.1.18)$$

Graph (e) gives a higher-order correction to the vertex,

$$F^{(0)}(x,y)_{(e)} \sim \langle 0|T\phi^3(x)\phi^4(u)\phi(y)|0\rangle = \Gamma(x-u)\Delta_F(u-y) \quad (5.1.19)$$

where again

$$\Gamma(x-u) = \Delta_F^3(x-u).$$

Finally, graphs (f) and (g) are counterparts of (d) and (e),

$$F^{(0)}(x,y)_{(f)} \sim \langle 0|T\phi(x)\phi^3(y)|0\rangle \sim M^2 G(M)\Delta_F(x-y) \quad (5.1.20)$$

$$F^{(0)}(x,y)_{(g)} \sim \langle 0|T\phi(x)\phi^4(u)\phi^3(y)|0\rangle \sim \Delta_F(x-u)\Gamma(u-y). \quad (5.1.21)$$

Next consider the *thermal* case, where we find the contributions of the same graphs to the ij-th component of the two-point function

$$F_{ij}(x,y) \sim \langle T J_i(x)J_j(y)\rangle \quad (5.1.22)$$

again in coordinate space. The individual graphs are evaluated as

$$F_{ij}(x,y)_{(a)} \sim \langle T\phi_i(x)\phi_j(y)|0\rangle = D_{ij}(x-y) \quad (5.1.23)$$

$$F_{ij}(x,y)_{(b)}$$
$$\sim \langle T\phi_i(x)\{\phi_1^4(u) - \phi_2^4(u)\}\phi_j(y)\rangle$$
$$\sim \overline{G}(M)\{D_{i1}(x-u)D_{1j}(u-y) - D_{i2}(x-u)D_{2j}(u-y)\}$$
$$= \overline{G}(M)\{\boldsymbol{D}(x-u)\boldsymbol{\tau}\boldsymbol{D}(u-y)\}_{ij} \tag{5.1.24}$$

$$F_{ij}(x,y)_{(c)}$$
$$\sim \langle T\phi_i(x)\{\phi_1^4(u) - \phi_2^4(u)\}\{\phi_1^4(v) - \phi_2^4(v)\}\phi_j(y)\rangle$$
$$\sim D_{i1}(x-u)D_{11}^3(u-v)D_{1j}(v-y) - D_{i1}(x-u)D_{12}^3(u-v)D_{2j}(v-y)$$
$$\quad - D_{i2}(x-u)D_{21}^3(u-v)D_{1j}(v-y) + D_{i2}(x-u)D_{22}^3(u-v)D_{2j}(v-y)$$
$$= \{\boldsymbol{D}(x-u)\boldsymbol{\Sigma}(u-v)\boldsymbol{D}(v-y)\}_{ij}, \tag{5.1.25}$$

where the self-energy matrix is

$$\boldsymbol{\Sigma} = \begin{pmatrix} s_{11} & -s_{12} \\ -s_{21} & s_{22} \end{pmatrix}, \quad s_{ij} = D_{ij}^3. \tag{5.1.26}$$

The expressions (5.1.24) and (5.1.25) may be compared to illustrate the use of $\boldsymbol{\tau}$ in absorbing the minus sign arising from the interaction Lagrangian. In (5.1.24) it involves the propagator matrix, which is already defined, requiring a $\boldsymbol{\tau}$ to be inserted explicitly, while in (5.1.25) we have a new structure $\boldsymbol{\Sigma}$, which can be defined to absorb the sign.

Continuing with the evaluation of the remaining graphs, we get

$$F_{ij}(x,y)_{(d)} \sim \langle T\phi_i^3(x)\phi_j(y)\rangle \sim M^2\overline{G}(M)D_{ij}(x-y) \tag{5.1.27}$$

$$F_{ij}(x,y)_{(e)}$$
$$\sim \langle T\phi_i^3(x)\{\phi_1^4(u) - \phi_2^4(u)\}\phi_j(y)\rangle$$
$$\sim D_{i1}^3(x-u)D_{1j}(u-y) - D_{i2}^3(x-u)D_{2j}(u-y)$$
$$\sim \{\boldsymbol{\Gamma}^{(1)}(x-u)\boldsymbol{D}(u-y)\}_{ij}, \tag{5.1.28}$$

where

$$\boldsymbol{\Gamma}^{(1)} = \begin{pmatrix} \gamma_{11} & -\gamma_{12} \\ \gamma_{21} & -\gamma_{22} \end{pmatrix}, \quad \gamma_{ij} = D_{ij}^3 \tag{5.1.29}$$

$$F_{ij}(x,y)_{(f)} \sim \langle T\phi_i(x)\phi_j^3(y)\rangle \sim M^2\overline{G}(M)D_{ij}(x-y) \tag{5.1.30}$$

$$F_{ij}(x,y)_{(g)}$$
$$\sim \langle T\phi_i(x)\{\phi_1^4(u) - \phi_2^4(u)\}\phi_j^3(y)\rangle$$
$$\sim D_{i1}(x-u)D_{1j}^3(u-y) - D_{i2}(x-u)D_{2j}^3(u-y)$$
$$= \{\boldsymbol{D}(x-u)\boldsymbol{\Gamma}^{(2)}(u-y)\}_{ij}, \tag{5.1.31}$$

where

$$\mathbf{\Gamma}^{(2)} = \begin{pmatrix} \gamma_{11} & \gamma_{12} \\ -\gamma_{21} & -\gamma_{22} \end{pmatrix}. \tag{5.1.32}$$

Comparing the evaluations at finite temperature with those in vacuum, we see that the thermal matrix amplitudes may simply be obtained from the vacuum amplitudes by replacing propagators, self-energies and vertices with the corresponding matrices

$$\Delta_F \to \mathbf{D}, \quad \Sigma \to \mathbf{\Sigma}, \quad \Gamma \to \mathbf{\Gamma}^{(1,2)}, \quad G \to \overline{G}. \tag{5.1.33}$$

When two propagators occur consecutively in an expression, a $\boldsymbol{\tau}$ is to be inserted between them,

$$\Delta^2 \to \mathbf{D}\boldsymbol{\tau}\mathbf{D}. \tag{5.1.34}$$

Such products of thermal propagators have a simple structure, resembling the one for the vacuum propagators, namely

$$\Delta^{r+1} = \frac{(-1)^r}{r!} \left(\frac{\partial}{\partial m^2} \right)^r \Delta. \tag{5.1.35}$$

Taking $\mathbf{D}$ in the form (4.2.18), each time it is multiplied by $\boldsymbol{\tau}\mathbf{D}$ the powers of Δ and Δ^* in $\mathbf{D}$ are raised by one, keeping their coefficients unaltered. Then using (5.1.35) we get

$$(\mathbf{D}\boldsymbol{\tau})^{r+1} = \frac{(-1)^r}{r!} \left(\frac{\partial}{\partial m^2} \right)^r \mathbf{D}\boldsymbol{\tau} \tag{5.1.36}$$

which is the so-called mass derivative formula for the thermal propagator [2].

5.2 Diagonalisation

Having obtained the thermal matrix amplitudes from the corresponding vacuum amplitudes, we recall the result of the previous chapter, namely that any two-point function, including the free and complete propagators, can be diagonalised to give its diagonal elements by a single analytic function. Then, as we show below, other thermal matrices like $\mathbf{\Sigma}$, $\mathbf{\Gamma}^{(1,2)}$ can also be diagonalised by the same diagonalising matrix [3]. As a result, a matrix equation involving these matrices collapses to a single equation of ordinary functions.

The Dyson–Schwinger equation for the scalar propagator reads

$$\mathbf{D}'(k) = \mathbf{D}(k) + \mathbf{D}(k)\mathbf{\Sigma}(k)\mathbf{D}'(k). \tag{5.2.1}$$

Let us display once again the diagonalised forms of $\mathbf{D}$ and $\mathbf{D}'$ as given by (4.2.19) and (4.4.23),

$$\mathbf{D}(k) = \mathbf{U}(|k_o|) \begin{pmatrix} \Delta(k) & 0 \\ 0 & -\Delta^*(k) \end{pmatrix} \mathbf{U}(|k_0|), \tag{5.2.2}$$

$$D'(k) = U(|k_0|) \begin{pmatrix} \overline{D}(k) & 0 \\ 0 & -\overline{D}^*(k) \end{pmatrix} U(|k_0|) \tag{5.2.3}$$

where the diagonalising matrix U involves only the distribution function and is given by (4.2.20). Inserting these in (5.2.1), we see that $\Sigma(k)$ must be of the form

$$\Sigma(k) = U^{-1}(k) \begin{pmatrix} \overline{\Sigma}(k) & 0 \\ 0 & -\overline{\Sigma}^*(k) \end{pmatrix} U^{-1}(k) \tag{5.2.4}$$

so that the matrix equation reduces to an ordinary equation

$$\overline{D} = \Delta + \Delta\overline{\Sigma}\,\overline{D} \tag{5.2.5}$$

having the solution

$$\overline{D} = \frac{-1}{k^2 - m^2 + \overline{\Sigma}}. \tag{5.2.6}$$

It immediately follows from (5.2.4) that

$$\Sigma_{22} = -\Sigma_{11}^*, \qquad \Sigma_{21} = \Sigma_{12}. \tag{5.2.7}$$

Further, we can get the function $\overline{\Sigma}$ entirely from any one component, say Σ_{11},

$$\mathrm{Re}\,\overline{\Sigma} = \mathrm{Re}\,\Sigma_{11},$$

$$\mathrm{Im}\,\overline{\Sigma} = \frac{1}{1+2n}\mathrm{Im}\,\Sigma_{11} = \epsilon(k_0)\tanh(\beta k_0/2)\mathrm{Im}\,\Sigma_{11} \quad \text{(bosonic)} \tag{5.2.8}$$

the relations being identical to those between $\overline{D}$ and D'_{11} given by (4.4.27). The above treatment actually holds for any bosonic propagator irrespective of its spin. We shall illustrate one such example in Chapter 6, where the ρ meson propagator will be treated in this way.

We now turn to the Dirac propagator, whose Dyson–Schwinger equation,

$$S'(p) = S(p) + S(p)\Sigma(p)S'(p), \tag{5.2.9}$$

can be solved in a similar way. Here, as usual, the Dirac indices are suppressed. Recalling (4.3.17) and (4.4.46) both S and S' are again diagonalised by the same matrix, namely

$$S(p) = V(|p_0|) \begin{pmatrix} S(p) & 0 \\ 0 & -S^*(p) \end{pmatrix} V(|p_0|),$$

$$S'(p) = V(|p_0|) \begin{pmatrix} \overline{S}(p) & 0 \\ 0 & -\overline{S}^*(p) \end{pmatrix} V(|p_0|) \tag{5.2.10}$$

where V again depends only on the distribution function and is given by (4.3.18). The $*$ in the above equations is only implied to change the sign of $i\epsilon$; in particular, it does not act on γ-matrices. Inserting these decompositions in (5.2.9), we see that $\Sigma(p)$ must be of the form

$$\Sigma(p_0, \boldsymbol{p}) = V^{-1} \begin{pmatrix} \overline{\Sigma} & 0 \\ 0 & -\overline{\Sigma}^* \end{pmatrix} V^{-1}. \tag{5.2.11}$$

Then the thermal matrix equation (5.2.9) becomes

$$\overline{S} = S + S\overline{\Sigma}\,\overline{S} \tag{5.2.12}$$

which gives

$$\overline{S} = \frac{-1}{\not{p} - m + \overline{\Sigma}}. \tag{5.2.13}$$

Two relations among the elements of Σ follow from (5.2.11),

$$\Sigma_{12} = -e^{\beta\mu}\Sigma_{21}, \qquad \Sigma_{22} = -\Sigma_{11}^{*}. \tag{5.2.14}$$

Also (5.2.11) relates the real and the imaginary parts of $\overline{\Sigma}$ to those of Σ_{11},

$$\mathrm{Re}\overline{\Sigma} = \mathrm{Re}\Sigma_{11}$$
$$\mathrm{Im}\overline{\Sigma} = \frac{1}{1 - 2n}\mathrm{Im}\Sigma_{11} = \epsilon(p_0)\coth[\beta(p_0 - \mu)/2]\mathrm{Im}\Sigma_{11}. \quad \text{(fermionic)} \tag{5.2.15}$$

Again, these relations are identical to those for the fermion two-point function obtained in (4.4.50). As with bosons, the same treatment holds for any fermionic propagator.

The vertex structures $\Gamma^{(1,2)}$ can be treated in a way similar to the self-energies. Let us rewrite (5.1.28) and (5.1.31) as matrices

$$\boldsymbol{F}^{(1)} = \boldsymbol{\Gamma}^{(1)}\boldsymbol{D}, \qquad \boldsymbol{F}^{(2)} = \boldsymbol{D}\boldsymbol{\Gamma}^{(2)}. \tag{5.2.16}$$

Being two-point functions, $\boldsymbol{F}^{(1,2)}$ must be diagonalisable in the same way as $\boldsymbol{D}$, so that we have the factorisations

$$\boldsymbol{\Gamma}^{(1)} = \boldsymbol{U}\begin{pmatrix} \overline{\Gamma} & 0 \\ 0 & \overline{\Gamma}^{*} \end{pmatrix}\boldsymbol{U}^{-1}, \qquad \boldsymbol{\Gamma}^{(2)} = \boldsymbol{U}^{-1}\begin{pmatrix} \overline{\Gamma} & 0 \\ 0 & \overline{\Gamma}^{*} \end{pmatrix}\boldsymbol{U}. \tag{5.2.17}$$

Expanding out the matrix products we see that the vertices $\boldsymbol{\Gamma}^{(1)}$ and $\boldsymbol{\Gamma}^{(2)}$ differ only by a change of sign of the off-diagonal elements, in agreement with the examples in (5.1.29) and (5.1.32). For both types of vertices, we have

$$\boldsymbol{\Gamma}_{22} = \boldsymbol{\Gamma}_{11}^{*}, \qquad \boldsymbol{\Gamma}_{21} = -\boldsymbol{\Gamma}_{12}. \tag{5.2.18}$$

Again any element, say the 11-element, determines the function $\overline{\Gamma}$ completely,

$$\mathrm{Re}\overline{\Gamma} = \mathrm{Re}\boldsymbol{\Gamma}_{11}, \qquad \mathrm{Im}\overline{\Gamma} = \frac{1}{1 + 2n}\mathrm{Im}\boldsymbol{\Gamma}_{11}. \tag{5.2.19}$$

To summarise, the replacements (5.1.33) and (5.1.34) transform the vacuum amplitudes to the corresponding thermal matrix amplitudes. A matrix equation, when diagonalised, reduces to a single ordinary equation involving functions like $\Delta, S, \overline{D}, \overline{S}, \overline{G}, \overline{\Gamma}, \overline{\Sigma}$. As the diagonalising matrix (one for bosonic systems and another for fermionic systems) is a scalar, the spin and isospin structure of the diagonalised elements will remain the same as for the vacuum amplitudes. Finally, the real and imaginary parts of these elements are respectively equal and related by a trigonometric function to those of the 11-element of the matrix amplitudes. Accordingly we may dispense with writing matrix amplitudes and their diagonalisation in the intermediate stage and work directly with the 11-elements.

5.3 One-Loop Self-Energy (Scalar Fields)

Knowing the matrix structure of thermal self-energies, we turn to find their imaginary parts to one loop. These were first derived by Weldon [4] for scalar and Dirac fields in the imaginary time formulation. These were rederived later in the real time formulation [3]. Using the spectral representation for propagators, we can easily generalise these results to particles of arbitrary spin. However, before we deal with general fields, it will be useful to describe the procedure first with all scalar fields and interpret the result physically in this simple context.

We calculate the self-energy of a real scalar field Φ interacting with two other real scalar fields ϕ_1 and ϕ_2 through a trilinear coupling

$$\mathscr{L}_{\text{int}} = g\Phi\phi_1\phi_2$$

where g is a coupling constant (Figure 5.3). As we have just seen, we need to calculate any one component of the self-energy matrix, say the 11, to get the diagonalised matrix, which in turn requires the 11-components of the propagators. Defining the self-energy as in (5.2.1) we get

$$\Sigma_{11}(q) = -ig^2 \int \frac{d^4k}{(2\pi)^4} D'_{11}(k, m_1) D'_{11}(q - k, m_2) \tag{5.3.1}$$

where the (complete) propagators for the two fields ϕ_1 and ϕ_2 are distinguished by their masses. Their spectral representations are given by (4.4.18)

$$D'(k_0, \vec{k})_{11} = \int_{-\infty}^{\infty} \frac{dk'_0}{2\pi} \rho(k'_0, \vec{k}) \left(\frac{1 + f(k'_0)}{k'_0 - k_0 - i\eta} - \frac{f(k'_0)}{k'_0 - k_0 + i\eta} \right). \tag{5.3.2}$$

Inserting these propagators in the self-energy, we want to work out the k_0 integral

$$\Sigma_{11}(q) = g^2 \int \frac{d^3k}{(2\pi)^3} \int \frac{dk'_0}{2\pi} \rho(k'_0, \boldsymbol{k}) \int \frac{dk''_0}{2\pi} \rho(k''_0, \boldsymbol{q} - \boldsymbol{k}) \cdot I \tag{5.3.3}$$

where

$$I = -i \int \frac{dk_0}{(2\pi)} \left(\frac{1 + f'}{k'_0 - k_0 - i\eta} - \frac{f'}{k'_0 - k_0 + i\eta} \right)$$

$$\times \left(\frac{1 + f''}{k''_0 - (q_0 - k_0) - i\eta} - \frac{f''}{k''_0 - (q_0 - k_0) + i\eta} \right). \tag{5.3.4}$$

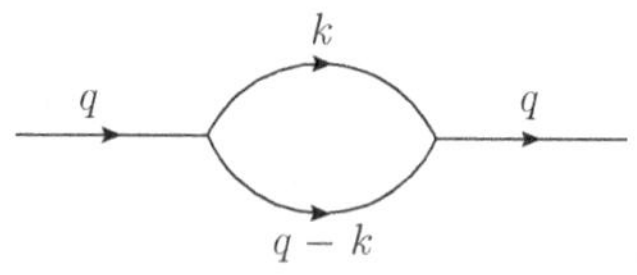

Figure 5.3 One-loop self-energy diagram.

Here $f' = (e^{\beta k_0'} - 1)^{-1}$ and $f'' = (e^{\beta k_0''} - 1)^{-1}$. Multiplying out the two factors in the integrand, I splits into four terms. On closing the integration contour in the upper- or lower-half k_0 plane, we see that only the two terms, having poles in *both* the half-planes, are non-zero, giving

$$I = -\left\{ \frac{(1 + f')(1 + f'')}{q_0 - k_0' - k_0'' + i\eta} - \frac{f'f''}{q_0 - k_0' - k_0'' - i\eta} \right\}. \tag{5.3.5}$$

We now look for the imaginary part of the self-energy. From (5.3.5) we can read off

$$\mathrm{Im}I = \pi\{(1 + f')(1 + f'') + f'f''\}\delta(q_0 - k_0' - k_0'') \tag{5.3.6}$$

giving $\mathrm{Im}\Sigma_{11}(q)$. Referring to (5.2.8) we find a trigonometric factor between $\mathrm{Im}\Sigma_{11}$ and $\mathrm{Im}\overline{\Sigma}$. If we note $\{1 + f(k_0)\}/f(k_0) = e^{\beta k_0}$, this factor can indeed be extracted by changing the sign between the two terms within $\{\cdots\}$ in (5.3.6)

$$(1 + f')(1 + f'') + f'f'' = \{(1 + f')(1 + f'') - f'f''\}\coth(\beta q_0/2) \tag{5.3.7}$$

on using the delta function in (5.3.6). (We do not cancel out the quadratic terms in f for proper interpretation, as we show below.) We then get

$$\mathrm{Im}\overline{\Sigma}(q) = \epsilon(q_0)\pi g^2 \int \frac{d^3k}{(2\pi)^3} \int \frac{dk_0'}{2\pi}\rho(k_0', \boldsymbol{k}, m_1) \int \frac{dk_0''}{2\pi}\rho(k_0'', \boldsymbol{q} - \boldsymbol{k}, m_2)$$
$$\times\{(1 + f')(1 + f'') - f'f''\}\delta(q_0 - k_0' - k_0''). \tag{5.3.8}$$

So far we have retained complete propagators in the loop. Let us now evaluate $\mathrm{Im}\overline{\Sigma}$ with free propagators, for which the spectral function is

$$\rho_0(k) = 2\pi\epsilon(k_0)\delta(k^2 - m^2) = \frac{2\pi}{2\omega}\epsilon(k_0)\{\delta(k_0 - \omega) + \delta(k_0 + \omega)\}. \tag{5.3.9}$$

It is now simple to integrate over k_0' and k_0'' with the δ-functions in the spectral functions, raising the internal lines to their mass-shells with positive and negative energies, corresponding to particles and antiparticles[2]. Then changing over to the notation

$$\omega_1 = \sqrt{\boldsymbol{k}^2 + m_1^2}, \quad \omega_2 = \sqrt{(\boldsymbol{q} - \boldsymbol{k})^2 + m_2^2}, \quad n_1 = f'(\omega_1), \quad n_2 = f''(\omega_2)$$

we get [4]

$$\mathrm{Im}\overline{\Sigma}(q_0, \boldsymbol{q}) = \pi g^2 \epsilon(q_0) \int \frac{d^3k}{(2\pi)^3 2\omega_1 2\omega_2}$$
$$[\{(1 + n_1)(1 + n_2) - n_1 n_2\}\delta(q_0 - \omega_1 - \omega_2)$$
$$+\{n_1(1 + n_2) - n_2(1 + n_1)\}\delta(q_0 + \omega_1 - \omega_2)$$
$$+\{n_2(1 + n_1) - n_1(1 + n_2)\}\delta(q_0 - \omega_1 + \omega_2)$$
$$+\{n_1 n_2 - (1 + n_1)(1 + n_2)\}\delta(q_0 + \omega_1 + \omega_2)]. \tag{5.3.10}$$

---

[2] When the function f gets a negative argument, it is $f(-\omega) = -(1 + f(\omega))$. Also to note is the overall sign from $\epsilon(k_0)$.

The regions of q_0 (at fixed $\boldsymbol{q}$) where the different terms in (5.3.10) can contribute are controlled by the delta functions. In Chapter 8 we shall study the resulting cut structure of $\overline{\Sigma}$ and give simpler analytic expressions for $\mathrm{Im}\overline{\Sigma}$ in the general case.

Let us check directly that (5.3.10) includes the full vacuum contribution by evaluating the vacuum self-energy

$$\Sigma^{(0)}(q) = -ig^2 \int \frac{d^4 k}{(2\pi)^4} \frac{1}{(k^2 - m_1^2 + i\eta)((q-k)^2 - m_2^2 + i\eta)}. \tag{5.3.11}$$

Splitting the propagators into partial fractions and integrating over k_0 in the same way as we did above, we can read off the imaginary part

$$\mathrm{Im}\Sigma^{(0)}(q) = \pi g^2 \int \frac{d^3 k}{(2\pi)^3 2\omega_1 2\omega_2} \{\delta(q_0 - \omega_1 - \omega_2) + \delta(q_0 + \omega_1 + \omega_2)\} \tag{5.3.12}$$

which coincides with $\mathrm{Im}\overline{\Sigma}(q)$ after putting $n_1 = n_2 = 0$ in vacuum. In medium the vacuum contributions are thus altered by terms linear in n_1 and n_2 (to one loop); in addition, two new contributions arise, again linear in n_1 and n_2, due to the possibility of Φ interacting with ϕ_1 and ϕ_2 from medium.

Interpretation

The physical meaning of the imaginary part of self-energy in *medium* differs from that in *vacuum*. Let us first obtain the familiar result in vacuum, namely $\mathrm{Im}\Sigma^{(0)}$ of a resonance is proportional to its decay rate Γ. In the present example, it relates to the Φ particle decaying as $\Phi \to \phi_1 + \phi_2$. The decay rate is obtained from (E.11)

$$\Gamma(q) = \frac{1}{2\omega} \int \frac{d^3 k}{(2\pi)^3 2\omega_1} \frac{d^3 k'}{(2\pi)^3 2\omega_2} (2\pi)^4 \delta^4(q - k - k')|M_{\beta\alpha}|^2 \tag{5.3.13}$$

with $|M_{\beta\alpha}|^2 = g^2$. It immediately reduces to (5.3.12) for $q_0 = \omega > 0$, giving

$$\Gamma = \frac{\mathrm{Im}\Sigma^{(0)}}{\omega}. \tag{5.3.14}$$

Alternatively, we can see this interpretation by looking at the complete (vacuum) propagator for the Φ particle [5]

$$\Delta'_F(q) = \frac{-1}{q^2 - m^2 + \Sigma^{(0)}(q)}. \tag{5.3.15}$$

The physical mass m_R is determined by the condition

$$m_R^2 - m^2 + \mathrm{Re}\Sigma^{(0)}(m_R) = 0.$$

For q^2 near m_R^2, (5.3.15) may be written as

$$\Delta'_F(q) = \frac{-1}{q^2 - m_R^2 + i\mathrm{Im}\Sigma^{(0)}(q)}. \tag{5.3.16}$$

For $\mathrm{Im}\Sigma^{(0)} \neq 0$, it shifts the pole from the real axis. Writing $q_0 = E$ for the off-shell propagator and assuming $\mathrm{Im}\,\Sigma^{(0)}/\omega$ to be small, we can put (5.3.16) in the form

$$\Delta'_F(E, \boldsymbol{q}) = \frac{-1}{E^2 - \{\omega - i\mathrm{Im}\Sigma^{(0)}/(2\omega)\}^2} \qquad \omega = \sqrt{m_R^2 + \boldsymbol{q}^2}$$

$$\rightarrow \frac{1}{2\omega} \frac{-1}{E - \omega + i\Gamma/2} \tag{5.3.17}$$

near the Φ particle pole with Γ defined by (5.3.14). The pole at $E = \omega - i\Gamma/2$ arises in the scattering *amplitude* for particles ϕ_1 and ϕ_2. Its Fourier transform behaves as

$$\exp(-iEt) = \exp(-i\omega t - \Gamma t/2) \tag{5.3.18}$$

corresponding to a resonant state, whose *probability* decays as $\exp(-\Gamma t)$, establishing again that $\mathrm{Im}\Sigma^{(0)}/\omega$ is the decay rate of the resonance in vacuum.

In a medium a reaction involving Φ cannot simply deplete its particles. Any reaction decreasing their number will be accompanied by the inverse reaction trying to replenish them. Indeed, Weldon [4] recognised the terms in (5.3.10) to be proportional to rates for decay and inverse decay of the Φ particle. As shown in Appendix E, differential rates in vacuum have to be multiplied with certain statistical weights to become the same in medium. The weights which arise naturally in the derivation of (5.3.10) are exactly the ones needed for this purpose. At this point we note that time reversal invariance requires the amplitudes for direct and inverse reactions to be equal. What is more, all amplitudes are equal in the present example. Thus, the first term on the right in this equation is, up to a factor of ω, the decay rate for $\Phi \rightarrow \phi_1\phi_2$ with weight $(1 + n_1)(1 + n_2)$ for stimulated emission *minus* the rate for the inverse decay $\phi_1\phi_2 \rightarrow \Phi$ with weight $n_1 n_2$ for absorption. Similarly the second term gives the decay rate for $\Phi\phi_1 \rightarrow \phi_2$ with weight $n_1(1 + n_2)$ *minus* the rate for the inverse reaction $\phi_2 \rightarrow \Phi\phi_1$ with weight $n_2(1 + n_1)$. (All the eight amplitudes involved are shown in Figure 5.4.) Thus (5.3.10) gives us the relation

$$\mathrm{Im}\overline{\Sigma}(q_0, \boldsymbol{q}) = \omega(\Gamma_d - \Gamma_i) \tag{5.3.19}$$

where Γ_d and Γ_i are respectively the rates for decay and inverse decay of Φ particles in the medium. The contributing amplitude is determined by the delta functions. (The relation (5.3.19) is also valid for multiloop self-energy graphs. See Problem 7.2.) The rates Γ_d and Γ_i are related to each other, as can be seen by expressing the distribution function as

$$n(\omega) = \frac{e^{-\beta\omega/2}}{2\sinh(\beta\omega/2)} \qquad 1 + n(\omega) = \frac{e^{\beta\omega/2}}{2\sinh(\beta\omega/2)}.$$

Also, using the respective delta functions, we get from each term in (5.3.10),

$$\Gamma_d/\Gamma_i = e^{\beta q_0}. \tag{5.3.20}$$

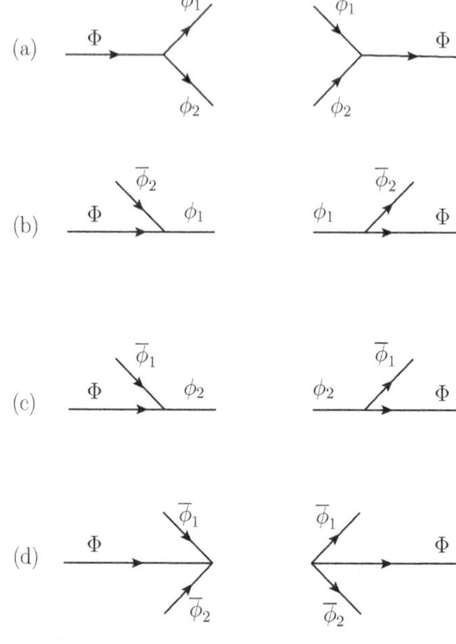

Figure 5.4 Eight amplitudes destroying and creating the Φ particles, whose probabilities appear in the imaginary part of its self-energy.

So far we have studied the propagation of a Φ particle through a medium populated by equilibrium distributions of ϕ_1 and ϕ_2 particles. Let us now assume that there is also a distribution of Φ particles in the medium. Let it be given by an arbitrary (non-equilibrium) distribution $N(\omega, t_0)$ at time t_0. This distribution decreases at the rate $N\Gamma_d$ and increases at the rate $(1 + N)\Gamma_i$. So it admits an equation of the Boltzmann type

$$\frac{dN}{dt} = (1 + N)\Gamma_i - N\Gamma_d. \tag{5.3.21}$$

This inhomogeneous first-order differential equation has the solution ($\omega = q_0$)

$$N(\omega, t) = \frac{\Gamma_i}{\Gamma_d - \Gamma_i} + c(\omega)e^{-(\Gamma_d - \Gamma_i)t} \tag{5.3.22}$$

where $c(\omega)$ is an arbitrary function. Using (5.3.20) the solution may be written as

$$N(\omega, t) = \frac{1}{e^{\beta\omega} - 1} + c(\omega)e^{-\Gamma(\omega)t}, \tag{5.3.23}$$

where

$$\Gamma = \Gamma_d - \Gamma_i = \mathrm{Im}\overline{\Sigma}/\omega. \tag{5.3.24}$$

Thus independently of the initial distribution, $N(\omega, t)$ approaches the equilibrium distribution as t becomes large. More specifically, $\Gamma(\omega)$ gives the inverse time for the distribution $N(\omega)$ to relax to equilibrium.

Summarising, $\mathrm{Im}\Sigma^{(0)}/\omega$ denotes the decay rate in vacuum of the resonant state, while $\mathrm{Im}\overline{\Sigma}/\omega$ denotes the decay rate of non-equilibrium excitations in an arbitrary distribution of the particles.

$$* \quad * \quad *$$

In the above analysis we considered the self-energy of a (scalar) boson. As we shall see in the next section, for a fermion it is given by (5.3.19) with a plus sign between Γ_d and Γ_i. So we may generally write

$$\mathrm{Im}\overline{\Sigma}(q_0, \boldsymbol{q}) = \omega(\Gamma_d - \sigma\Gamma_i) \tag{5.3.25}$$

where $\sigma = 1$ for a boson and $\sigma = -1$ for a fermion. Then the equation for the non-equilibrium distribution function (5.3.21) generalises to

$$\frac{dN}{dt} = (1 + \sigma N)\Gamma_i - N\Gamma_d \tag{5.3.26}$$

with the solution

$$N(\omega, t) = \frac{1}{e^{\beta\omega} - \sigma} + c(\omega)e^{-\Gamma(\omega)t}, \tag{5.3.27}$$

where

$$\Gamma = \Gamma_d - \sigma\Gamma_i = \mathrm{Im}\overline{\Sigma}/\omega. \tag{5.3.28}$$

5.4 One-Loop Self-Energy (General Fields)

Consider the three field vertex

$$V_{lmn}\Psi_l\psi_{1,m}\psi_{2,n}^\dagger + h.c.$$

where $\Psi_l(x)$, $\psi_{1,m}(x)$ and $\psi_{2,n}(x)$ are general fields of arbitrary spin and, in particular, may be bosonic or fermionic (see Appendix A). The coupling V may contain derivatives. As before, the 11-component of the self-energy of Ψ is given by (see Figure 5.5)

$$\Sigma(q)_{ll',11} = \mp i \int \frac{d^4p}{(2\pi)^4} V_{lmn}V^*_{l'm'n'}R_{mm'}(p, m_1)_{11}R_{n'n}(p - q, m_2)_{11} \tag{5.4.1}$$

where the lower sign applies only when both ψ_1 and ψ_2 are fermion fields. The propagators R having masses m_1 and m_2 are given by (4.4.51). On inserting it in (5.4.1), we get

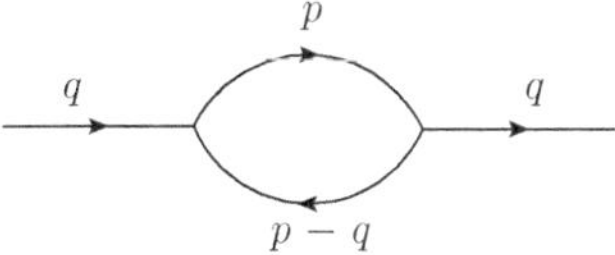

Figure 5.5 One-loop self-energy graph with momentum direction convenient for including fermion loop.

$$\Sigma(q)_{ll',11} = \int \frac{d^3p}{(2\pi)^3} \int \frac{dp_0'}{2\pi} \rho_{mm'}(p_0', \boldsymbol{p}, m_1) \int \frac{dp_0''}{2\pi} \rho_{n'n}(p_0'', \boldsymbol{p} - \boldsymbol{q}, m_2)$$
$$\times V_{lmn} V^*_{l'm'n'} \cdot J \tag{5.4.2}$$

where J is an integral over p_0

$$J = \mp i \int \frac{dp_0}{(2\pi)} \left(\frac{1+g'}{p_0' - p_0 - i\eta} - \frac{g'}{p_0' - p_0 + i\eta} \right)$$
$$\times \left(\frac{1+g''}{p_0'' - (p_0 - q_0) - i\eta} - \frac{g''}{p_0'' - (p_0 - q_0) + i\eta} \right) \tag{5.4.3}$$

with $g' = g(p_0')$ and $g'' = g(p_0'')$. Since we are interested in the imaginary part of J, any p_0 dependence of V and V^* has been fixed at the poles of the integrand, allowing them to be pulled outside the p_0 integral. As in the scalar case, we get

$$J = \pm \left(\frac{(1+g')g''}{p_0' - p_0'' - q_0 - i\eta} - \frac{g'(1+g'')}{p_0' - p_0'' - q_0 + i\eta} \right)$$

giving the imaginary part

$$\mathrm{Im} J = \pm \pi \{ (1+g')g'' + g'(1+g'') \} \delta(q_0 - p_0' + p_0'').$$

Recalling that the function $g(p_0)$ is $f(p_0)$ or $-\widetilde{f}(p_0)$, according to whether the associated field is bosonic or fermionic, we get $(1+g)/g = \pm e^{\beta p_0}$ for the two types of fields. We can now find the ratio r

$$r = \{ (1+g')g'' + g'(1+g'') \} / \{ (1+g')g'' - g'(1+g'') \}.$$

If both ψ_1 and ψ_2 are bosonic or fermionic, $r = \coth(\beta q_0/2)$. If, on the other hand, they belong to different statistics, $r = \tanh(\beta q_0/2)$. Referring to (5.2.8) and (5.2.15) we have the correct trigonometric function to extract the imaginary part of the diagonalised self-energy as

$$\mathrm{Im} \overline{\Sigma}(q)_{ll'} = \epsilon(q_0)\pi \int \frac{d^3p}{(2\pi)^3} \int \frac{dp_0'}{2\pi} \rho_{mm'}(p_0', \boldsymbol{p}) \int \frac{dp_0''}{2\pi} \rho_{n'n}(p_0'', \boldsymbol{p} - \boldsymbol{q})$$
$$\times V_{lmn} V^*_{l'm'n'} K \delta(q_0 - p_0' + p_0'') \tag{5.4.4}$$

where

$$K = \pm \{ (1+g')g'' - g'(1+g'') \}$$

with the quadratic terms in distribution functions cancelling out. Let us write K explicitly for the three cases of particles in the loop,

$$\begin{aligned}
f(p_0'') - f(p_0'), &\quad \text{(boson} - \text{boson)} \\
\widetilde{f}(p_0'') - \widetilde{f}(p_0'), &\quad \text{(fermion} - \text{fermion)} \\
f(p_0'') + \widetilde{f}(p_0'). &\quad \text{(fermion} - \text{boson)}
\end{aligned}$$

Observe that the distribution functions add up in the case of fermionic self-energy.

So far the propagators in the loop are exact. Let us now work with free propagators, for which ρ is of the form

$$\rho_{mm'}(p) = P_{mm'}(p)\rho_0(p)\,, \qquad \rho_0(p) = 2\pi\epsilon(p_0)\delta(p^2 - m^2) \tag{5.4.5}$$

where $P_{mm'}(p)$ is defined in (1.6.1) in terms of the coefficient functions in the free field expansion of $\psi_1(x)$ and $\psi_2(x)$. In calculating correction to the Ψ propagator, the self-energy Σ appears between propagators. But to interpret as probabilities, we take its matrix element between the polarisation tensors (spinors) $U_l(s)^*$ and $U_{l'}(s)$ of the Ψ particle, with s denoting the spin projection. Also it is convenient to sum over s. We collect quantities with indices in (5.4.4) to get the spin sum of the squared vertex amplitude M. Omitting momentum variables, we have

$$\sum_s U_l^*(s)V_{lmn}P_{mm'}P_{n'n}V_{l'm'n'}^*U_{l'}(s) \equiv |M|^2. \tag{5.4.6}$$

Then (5.4.4) can be written as

$$\mathrm{Im}\sum_s U(s)^*\overline{\Sigma}(q)U(s) = \epsilon(q_0)\pi \int \frac{d^3p}{(2\pi)^3} \int \frac{dp_0'}{2\pi}\rho_0(p_0', \boldsymbol{p}) \int \frac{dp_0''}{2\pi}\rho_0(p_0'', \boldsymbol{p} - \boldsymbol{q})$$
$$|M|^2 K\delta(q_0 - p_0' + p_0''). \tag{5.4.7}$$

As in the scalar case, we integrate p_0' and p_0'' over the delta functions (5.3.9) in the spectral functions. We can then interpret $|M|^2$ physically as probabilities for emission and absorption processes at the vertices of the self-energy graph. As an example, let us find out the contribution to the self-energy of a bosonic field Φ from two-boson (ϕ_1, ϕ_2) intermediate state with $p_0' = \omega_1$, $p_0'' = -\omega_2$. Inserting (1.6.1) for the projection matrices, (5.4.6) becomes

$$|M|^2 \rightarrow \sum_s U_l^*(s)V_{lmn}\sum_\sigma u_m u_{m'}^*\sum_{\sigma'} v_{n'}v_n^* V_{l'm'n'}^* U_{l'}(s)$$
$$= \sum_{s,\sigma,\sigma'} |U_l^*V_{lmn}u_m(\sigma)v_n^*(\sigma')|^2 \equiv |M(\Phi \rightarrow \phi_1\phi_2)|^2. \tag{5.4.8}$$

Then this contribution to (5.4.7) becomes

$$\mathrm{Im}U^*\overline{\Sigma}(q)U \rightarrow \pi \int \frac{d^3p}{(2\pi)^3 4\omega_1\omega_2}|M(\Phi \rightarrow \phi_1\phi_2)|^2(1 + n_1 + n_2)\delta(q_0 - \omega_1 - \omega_2) \tag{5.4.9}$$

in the notation used in Section 5.3.

Below we write the results explicitly for the three different cases. We denote the bosonic fields by Φ, ϕ and the fermionic fields by Ψ, ψ, χ, where the boldface letters denote the field for which we find the self-energy. (We previously used $f(-\omega) = -(1 + n(\omega))$. We now also use $\widetilde{f}(-\omega) = 1 - \widetilde{n}(\omega)$.)

i) Boson (Φ) self-energy from bosons (ϕ_1, ϕ_2)

$$\mathrm{Im}\,U^*\overline{\Sigma}(q)U = \pi\epsilon(q_0)\int\frac{d^3k}{(2\pi)^3 4\omega_1\omega_2}\times$$

$$[\,|M(\Phi\to\phi_1\phi_2)|^2\{(1+n_1)(1+n_2)-n_1n_2\}\delta(q_0-\omega_1-\omega_2)$$

$$+\,|M(\Phi\overline{\phi}_1\to\phi_2)|^2\{n_1(1+n_2)-n_2(1+n_1)\}\delta(q_0+\omega_1-\omega_2)$$

$$+\,|M(\Phi\overline{\phi}_2\to\phi_1)|^2\{n_2(1+n_1)-n_1(1+n_2)\}\delta(q_0-\omega_1+\omega_2)$$

$$+\,|M(\Phi\phi_1\phi_2\to0)|^2\{n_1n_2-(1+n_1)(1+n_2)\}\delta(q_0+\omega_1+\omega_2)\,]. \quad (5.4.10)$$

It agrees with (5.3.10) for the scalar case with the constant vertex replaced by momentum and spin-dependent vertices.

ii) Boson (Φ) self-energy from fermions (ψ, χ)

$$\mathrm{Im}\,U^*\overline{\Sigma}(q)U = \pi\epsilon(q_0)\int\frac{d^3k}{(2\pi)^3 4\omega_1\omega_2}\times$$

$$[\,|M(\Phi\to\psi\chi)|^2\{(1-\widetilde{n}_1)(1-\widetilde{n}_2)-\widetilde{n}_1\widetilde{n}_2\}\delta(q_0-\omega_1-\omega_2)$$

$$+\,|M(\Phi\overline{\psi}\to\chi)|^2\{\widetilde{n}_1(1-\widetilde{n}_2)-\widetilde{n}_2(1-\widetilde{n}_1)\}\delta(q_0+\omega_1-\omega_2)$$

$$+\,|M(\Phi\overline{\chi}\to\psi)|^2\{\widetilde{n}_2(1-\widetilde{n}_1)-\widetilde{n}_1(1-\widetilde{n}_2)\}\delta(q_0-\omega_1+\omega_2)$$

$$+\,|M(\Phi\chi\psi\to0)|^2\{\widetilde{n}_1\widetilde{n}_2-(1-\widetilde{n}_1)(1-\widetilde{n}_2)\}\delta(q_0+\omega_1+\omega_2)\,]. \quad (5.4.11)$$

iii) Fermion (Ψ) self-energy from fermion (ψ) and boson (ϕ)

$$U^*\mathrm{Im}\,\overline{\Sigma}(q)U = \pi\epsilon(q_0)\int\frac{d^3k}{(2\pi)^3 4\omega_1\omega_2}\times$$

$$[\,|M(\Psi\to\phi\psi)|^2\{(1-\widetilde{n}_1)(1+n_2)+\widetilde{n}_1n_2\}\delta(q_0-\omega_1-\omega_2)$$

$$+\,|M(\Psi\overline{\psi}\to\phi)|^2\{\widetilde{n}_1(1+n_2)+(1-\widetilde{n}_1)n_2\}\delta(q_0+\omega_1-\omega_2)$$

$$+\,|M(\Psi\overline{\phi}\to\psi)|^2\{n_2(1-\widetilde{n}_1)+(1+n_2)\widetilde{n}_1\}\delta(q_0-\omega_1+\omega_2)$$

$$+\,|M(\Psi\overline{\phi}\,\overline{\psi}\to0)|^2\{\widetilde{n}_1n_2+(1-\widetilde{n}_1)(1+n_2)\}\delta(q_0+\omega_1+\omega_2)\,]. \quad (5.4.12)$$

As noted above, for bosonic self-energy, the factor involving the distribution functions is a *difference* of terms for decay and inverse decay of Φ, while for fermionic self-energy, it is a *sum* of these terms for Ψ.

Problems

Problem 5.1: Find the imaginary part of the vacuum self-energy of a scalar particle from the unitarity of the S-matrix.

Solution: We have $S^\dagger S = 1$. Writing $S = 1 + iT$, it gives $-i(T - T^\dagger) = T^\dagger T$. Stretching the notion of scattering, we take its matrix element between states of one Φ particle. Then, to evaluate the right-hand side, insert a complete set of states between $T^\dagger$ and T and isolate the lowest order contribution, arising from two-particle states of ϕ_1 and ϕ_2. We thus get

$$- i\langle q'|T - T^\dagger|q\rangle = \int \frac{d^3k}{(2\pi)^3 2\omega} \frac{d^3k'}{(2\pi)^3 2\omega'} \langle kk'|T|q'\rangle^* \langle kk'|T|q\rangle. \qquad (P5.1)$$

The two matrix elements appearing in this unitarity equation are, to lowest order

$$\langle q'|T|q\rangle = (2\pi)^4 \delta^4(q' - q)\Sigma^{(0)}(q)$$
$$\langle q'|T|kk'\rangle = (2\pi)^4 \delta^4(q - k - k')g\,,$$

Inserting these in (P5.1) and cancelling the overall delta function $(2\pi)^4\delta^4(q - k - k')$ from both sides, we get

$$\mathrm{Im}\Sigma^{(0)}(q) = \pi g^2 \int \frac{d^3k}{(2\pi)^3 4\omega\omega'}\delta(q_0 - k_0 - k_0') \qquad (P5.2)$$

to be compared with (5.3.12), obtained by evaluating the one-loop self-energy graph. Observe that the second term there, contributing in the negative energy region, is absent here, as unitarity concerns physical amplitudes.

Problem 5.2: Find the imaginary part of the thermal self-energy of a scalar particle without using the spectral form of propagators.

Solution: On using

$$D_{11}(k, m) = \frac{-1}{k^2 - m^2 + i\eta} + 2\pi in(\omega)\delta(k^2 - m^2)$$

the self-energy function (5.3.1) splits into

$$\begin{aligned}
\Sigma_{11}(q) = -ig^2 \int \frac{d^4k}{(2\pi)^4} \Bigg\{ &\frac{1}{(k^2 - m^2 + i\eta)\{(q - k)^2 - m'^2 + i\eta\}} \\
&-2\pi i\left(\frac{n\delta(k^2 - m^2)}{(q - k)^2 - m'^2 + i\eta} + \frac{n'\delta((q - k)^2 - m'^2)}{k^2 - m^2 + i\eta}\right) \\
&-(2\pi)^2 nn'\delta(k^2 - m^2)\delta((q - k)^2 - m'^2) \Bigg\}
\end{aligned} \qquad (P5.3)$$

where $n = n(\omega)$, $n' = n(\omega')$, $\omega = \sqrt{k^2 + m^2}$, $\omega' = \sqrt{(q - k)^2 + m'^2}$. Here the first term is the vacuum contribution, and the second and third are thermal ones, being linear and quadratic in the distribution functions. Note that the third term is purely imaginary and the imaginary part of the second term can be obtained immediately. So these two terms together give

$$\mathrm{Im}\Sigma_{11}(q)|_{(3+2)} = \pi g^2 \int \frac{d^3k}{(2\pi)^3}(n + n' + 2nn') \int dk_0\delta(k^2 - m^2)\delta((q - k)^2 - m'^2).$$
$$(P5.4)$$

The imaginary part of the first term is already obtained in (5.3.12). So adding it to (P5.4) we get the total imaginary part as

$$\mathrm{Im}\Sigma_{11}(q)|_{(3+2+1)} = \pi g^2 \int \frac{d^3k}{(2\pi)^3 4\omega\omega'}$$
$$[\{(1+n)(1+n') + nn'\}\{\delta(q_0 - \omega - \omega') + \delta(q_0 + \omega + \omega')\}$$
$$+ \{n(1+n') + n'(1+n)\}\{\delta(q_0 - \omega + \omega') + \delta(q_0 + \omega - \omega')\}].$$
$$(\mathrm{P5.5})$$

Using the respective delta functions, the factors with distribution functions can be written as

$$(1+n)(1+n') + nn' = \{(1+n)(1+n') - nn'\}\coth\{\beta(\omega + \omega')/2\},$$
$$n(1+n') + n'(1+n) = \{n(1+n') - n'(1+n)\}\coth\{\beta(\omega' - \omega)/2\}.$$

Noting (5.2.8) we reproduce (5.3.10).

References

[1] A.J. Niemi and G.W. Semenoff, Ann. Phys. **152**, 105 (1984); Nucl. Phys. **B230**, 181 (1984).

[2] Y. Fujimoto, H. Matsumoto, H. Umezawa and I. Ozima, Phys. Rev. **D 30**, 1400 (1984).

[3] R.L. Kobes and G.W. Semenoff, Nucl. Phys. **B 260**, 714 (1985).

[4] H.A. Weldon, Phys. Rev., **D 28**, 2007 (1983).

[5] M.E. Peskin and D.V. Schroeder, *An Introduction to Quantum Field Theory*, Westview (1995).

6

Thermal Parameters

We are now in a position to apply chiral perturbation theory to find the changes in hadronic parameters at non-zero temperature and density. This topic was initiated by Leutwyler and collaborators, who calculated in this theory the thermal properties of pion gas, including the quark condensate and parameters of hadrons like the pion and nucleon [1–4]. This chapter and the next one review some of their works. We also describe results for other hadrons and consider the medium consisting of (infinite) nuclear matter.

Earlier in Sections 3.10 to 3.13 we used the method of external fields to evaluate the one- and two-point functions of densities and currents in vacuum. As an example of one-point functions, we found the renormalised quark condensate. The two-point functions were obtained in the neighbourhood of poles of particles communicating with the currents. The pole positions and residues gave the renormalised hadronic couplings and masses of particles. Essentially the same method can be applied to find how such parameters depend on temperature and chemical potential. Instead of their vacuum expectation values, we now have to work with the ensemble averages. The same vertices and graphs appear in the evaluation, but with vacuum propagators replaced by thermal ones, a procedure that we explained in detail in the last chapter. The additional pieces in thermal propagators give rise to temperature and chemical potential dependence of the parameters. In this chapter we describe results to one loop, extending some of these to two loops in the next chapter.

As mentioned earlier, describing a particle in medium involves the four-velocity u^μ of the medium, in addition to its four-momentum. Because we work with the medium at rest $[u^\mu = (1,0)]$, we do not expect relativistic covariance in calculations [5]. In this chapter the one-loop calculations, however, do not show up this non-covariance, except for pion parameters in nuclear matter, leading to two pion decay constants [6]. In the next chapter the two-loop calculation of pion parameters in pion medium will show this behaviour [7]. This situation is also known to arise in non-relativistic systems [8].

6.1 Quark Condensate at Finite Temperature

It is simple to extend the case of vacuum condensate to its thermal counterpart. We recall from Section 5.1 that the thermal condensate is a two-component object, having the same value for both components. So in its definition we can restrict the time contour to the real axis only, the matrix structure arising in higher-order perturbations only. Then we can rewrite (3.10.28) for the ensemble averaged generating functionals:

$$\langle T \exp \left(i \int d^4x (-) \delta M^2(x) \frac{1}{2B} \bar{q}q \right) \rangle = \langle T \exp \left(i \int d^4x \delta M^2(x) \mathcal{N} \right) \rangle \quad (6.1.1)$$

where $\mathcal{N}$ is already obtained in (3.10.27). As before, we differentiate both sides functionally with respect to δM^2 to get the thermal equivalent of (3.10.29)

$$-\frac{1}{2B}\langle \bar{q}q \rangle = \langle \mathcal{N} \rangle \quad (6.1.2)$$

$$= F^2 \left(1 - \frac{1}{2F^2}\langle \vec{\phi} \cdot \vec{\phi} \rangle + 2(l_3 + h_1)\frac{M^2}{F^2} \right) \quad (6.1.3)$$

where we do not write the thermal index anymore. To evaluate (6.1.3) we require the thermal propagator at the origin of coordinates,

$$\langle \vec{\phi} \cdot \vec{\phi} \rangle = \langle 0|\vec{\phi} \cdot \vec{\phi}|0 \rangle + 3N(T)$$

which is evaluated in Appendix D. As we saw in Section 3.10, the divergent piece of the vacuum propagator cancels with those of l_3 and h_1. Using (3.10.33) in (6.1.3) we get

$$\langle \bar{q}q \rangle = \langle 0|\bar{q}q|0 \rangle \left\{ 1 - \frac{3N(T)}{2F^2} \left(1 + \frac{M^2}{32\pi^2 F^2}(4\bar{h}_1 - \bar{l}_3) \right)^{-1} \right\} \quad (6.1.4)$$

where the vacuum condensate is now correct to one loop. The correction terms from $\bar{h}_1$ and $\bar{l}_3$ are of higher order and can be significant only if we extend the calculation to two loops. We thus get to one loop [9]

$$\langle \bar{q}q \rangle = \langle 0|\bar{q}q|0 \rangle \left(1 - \frac{T^2}{8F^2} \right). \quad (6.1.5)$$

In Appendices H and I we rederive this formula directly from the partition function for pion gas and from density expansion of the operator $\bar{q}q$.

6.2 Pion at Non-Zero Temperature

Here we shall calculate the thermal two-point function of the axial-vector current. We begin by describing the general procedure to be followed for other cases in this chapter. Like the vacuum case (3.10.8) we equate ensemble averaged generating functionals of the QCD and effective theories,

$$\langle T \exp\left(i \int d^3x \int_C d\tau \vec{a}^\mu(x) \cdot \vec{A}_\mu(x)\right)\rangle = \langle T \exp\left(i \int d^3x \int_C d\tau \vec{a}^\mu(x) \cdot \vec{\mathscr{A}}_\mu(x)\right)\rangle \tag{6.2.1}$$

Again taking two functional derivatives with respect to the external axial-vector field, we can express the equality of the thermal two-point functions in the two theories. We then rearrange the left side of this equality in the form of the matrix

$$i \begin{pmatrix} \langle T A^a_\mu(\boldsymbol{x},t) A^b_\nu(\boldsymbol{x}',t') \rangle & \langle A^b_\nu(\boldsymbol{x}',t'-i\beta/2) A^a_\mu(\boldsymbol{x},t) \rangle \\ \langle A^a_\mu(\boldsymbol{x},t-i\beta/2) A^b_\nu(\boldsymbol{x}',t') \rangle & \langle \widetilde{T} A^a_\mu(\boldsymbol{x},t-i\beta/2) A^b_\nu(\boldsymbol{x}',t'-i\beta/2) \rangle \end{pmatrix} \tag{6.2.2}$$

as we did for the scalar propagator in (4.2.29). On the right we can get the corresponding matrix by the replacements (5.1.33) and (5.1.34) in the vacuum amplitudes of Section 3.10. But, as explained at the end of Section 5.2, we can avoid this step by finding directly the 11-components of the matrices. However, for illustration, we shall write the matrices in this example.

The vertices and self-energies in this case are momentum independent, being integrals over the 11- or 22-component of the thermal propagator at the origin. Recalling the replacements, the thermal matrix amplitude can be obtained directly from (3.10.16),

$$\boldsymbol{T}(k) = F^2\{1 + \eta(\overline{\gamma} + \overline{\sigma}')\}\{\boldsymbol{D}(k) + M^2\eta\overline{\sigma}\,\boldsymbol{D}\boldsymbol{\tau}\boldsymbol{D} + \cdots\} \tag{6.2.3}$$

where the quantities with a bar are given by (3.10.15) after replacing G by $\overline{G}$,

$$\overline{\sigma} = -\frac{1}{2}(4l_3 + \overline{G}), \qquad \overline{\sigma}' = -\frac{2}{3}(3l_4 - \overline{G}), \qquad \overline{\gamma} = \frac{4}{3}(3l_4 - 2\overline{G}), \tag{6.2.4}$$

We now multiply (6.2.3) by $\boldsymbol{U}(|k_0|)^{-1}$ on the left and right to diagonalise the amplitude, getting the diagonal element

$$\begin{aligned} \overline{T}(k) &= F^2\{1 + \eta(\overline{\gamma} + \overline{\sigma}')\}\{\Delta + M^2\eta\overline{\sigma}\Delta^2 + \cdots\} \\ &= \frac{-F_\pi^2(T)}{k^2 - M_\pi^2(T)} \end{aligned} \tag{6.2.5}$$

where

$$M_\pi^2(T) = M^2\{1 + \eta(\frac{1}{2}\overline{G} + 2l_3)\} \tag{6.2.6}$$

$$F_\pi(T) = F\{1 + \eta(-\overline{G} + l_4)\}. \tag{6.2.7}$$

The renormalisation proceeds as in the vacuum case. The divergence of the vacuum part of $\overline{G}$ cancels with those of l_3 and l_4, giving the renormalised parameters M_π and F_π. The thermal part is left over as correction.

$$M_\pi^2(T) = M_\pi^2 + \frac{M^2 N(T)}{2F^2}, \qquad F_\pi(T) = F_\pi - F\frac{N(T)}{F^2}.$$

In the correction terms we may replace M and F by their renormalised (physical) values, as it affects the result only to two loops. Inserting the evaluation of $N(T)$ in the limit $M \to 0$, we finally get [1]

$$M_\pi^2(T) = M_\pi^2 \left(1 + \frac{T^2}{24F_\pi^2} \right) \tag{6.2.8}$$

$$F_\pi(T) = F_\pi \left(1 - \frac{T^2}{12F_\pi^2} \right). \tag{6.2.9}$$

$$* \quad * \quad *$$

This example of pion is somewhat special in that both the self-energy and vertex loops are (momentum independent) parameters. So the only diagonalisation needed here is that of the pion propagator. In the examples below, we shall have non-trivial loop contributions. In their analysis, it will be useful to note the following points:

a) As we are interested in thermal parameters like mass and coupling, we need calculate only the real part of loop graphs.
b) Being interested in the real part of graphs, it is not convenient to use the spectral form of propagators. Instead we shall use the form in which the vacuum and thermal parts appear separated.
c) Of the two particles circulating in a loop, if one is heavy (with zero chemical potential), it will be dilute enough to justify omitting the thermal part from its propagator.
d) To avoid new symbols, we shall continue to denote the diagonalised quantities, like self-energy, by a bar, even after removing their vacuum parts.

6.3 Nucleon at Finite Temperature

The vacuum two-point function (3.12.6) of nucleon current generalises to the thermal matrix

$$i \begin{pmatrix} \langle T\eta(\boldsymbol{x},t)\overline{\eta}(\boldsymbol{x}',t')\rangle & \langle \overline{\eta}(\boldsymbol{x}',t'-i\beta/2)\eta(\boldsymbol{x},t)\rangle \\ \langle \eta(\boldsymbol{x},t-i\beta/2)\overline{\eta}(\boldsymbol{x}',t')\rangle & \langle \widetilde{T}\eta(\boldsymbol{x},t-i\beta/2)\overline{\eta}(\boldsymbol{x}',t'-i\beta/2)\rangle \end{pmatrix}. \tag{6.3.1}$$

We find the diagonalised thermal amplitudes in the effective theory for the graphs of Figure 3.4. We calculate for $\boldsymbol{p} = 0$, when the free vacuum propagator (3.12.7) of graph (a) becomes

$$\frac{-\lambda^2}{E - m_N + i\epsilon} \frac{1}{2}(1 + \gamma_0) \tag{6.3.2}$$

which is also the diagonalised element of the free thermal matrix. We want to find the changes in the coupling λ and mass m_N of nucleon at finite temperature.

From the expression (3.12.9) of the vacuum self-energy, we get the diagonalised thermal one

$$\overline{\Sigma}(p) = -\frac{3}{4}\left(\frac{g_A}{F}\right)^2 \int \frac{d^4k}{(2\pi)^3} n(\omega)\delta(k^2 - M^2)\not{k}\gamma^5 \frac{1}{\not{p} - \not{k} - m_N}\not{k}\gamma^5. \qquad (6.3.3)$$

For $\boldsymbol{p} = 0$, it becomes in the chiral limit

$$\overline{\Sigma}(E) = -\frac{3}{2}\left(\frac{g_A}{F}\right)^2 E(E^2 - m_N^2) \int \frac{d^3k\, n(\omega)}{(2\pi)^3\omega} \frac{\omega^2}{(E^2 - m_N^2)^2 - 4E^2\omega^2}\gamma_0 \qquad (6.3.4)$$

where now $\omega = |\boldsymbol{k}|$. In the neighbourhood of the nucleon pole, it gives

$$\overline{\Sigma}(E) = a(E - m_N)\gamma_0 + O((E - m_N)^2), \qquad a = \frac{3}{4}\left(\frac{g_A}{F}\right)^2 N(T). \qquad (6.3.5)$$

As $\overline{\Sigma}(m_N) = 0$, there is no mass shift of nucleon to $O(T^2)$.[1] This is because the chiral coupling has one derivative on the pion field in the interaction term, giving the factor ω^2 in the integrand of the above self-energy. Thus it is the chiral symmetry which prevents nucleon from acquiring an effective mass of $O(T^2)$ at low temperature.

From (3.12.11) we get the thermal contribution of graphs from (a) to (d) as

$$\overline{W}_{(a+b+c+d)} = \lambda^2\left(1 - \frac{3N(T)}{4F^2}\right)\frac{-1}{\gamma_0 E - m_N + a(E - m_N)\gamma_0}$$

$$= \lambda^2\left(1 - \frac{3(1 + g_A^2)N(T)}{4F^2}\right)\frac{-1}{E - m_N + i\epsilon}\frac{1}{2}(1 + \gamma_0). \qquad (6.3.6)$$

The thermal part from the momentum-dependent vertex (3.12.12) is

$$\overline{W}_{(e+f)} = \frac{3\lambda^2}{4F^2}g_A^2 \int \frac{d^4k}{(2\pi)^3}\frac{n(|k_0|)\delta(k^2 - M^2)P(p,k)}{(p-k)^2 - m^2} \cdot \Delta_N(p) \qquad (6.3.7)$$

with Δ_N standing for the scalar propagator with nucleon mass and

$$P(p,k) = (\not{p} + m_N)\not{k}(\not{p} - \not{k} - m_N) + (\not{p} - \not{k} - m_N)\not{k}(\not{p} + m_N).$$

When evaluated at $\boldsymbol{p} = 0$, P reduces to $P = 2(E^2 - m_N^2)k_0\gamma_0$ in the chiral limit, so that nucleon pole cancels out in (6.3.7). Finally the thermal part of graph (g) is also similar to that of (e+f), with no contribution to nucleon pole.

We thus read off from (6.3.6) temperature-dependent coupling $\lambda(T)$ between the nucleon current and nucleon field as [4]

$$\lambda(T) = \lambda\left(1 - \frac{g_A^2 + 1}{32}\frac{T^2}{F^2}\right) \qquad (6.3.8)$$

and, as we already observed above, the nucleon mass does not shift to order $O(T^2)$.

---

[1] If we take an intermediate state like $\pi\Delta(1232)$, where the baryon mass differs from that of nucleon, we would get a non-zero value for $\overline{\Sigma}(m_N)$ but of $O(T^4)$.

6.4 ρ at Finite Temperature

Here we consider the matrix of the thermal two-point function of vector currents,

$$
i \begin{pmatrix} \langle TV_\mu^a(\boldsymbol{x},t)V_\nu^b(\boldsymbol{x}',t')\rangle & \langle V_\nu^b(\boldsymbol{x}',t'-i\beta/2)V_\mu^a(\boldsymbol{x},t)\rangle \\ \langle V_\mu^a(\boldsymbol{x},t-i\beta/2)V_\nu^b(\boldsymbol{x}',t')\rangle & \langle \widetilde{T}V_\mu^a(\boldsymbol{x},t-i\beta/2)V_\nu^b(\boldsymbol{x}',t'-i\beta/2)\rangle \end{pmatrix} . \tag{6.4.1}
$$

As before, we infer from the vacuum result (3.13.19) the corresponding thermal diagonalised element as

$$
\overline{M}_{\mu\nu}^{ab}(q) = \delta^{ab} \left(\frac{F_\rho q^2}{m_\rho}\right)^2 \left(1 - \frac{M^2 \overline{G}}{F^2}\right) \overline{D}_{\mu\nu} + \cdots \tag{6.4.2}
$$

where the dots are contributions from momentum dependent vertices and intermediate states. The thermal complete propagator satisfies

$$
\overline{D}_{\mu\nu} = \widetilde{\Delta}_{\mu\nu} + \widetilde{\Delta}_{\mu\rho}\overline{\Pi}^{\rho\sigma}\overline{D}_{\sigma\nu}. \tag{6.4.3}
$$

Having eliminated the thermal matrix structure, we analyse the Lorentz indices in medium to extract relevant scalar amplitudes.

Current Conservation

We first obtain the condition of current conservation. For this we go back to the representation (6.4.1) in terms of thermal matrix elements of current operators. In deriving this condition for the vacuum two-point function, we actually worked with operators, without involving the (vacuum) state. So we can proceed for the thermal case in the same way. Denoting the Fourier transform of the matrix elements in (6.4.1) by $(M(q))_{ij}$, we get for diagonal elements

$$
\left(q^\mu M_{\mu\nu}^{ab}\right)_{11} = - \left(q^\mu M_{\mu\nu}^{ab}\right)_{22} = -i\epsilon^{abc}\langle V_\nu^c(0)\rangle
$$

and for off-diagonal elements

$$
\left(q^\mu M_{\mu\nu}^{ab}\right)_{12} = \left(q^\mu M_{\mu\nu}^{ab}\right)_{21} = 0.
$$

In the vacuum case we gave two arguments requiring $\langle 0|V_\nu^c|0\rangle = 0$; that there is no four-vector in vacuum and that we assume isospin symmetry. In the medium we do have a four-vector, namely its four-velocity u^μ, so that $\langle 0|V_\nu^c|0\rangle$ may be proportional to u_ν. But our assumption of isospin symmetry makes it zero. So for all elements of the thermal matrix

$$
\left(q^\mu M_{\mu\nu}^{ab}\right)_{ij} = 0
$$

and hence also for the diagonalised element

$$
q^\mu \overline{M}_{\mu\nu} = q^\nu \overline{M}_{\mu\nu} = 0. \tag{6.4.4}
$$

Lorentz Structure

We can now analyse the tensor structure into independent covariants satisfying current conservation [10, 11]. With the additional four-vector u^μ, one can form two scalar variables, $\omega \equiv u \cdot q$ and $\bar{q} \equiv \sqrt{\omega^2 - q^2}$. It is convenient here to replace u^μ by the transverse variable $\tilde{u}^\mu \equiv u^\mu - (\omega/q^2)q^\mu$, $q_\mu \tilde{u}^\mu = 0$. Then we have four elementary tensors, namely

$$g_{\mu\nu}, \qquad q_\mu q_\nu, \qquad \tilde{u}_\mu \tilde{u}_\nu \quad \text{and} \quad q_\mu \tilde{u}_\nu + q_\nu \tilde{u}_\mu$$

in terms of which the symmetric, second rank tensor $\overline{M}$ can be expanded with four scalar functions as coefficients. Current conservation reduces them to two independent tensors, say $g_{\mu\nu} - q_\mu q_\nu/q^2$ and $\tilde{u}_\mu \tilde{u}_\nu$. In choosing their actual forms, we want to remove any kinematical pole at $q^2 = 0$, as dynamical singularities at finite temperature can extend to this point (see Section 8.2). So we choose

$$P_{\mu\nu} = -g_{\mu\nu} + \frac{q_\mu q_\nu}{q^2} - \frac{q^2}{\bar{q}^2}\tilde{u}_\mu \tilde{u}_\nu, \qquad Q_{\mu\nu} = \frac{(q^2)^2}{\bar{q}^2}\tilde{u}_\mu \tilde{u}_\nu, \tag{6.4.5}$$

which are orthogonal and projective,

$$P \cdot Q = 0, \qquad P \cdot P = -P, \qquad Q \cdot Q = -q^2 Q. \tag{6.4.6}$$

We then have

$$\overline{M}_{\mu\nu} = \overline{M}_t P_{\mu\nu} + \overline{M}_l Q_{\mu\nu} \tag{6.4.7}$$

with $\overline{M}_{t,l}$ as functions of the scalar variables, ω and $\bar{q}$. As the indices of the self-energy tensor $\overline{\Pi}$ in (6.4.3) are contracted with transverse propagators, it can also be decomposed as (6.4.7).

That these covariants are indeed free from singularities at $q^2 = 0$ can be seen clearly, if we write their space and time components in the rest frame of the medium $u^\mu = (1, 0)$.

$$P_{00} = 0, \qquad P_{0i} = P_{i0} = 0, \qquad P_{ij} = \delta_{ij} - \frac{q_i q_j}{|\boldsymbol{q}|^2}$$

$$Q_{00} = |\boldsymbol{q}|^2, \qquad Q_{0i} = Q_{i0} = q_0 q_i, \qquad Q_{ij} = q_0^2 \frac{q_i q_j}{|\boldsymbol{q}|^2}$$

(Observe that P is transverse not only four-dimensionally, but also three-dimensionally.) There is still an ambiguity in the space components. To see this, let $\boldsymbol{n}$ be the unit vector along $\boldsymbol{q}$, so that $q_i = |\boldsymbol{q}|n_i$. Then we can write

$$P_{ij} = \delta_{ij} - n_i n_j, \qquad Q_{ij} = q_0^2 n_i n_j, \tag{6.4.8}$$

when (6.4.7) gives

$$\overline{M}_{ij} = \overline{M}_t \delta_{ij} - n_i n_j (\overline{M}_t - q_0^2 \overline{M}_l). \tag{6.4.9}$$

So the amplitude appears to depend on the direction of $\boldsymbol{q}$ even when it is zero. This dependence may be eliminated by requiring [10]

$$\overline{M}_t(q_0, \boldsymbol{q} = 0) = q_0^2 \overline{M}_l(q_0, \boldsymbol{q} = 0) \tag{6.4.10}$$

which relates the invariant amplitudes at $\boldsymbol{q} = 0$. To extract $\overline{M}_t$ and $\overline{M}_l$ from $\overline{M}_{\mu\nu}$, it is convenient to obtain first

$$\overline{M}{}^{\mu}_{\ \mu} \equiv g_{\mu\nu}\overline{M}^{\mu\nu} = -2\overline{M}_t - q^2\overline{M}_l \tag{6.4.11}$$

$$\overline{M}_{00} = u^{\mu}u^{\nu}\overline{M}_{\mu\nu} = |\boldsymbol{q}|^2\overline{M}_l. \tag{6.4.12}$$

Then

$$\overline{M}_t = -\frac{1}{2}(\overline{M}{}^{\mu}_{\ \mu} + \frac{q^2}{\overline{q}^2}\overline{M}_{00}), \quad \overline{M}_l = \frac{1}{\overline{q}^2}\overline{M}_{00} \tag{6.4.13}$$

Similar relations hold for $\overline{\Pi}_{t,l}$.

ρ *Meson Pole*

We can now solve (6.4.3) for the thermal propagator. The tensors in it can be decomposed as

$$\widetilde{\Delta}_{\mu\nu} = (P_{\mu\nu} + \frac{1}{q^2}Q_{\mu\nu})\Delta_F$$

$$\overline{D}_{\mu\nu} = P_{\mu\nu}\overline{D}_t + Q_{\mu\nu}\overline{D}_l$$

$$\overline{\Pi}_{\mu\nu} = P_{\mu\nu}\overline{\Pi}_t + Q_{\mu\nu}\overline{\Pi}_l \tag{6.4.14}$$

giving the solution

$$\overline{D}_t = \frac{-1}{q^2 - m_V^2 + \overline{\Pi}_t}, \quad \overline{D}_l = \frac{1}{q^2}\frac{-1}{q^2 - m_V^2 + q^2\overline{\Pi}_l}. \tag{6.4.15}$$

We shall simplify calculations by taking $\boldsymbol{q} = 0$, when (6.4.10) shows that the invariant amplitudes $\overline{M}_t$ and $\overline{M}_l$ are related, so that we can work with either of these, say the longitudinal one, for which (6.4.2) gives

$$\overline{M}_l(q) = \frac{q^2}{m_V^2}F_\rho^2\left(1 - \frac{M^2\overline{G}}{F^2}\right)\frac{-1}{q^2 - m_V^2 + q^2\overline{\Pi}_l + i\epsilon} + \cdots \tag{6.4.16}$$

The vacuum self-energies $\Pi_{\mu\nu}$ for different loop graphs are given in (3.13.15–17), from which we can find the corresponding diagonalised elements $\overline{\Pi}_{\mu\nu}$. Here the $\pi\omega$ loop is singled out for dominant contribution at low temperature. As we shall see below, because of the closeness of ω and ρ meson masses, which we set to be equal (denoted by m_V), this loop alone can contribute to $O(T^2)$. It is

$$\overline{\Pi}^{(\pi\omega)}_{\alpha\beta}(q) = \left(\frac{2g_1}{F}\right)^2 \int \frac{d^4k}{(2\pi)^3}\frac{n(\omega)\,\delta(k^2 - M^2)}{(q-k)^2 - m_V^2}(B_{\alpha\beta} + M^2q^2A_{\alpha\beta}). \tag{6.4.17}$$

The longitudinal function is given by the 00-component of this tensor. As defined by (3.13.16), A_{00} and B_{00} are proportional to $\boldsymbol{q}^2$, which cancels the singularity at this point in the definition of $\overline{\Pi}_l$ defined in the same way as $\overline{M}_l$ in (6.4.13). For the case at hand

$$\frac{1}{\boldsymbol{q}^2}(B_{00} + q^2k^2A_{00}) = -\boldsymbol{k}^2(1 - \cos^2\theta)$$

where θ is the angle between $\boldsymbol{k}$ and $\boldsymbol{q}$. We can now take the limit $\boldsymbol{q} \to 0$ and do the angular integration. The k_0 integration over the delta function gives two terms, which can be combined to give in the chiral limit (with $q_0 \equiv E$)

$$\overline{\Pi}_l^{(\pi\omega)}(E) = -\frac{2}{3}\left(\frac{2g_1}{F}\right)^2 (E^2 - m_V^2) \int_0^\infty \frac{dk\, k\, n(k)}{2\pi^2} \frac{k^2}{(E^2 - m_V^2)^2 - 4E^2 k^2} \tag{6.4.18}$$

We now find the temperature dependence of (6.4.18) in much the same way as for the nucleon in the last section. Because of the pion distribution function, the integral is effectively cut off at $k \sim T$. As we are interested in the pole parameters, we evaluate the integral for E close to m_V. So the denominator of the integrand in (6.4.18) behaves as k^2, cancelling this factor in the numerator, getting[2]

$$\overline{\Pi}_l^{(\pi\omega)}(E) = \frac{2g_1^2 N(T)}{3F^2 m_V^2}(E^2 - m_V^2). \tag{6.4.19}$$

The crucial point here is the proximity of m_ω to m_ρ, which is not the case for loops involving particles like πa_1 or $\pi\pi$. (See Problem 6.1.)

So far we have investigated only the self-energy graph contributions to the two-point function. The other graphs involving momentum-dependent vertices and intermediate states can also be written and analysed in the same way, with the result that they do not have the ρ meson pole. Thus, as far as the thermal corrections to $O(T^2)$ are concerned, we are left with corrections from momentum independent vertex and $\pi\omega$ self-energy, giving the pole term as

$$\overline{M}_l(E) = E^2 \left(\frac{F_\rho^T}{m_V}\right)^2 \frac{-1}{E^2 - m_V^2} \tag{6.4.20}$$

with [12]

$$F_\rho^T = F_\rho \left\{ 1 - \left(1 + \frac{g_1^2}{3}\right) \frac{T^2}{12F_\pi^2} \right\}. \tag{6.4.21}$$

*   *   *

In Section 6.2 we evaluated the thermal two-point function of the axial-vector current in the neighbourhood of the pion pole. We can also find it near the a_1 meson pole to get results similar to ρ meson pole. The two evaluations run parallel, especially in the chiral limit, when the axial-vector current is also conserved. Like the $\rho\omega\pi$ vertex in the vector current case, the self-energy from the $a_1 f_1 \pi$

---

[2] In the above analysis we took the masses of ρ and ω to be equal. If we take them to be different ($m_\omega = m_\rho + \Delta m$), the denominator of the integrand in (6.4.18) becomes $(E^2 - m_\omega^2)^2 - 4E^2 k^2$, which at $E = m_\rho$ is proportional to $(\Delta m)^2 - k^2$. Experimentally $\Delta m = 10$MeV, so unless $T < 10$ MeV, we may ignore Δm and get $\overline{\Pi}_t^{(\omega)} \sim T^2$ as written in (6.4.20). But if the temperature is as low as a few MeV – a region not of any interest – then $\overline{\Pi}_t^{(\omega)} \sim T^4$.

vertex will dominate here, again because of the proximity of a_1 and f_1 masses. We then get the result to $O(T^2)$ that the a_1 mass remains unchanged, while the coupling F_{a_1} of axial-vector current to a_1 meson shifts to [12]

$$F_{a_1}^T = F_{a_1} \left\{ 1 - \left(1 + \frac{g_2^2}{3} \right) \frac{T^2}{12 F_\pi^2} \right\}. \tag{6.4.22}$$

$$*\quad*\quad*$$

We comment here briefly on the work of Ioffe et al. [13] on the thermal two-point functions of vector and axial-vector currents. Considering the vector current, they expand the two-point function in pion density. From (I.7) of Appendix I, we get

$$\langle T V_\mu^i(x) V_\nu^j(0) \rangle = \langle 0 | T V_\mu^i(x) V_\nu^j(0) | 0 \rangle$$
$$+ \int \frac{d^3 k\, n(k)}{(2\pi)^3\, 2k} \sum_m \langle \pi^m(k) | T V_\mu^i(x) V_\nu^j(0) | \pi^m(k) \rangle_{\mathrm{con}}. \tag{6.4.23}$$

They evaluate the pion matrix element in (6.4.23) using the old technique of PCAC and current algebra. Their result implies (6.4.21) without the g_1^2 term.

We now show in outline that the methods of current algebra would also yield the g_1^2 term, if applied carefully [14]. These methods try to find singular amplitudes as $k_\mu \to 0$ in the chiral limit. The amplitude to start with is

$$i \int d^4 x\, e^{ik' \cdot z} \langle 0 | T A_\alpha^d(z) V_\mu^a(x) V_\nu^b(0) | \pi^c(k) \rangle. \tag{6.4.24}$$

As described in Section 2.4 it can give rise to singular amplitudes in two ways. The axial-vector current can attach to a pion line, which terminates on the graph for the amplitude $\langle 0 | T V_\mu^i(x) V_\nu^j(0) | \pi^m(k) \rangle$. The resulting pion propagator produces the singularity as shown in graph (a) of Figure 6.1. It is this singularity which can be obtained naively by PCAC and current algebra. But

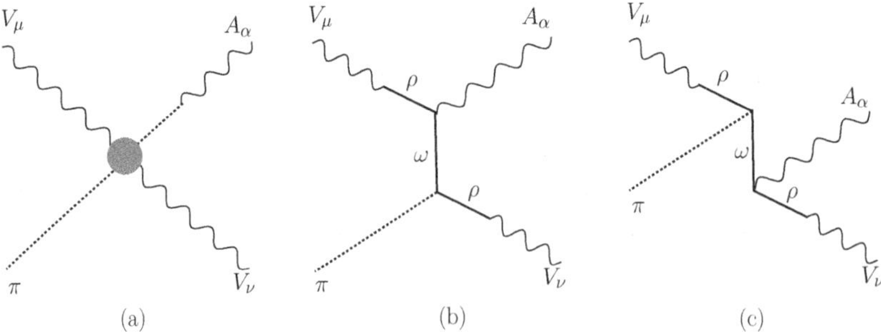

Figure 6.1 Graphs producing singular behaviour as the momenta carried by the currents go to zero. The axial-vector current can attach via the pion propagator (a) or directly (b,c) to the amplitude (6.4.24).

the axial-vector current can also attach directly to the amplitude producing an extra propagator, which can be singular. This is shown in graphs (b,c) of Figure 6.1, where the vector currents couple to ρ meson. Then the axial-vector current attaches to the ρ propagator generating the ω propagator. Because the masses of ρ and ω are very close, these graphs will produce a singular contribution proportional to g_1^2. (The details of the calculation can be found in [14].) It is this piece which is missing in [13].

6.5 Nuclear Matter

So far we have discussed the effect of pions in medium on the quark condensate and parameters of different hadrons. We now find the effect of nucleons in medium on these parameters. As we already pointed out, at finite temperature, pions usually dominate the medium, the heavy particles being exponentially suppressed. But we can introduce a chemical potential μ for the nucleon, when we have only pions, and nucleons and anti-nucleons in the medium, distributed as

$$n(\omega) = \frac{1}{e^{\beta\omega} - 1}, \quad \widetilde{n}_+(E) = \frac{1}{e^{\beta(E-\mu)} + 1}, \quad \widetilde{n}_-(E) = \frac{1}{e^{\beta(E+\mu)} + 1} \quad (6.5.1)$$

with $\omega = \sqrt{k^2 + M^2}$ and $E = \sqrt{p^2 + m_N^2}$. It simplifies to let the temperature go to zero, when only nucleons will be present

$$n(\omega) \to 0, \quad \widetilde{n}_+(E) \to \theta(\mu - E), \quad \widetilde{n}_-(E) \to 0, \quad (6.5.2)$$

so that the chemical potential is related to the Fermi momentum p_F by $\mu = \sqrt{p_F^2 + m_N^2}$. Thus, at zero temperature, all the nucleon states are filled up to momentum p_F.

We have here what is called nuclear matter, which we take to be symmetric, consisting of equal numbers of neutrons and protons. The single nucleon states are four-fold degenerate (two from spin and another two from isospin), giving the nucleon number density

$$\overline{n} = 4 \int \frac{d^3p}{(2\pi)^3} \widetilde{n}_+ = 4 \int_0^{p_F} \frac{d^3p}{(2\pi)^3} = \frac{2p_F^3}{3\pi^2}. \quad (6.5.3)$$

We note that the saturation density for nuclear matter is $\overline{n}_0 = (110\text{MeV})^3$ corresponding to Fermi momentum $p_F = 270$ MeV.

Quark Condensate

To calculate the quark condensate in nuclear matter in a way similar to that in pion medium, we have to introduce an external scalar field in the πN effective Lagrangian and expand the resulting Lagrangian in this field around the quark mass. However, we do not follow this procedure here. Instead, we calculate this condensate in two other ways, namely from the partition function of nuclear matter and the density expansion of the operator $\bar{q}q$. These are obtained in Appendices H and I. Both methods give the quark condensate in nuclear matter (denoted by superscript N) as

$$\langle\bar{q}q\rangle^N = \langle 0|\bar{q}q|0\rangle\left(1 - \frac{\sigma\bar{n}}{F^2 M^2}\right), \tag{6.5.4}$$

where σ is the nucleon sigma term defined by (H.10).

Pion Parameters

We next consider the two-point function of axial-vector current in nuclear matter. We first take the vacuum two-point function

$$\mathcal{T}_{\mu\nu}^{ij}(q) = i\int d^4x\, e^{iq\cdot x}\langle 0|T A_\mu^i(x)\, A_\nu^j(0)|0\rangle$$

and find corrections to pion propagation in vacuum from graphs with a nucleon loop (Figure 6.2). To get the medium effect, we follow the familiar steps, replacing the nucleon vacuum propagator with the 11-component of the thermal propagator given by (4.3.15). It is convenient here to extract the projection matrix $(\not{p} + m_N)$ from the propagator and define a scalar nucleon propagator, $\Delta_F^N(p) = -1/(p^2 - m_N^2 + i\epsilon)$. Then the replacement is

$$\Delta_F^N(p) \to \Delta_F^N(p) - 2\pi i\tilde{n}_+\delta(p^2 - m_N^2)\theta(p_0). \tag{6.5.5}$$

To find the vertices in the graphs, we look at (3.10.7) and the pion–nucleon Lagrangian given by (3.11.12) and (3.11.13). Expanding the matrices U and u_μ up to two powers of pion field and keeping terms linear in the external axial-vector field $a_\mu(x)$, we get

$$\mathcal{L}_{\pi N} = \mathcal{L}_\pi + \mathcal{L}_N^{(1)} + \mathcal{L}_N^{(2)}$$

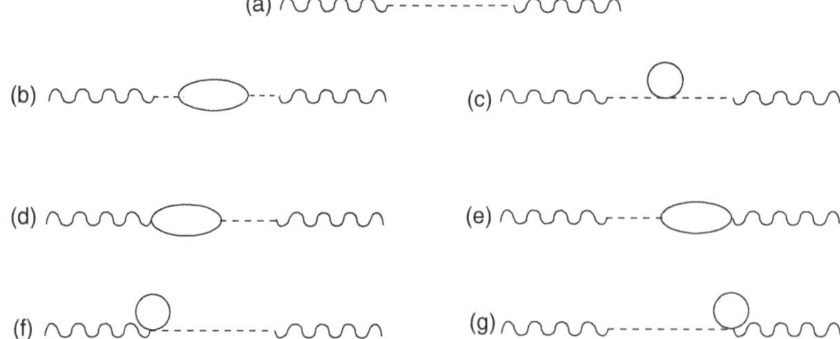

Figure 6.2 Feynman diagrams for the axial-vector two-point function to one loop. Only loops with nucleons are considered.

where[3]

$$\mathscr{L}_\pi = \mathscr{L}_\pi^{(0)} - F\partial_\mu\vec{\phi}\cdot\vec{a}^\mu,$$

$$\mathscr{L}_N^{(1)} = \mathscr{L}_N^{(0)} - \frac{g_A}{2F}\overline{\psi}\gamma^\mu\gamma^5\partial_\mu\vec{\phi}\cdot\vec{\tau}\psi + \frac{g_A}{2}\overline{\psi}\gamma^\mu\gamma^5\vec{a}^\mu\cdot\vec{\tau}\psi,$$

$$\mathscr{L}_N^{(2)} = -\frac{2M^2 c_1}{F^2}\vec{\phi}\cdot\vec{\phi}\,\overline{\psi}\psi$$

$$-\frac{c_2}{m_N^2}\left(\frac{1}{F^2}\partial_\mu\vec{\phi}\cdot\partial_\nu\vec{\phi} - \frac{2}{F}\partial_\mu\vec{\phi}\cdot\vec{a}^\nu\right)\overline{\psi}\partial^\mu\partial^\nu\psi$$

$$+c_3\left(\frac{1}{F^2}\partial_\mu\vec{\phi}\cdot\partial^\mu\vec{\phi} - \frac{2}{F}\partial_\mu\vec{\phi}\cdot\vec{a}^\mu\right)\overline{\psi}\psi. \tag{6.5.6}$$

The pion–nucleon system is described by the free Lagrangians $\mathscr{L}_\pi^{(0)}$ and $\mathscr{L}_N^{(0)}$ with interaction given by terms without $a_\mu(x)$. The sum of the coefficients of $a_\mu(x)$ defines the axial-vector current $\mathscr{A}_\mu$ for the system in the effective theory.

The free pion pole graph (a) gives the amplitude,

$$\overline{\mathscr{T}}_{\mu\nu}^{ij}(q)_{(a)} = \delta^{ij}q_\mu q_\nu F^2\Delta_F(q), \qquad \Delta_F(q) = -1/(q^2 - M^2 + i\epsilon). \tag{6.5.7}$$

Omitting the isospin indices from now on, the self-energy graph (b) gives

$$\overline{\mathscr{T}}_{\mu\nu}(q)_{(b)} = q_\mu q_\nu F^2\Delta(q)\overline{\Sigma}_1(q)\Delta(q). \tag{6.5.8}$$

To get $\overline{\Sigma}_1$, we first write the self-energy in vacuum

$$\Sigma_1(q) = \frac{ig_A^2}{2F^2}\int\frac{d^4p}{(2\pi)^4}\,\mathrm{tr}\left[\slashed{q}\gamma^5(\slashed{p}+m_N)\slashed{q}\gamma^5(\slashed{p}-\slashed{q}+m_N)\right]\Delta_F^N(p)\Delta_F^N(p-q). \tag{6.5.9}$$

Replacing these vacuum propagators by (6.5.5), we get the 11-component of the thermal self-energy matrix. Collecting terms linear in the distribution function, it gives the medium correction to the self-energy,

$$\overline{\Sigma}_1(q) = 8\frac{g_A^2}{F^2}m_N^2 q^4\int\frac{d^4p}{(2\pi)^3}\frac{\tilde{n}_+\delta(p^2 - m_N^2)\,\theta(p_0)}{q^4 - 4(p\cdot q)^2}. \tag{6.5.10}$$

Let us take the pion at rest, $\boldsymbol{q} = 0$, when (6.5.10) simplifies to

$$\overline{\Sigma}_1(q_0) = q_0^2\sigma_1, \qquad \sigma_1 = -\frac{g_A^2\overline{n}}{4F^2 m_N}. \tag{6.5.11}$$

The self-energy graph (c) receives contributions from all the interaction terms in $\mathscr{L}_N^{(2)}$

$$\overline{\mathscr{T}}_{\mu\nu}(q)_{(c)} = q_\mu q_\nu F^2\Delta_F(q)\overline{\Sigma}_2(q)\Delta_F(q) \tag{6.5.12}$$

with

$$\overline{\Sigma}_2 = M^2\sigma_2 + q_0^2\sigma_3, \qquad \sigma_2 = -4c_1\overline{n}/F^2, \qquad \sigma_3 = 2(c_2 + c_3)\overline{n}/F^2. \tag{6.5.13}$$

---

[3] The piece with c_4 in (3.11.13) is omitted, as it gives the nucleon loop with $\mathrm{tr}\,\tau^i$ which is zero.

From the vertex graphs (d) and (e) we get

$$\overline{\mathscr{T}}_{\mu\nu}(q)_{(d+e)} = q_\mu q_\nu F^2 \overline{\Gamma}_1 \Delta_F(q), \quad \overline{\Gamma}_1 = -\frac{g_A^2 \overline{n}}{2m_N F^2}. \tag{6.5.14}$$

Finally we evaluate the contribution from the constant vertex graphs (f) and (g),

$$\overline{\mathscr{T}}_{\mu\nu}(q)_{(f+g)} = \{q_\mu q_\nu 4c_3 \overline{n} + 2c_2(q_\mu K_\nu + q_\nu K_\mu)\}\Delta_F(q) \tag{6.5.15}$$

where

$$K_\nu(q) = \frac{8}{m_N} \int \frac{d^4p}{(2\pi)^3} \widetilde{n}_+ \{q_0 p_0 - \boldsymbol{q} \cdot \boldsymbol{p}\} p_\nu \delta(p^2 - m_N^2)\theta(p_0).$$

A special feature is presented by K_ν in that it is not proportional to q_ν. If we consider K_0, the second term within the curly bracket is zero, being odd in $\boldsymbol{p}$, getting $K_0 = q_0 \overline{n}$. But for K_i, the first term becomes zero for the same reason, making $K_i \sim q_i \overline{n}^{5/3}$, which we neglect, being of higher order in $\overline{n}$. So $K_\nu = (q_0 \overline{n}, 0)$ in the frame in which the medium is at rest. We can write it as $K_\nu = q \cdot u \overline{n} u_\nu$ in a general frame with u_ν as the four-velocity of the medium. We thus get

$$\overline{\mathscr{T}}_{\mu\nu}(q)_{(f+g)} = F^2 \{q_\mu q_\nu \overline{\Gamma}_2 + q \cdot u(q_\mu u_\nu + q_\nu u_\mu)\overline{\Gamma}_3\}\Delta(q) \tag{6.5.16}$$

where

$$\overline{\Gamma}_2 = 4c_3 \frac{\overline{n}}{F^2}, \qquad \overline{\Gamma}_3 = 2c_2 \frac{\overline{n}}{F^2}. \tag{6.5.17}$$

We now sum the results from all the graphs to get the pion pole correct to one loop. Including vacuum corrections to replace F and M by their physical values F_π and M_π, it is

$$\begin{aligned}
\overline{\mathscr{T}}_{\mu\nu}^{ij}(q) &= \delta^{ij} \frac{F_\pi^2 \{q_\mu q_\nu(1 - \sigma_1 - \sigma_3 + \overline{\Gamma}_1 + \overline{\Gamma}_2) + q \cdot u(q_\mu u_\nu + q_\nu u_\mu)\overline{\Gamma}_3\}}{q_0^2 - M_\pi^2(1 - \sigma_1 - \sigma_2 - \sigma_3)} \\
&\equiv \delta^{ij} \frac{f_\mu(q,\overline{n}) f_\nu(q,\overline{n})}{q_0^2 - (M_\pi^N)^2} \tag{6.5.18}
\end{aligned}$$

where

$$f_\mu(q,\overline{n}) = F_\pi[\{1 - (\sigma_1 + \sigma_3 - \overline{\Gamma}_1 - \overline{\Gamma}_2)/2\}q_\mu + q \cdot u\overline{\Gamma}_3 u_\mu] \tag{6.5.19}$$
$$(M_\pi^N)^2 = M_\pi^2(1 - \sigma_1 - \sigma_2 - \sigma_3). \tag{6.5.20}$$

It is useful to go to the rest frame and display the decay constants separately for the time and space components,

$$f_0(q,\overline{n}) = q_0 F_{\pi t}^N, \qquad f_i(q,\overline{n}) = q_i F_{\pi s}^N. \tag{6.5.21}$$

Inserting values for the σs and $\overline{\Gamma}$s from above, we finally get the correction to pion parameters in nuclear matter as [15, 16]

$$(M_\pi^N)^2 = M_\pi^2 \left\{ 1 + 2\left(2c_1 - c_2 - c_3 + \frac{g_A^2}{8m_N}\right)\frac{\overline{n}}{F_\pi^2} \right\},$$

$$F_{\pi\,t}^N = F_\pi \left\{ 1 + \left(c_2 + c_3 - \frac{g_A^2}{8 m_N} \right) \frac{\overline{n}}{F_\pi^2} \right\},$$

$$F_{\pi\,s}^N = F_\pi \left\{ 1 + \left(-c_2 + c_3 - \frac{g_A^2}{8 m_N} \right) \frac{\overline{n}}{F_\pi^2} \right\}. \tag{6.5.22}$$

Putting in values of the low-energy constants as quoted in Section 3.11, we get

$$(M_\pi^N)^2 = M_\pi^2 (1 + .16\overline{n}/\overline{n}_0)$$
$$F_{\pi\,t}^N = F_\pi (1 - .32\overline{n}/\overline{n}_0)$$
$$F_{\pi\,s}^N = F_\pi (1 - .88\overline{n}/\overline{n}_0). \tag{6.5.23}$$

Observe that the two decay constants have quite different density dependence. However, in questions of spontaneous symmetry breaking, it is the temporal one which is relevant [15]. The reason is that the axial-vector charge responsible for the breaking is given by (the spatial integral of) the time component of the axial-vector current.

6.6 Comments

Let us recall that QCD is a non-Abelian gauge theory of colour of quarks and gluons. Hadrons are assumed to be bound states of these elements confining their colour degrees of freedom. Though QCD describes the strong interactions of these hadrons, it is asymptotically free. It implies that at high temperature and/or density, the interactions of quarks and gluons become weak. So if the temperature/density is high enough, a phase transition is expected in a hadronic system liberating quarks and gluons.

In the absence of quark masses, QCD has also a non-Abelian global flavour symmetry, which is supposed to be spontaneously broken. (As the masses of quarks are small, the explicit symmetry breaking can be included perturbatively in the theory.) On this basis an effective theory can be constructed to describe the interactions of hadrons (Chapter 3). In particular, we can study the one- and two-point correlation functions of flavour densities and currents in vacuum. Using the real time method of thermal field theory (Chapter 4 and 5), we extend these results in the present chapter to the hadronic phase of matter.

We find from (6.1.5) and (6.5.4) that the quark condensate 'melts', as the temperature or density of the hadronic matter is increased. (Its thermal behaviour will be studied more thoroughly in the next chapter.) Hence the spontaneous breaking of flavour symmetry becomes weaker, the Nambu–Goldstone bosons (pions) gradually losing their right to exist. Although the calculations are not strictly valid at such temperature/density, it does support the expectation that at high enough values of these variables, chiral symmetry will be restored and quarks liberated, possibly simultaneously.

We also find the residues at the particle poles of different correlation functions, namely F_π^T, λ^T, F_ρ^T and $F_{a_1}^T$ given respectively by (6.2.9), (6.3.8), (6.4.21) and

(6.4.22), to decrease as T grows. In nuclear matter $F_{\pi,t}^N$ given by (6.5.23) shows the same behaviour as nuclear density increases. Also other residues calculated from QCD sum rules give similar results in nuclear matter [17, 18]. So the respective external sources become gradually weak in exciting the hadrons. In other words, the hadrons lose their prominence as the most likely configurations to be produced at finite temperature and density, again supporting the expectation of a phase transition in hadronic matter.

Problem

Problem 6.1: Find the low-temperature behaviour of vector meson self-energies from $\pi\pi$ and πa_1 loops.

Solution: From (3.13.15–17) we get the vacuum self-energies of different loop graphs. To get their thermal parts to leading order, we replace the vacuum propagator for pion with its thermal propagator and retain only the thermal part. Then the diagonalised self-energies of $\pi\pi$ and πa_1 loops are

$$\overline{\Pi}_{\alpha\beta}^{(\pi\pi)}(q) = -2\left(\frac{2G_\rho}{m_V F^2}\right)^2 \int \frac{d^4k}{(2\pi)^3} \frac{n(\omega)\,\delta(k^2 - M^2)}{(q-k)^2 - M^2}\, C_{\alpha\beta}, \qquad (P6.1)$$

$$\overline{\Pi}_{\alpha\beta}^{(\pi a_1)}(q) = 2\left(\frac{g_3}{F}\right)^2 \int \frac{d^4k}{(2\pi)^3} \frac{n(\omega)\,\delta(k^2 - M^2)}{(q-k)^2 - m_A^2}\, (B_{\alpha\beta} - C_{\alpha\beta}/m_A^2) \qquad (P6.2)$$

where the tensors $B_{\alpha\beta}$ and $C_{\alpha\beta}$ are defined in (3.13.16).

Consider first the $\pi\pi$ loop. For the longitudinal function, we have $C_{00} = C_1 q^2 k^2$, where C_1 depends on $q_0 \equiv E$ and the angle between $\boldsymbol{q}$ and $\boldsymbol{k}$. It removes the singularity at $\boldsymbol{q}^2 = 0$ from $\overline{\Pi}_l$, when we can take the limit $\boldsymbol{q} \to 0$. Doing the k_0 integration over the delta function we get in the chiral limit

$$\overline{\Pi}_l^{(\pi\pi)}(E) = -4\left(\frac{2G_\rho}{m_V F^2}\right)^2 E^2 \int \frac{d^3k\, n(k)}{(2\pi)^3 |\boldsymbol{k}|} \frac{C_1 k^2}{E^4 - 4E^2 k^2}. \qquad (P6.3)$$

In the same way using $B_{00} - C_{00}/m_{a_1}^2 = C_2 q^2 k^2$, we get

$$\overline{\Pi}_l^{(\pi a_1)}(E) = 4\left(\frac{g_3}{F}\right)^2 (E^2 - m_A^2) \int \frac{d^3k\, n(k)}{(2\pi)^3 |\boldsymbol{k}|} \frac{C_2 k^2}{(E^2 - m_A^2)^2 - 4E^2 k^2}, \qquad (P6.4)$$

Observe the difference in the behaviour of denominators of these integrands for E close to m_V compared to that of (6.4.18) in the text: there it behaves as k^2, cancelling k^2 in the numerator. Here the denominators are constants. Scaling the integration momentum by β, we get

$$\overline{\Pi}_l^{\pi\pi} \sim \overline{\Pi}_l^{\pi a_1} \sim O(T^4). \qquad (P6.5)$$

References

[1] J. Gasser and H. Leutwyler, Phys. Lett. **B 184**, 83 (1987).

[2] J. Gasser and H. Leutwyler, Phys. Lett. **B 188**, 477 (1987).

[3] P. Gerber and H. Leutwyler, Nucl. Phys. **B 321**, 387 (1989).

[4] H. Leutwyler and A. Smilga, Nucl. Phys. **B432**, 302 (1990).

[5] R.D. Pisarski and M. Tytgat, Phys. Rev. **D54**, 2989 (1996).

[6] V. Thorsson and A. Wirzba, Nucl. Phys. **A589**, 633 (1995).

[7] D. Toublan, Phys. Rev. **D 56**, 5629 (1997).

[8] H. Leutwyler, Phys. Rev. **D 49**, 3033 (1994).

[9] P. Binetruye and M. Gaillard, Phys. Rev. **D 32**, 931 (1985).

[10] E.S. Fradkin, Proc. P.N. Lebedev, Phys. Inst. **29** (1965). See also Bochkarev and M.E. Shaposhnikov, Nucl. Phys. **B 268**, 220 (1986).

[11] C. Gale and J.I. Kapusta, Nucl. Phys. **B 357**, 65 (1991).

[12] S. Mallik and S. Sarkar, Eur. Phys. J. **C 25**, 445 (2002).

[13] M. Dey, V.L. Eletsky, B.L. Ioffe, Phys. Lett. **B 252**, 620 (1990); V.L. Eletsky, B.L. Ioffe, Phys. Rev. **D 47**, 3083 (1993); ibid. **D 51**, 2371 (1995).

[14] S. Mallik, Eur. Phys. J. **C 45**, 777 (2006).

[15] U.-G. Meissner, J.A. Oller and A. Wirzba, Ann. Phys. (N.Y), **297**, 27 (2002).

[16] S. Mallik and S. Sarkar, Phys. Rev. **C 69**, 015204 (2004).

[17] T. Hatsuda, S.H. Lee and H. Shuomi, Phys. Rev. **C 52**, 3364 (1995).

[18] S. Mallik and S. Sarkar, Eur. Phys. J. **C 65**, 247 (2010).

7

Two-Loop Results

In this chapter we evaluate three cases of two-loop graphs. One is the *imaginary part* of pion self-energy, which arises first in two loops. The other two are extensions of one-loop results for the quark condensate, and pion mass and its coupling to the axial-vector current.

7.1 Quark Condensate

In Section 6.1 we evaluated the quark condensate at non-zero temperature to one loop. We now extend the result to two loops, again varying $M^2(x)$ in (3.10.1) and (3.10.2) around the pion mass M and retaining now pion fields up to fourth order in the effective Lagrangian,

$$\mathscr{L}_{\text{eff}} = C + \mathscr{L}_0 + \mathscr{L}_{int} + \delta M^2(x)\mathscr{N}(x) \tag{7.1.1}$$

where

$$\mathscr{L}_{int} = \frac{1}{6F^2}\{\vec{\phi}\cdot\partial^\mu\vec{\phi}\,\vec{\phi}\cdot\partial_\mu\vec{\phi} - \vec{\phi}\cdot\vec{\phi}\,\partial^\mu\vec{\phi}\cdot\partial_\mu\vec{\phi} + \frac{M^2}{4}(\vec{\phi}\cdot\vec{\phi})^2\}$$

$$+M^4(h_1+l_3) + l_4\frac{M^2}{F^2}\partial_\mu\vec{\phi}\cdot\partial^\mu\vec{\phi} - (l_3+l_4)\frac{M^4}{F^2}\vec{\phi}\cdot\vec{\phi} \tag{7.1.2}$$

and

$$\mathscr{N}(x) = F^2\left\{1 - \frac{1}{2F^2}\vec{\phi}\cdot\vec{\phi} + \frac{1}{24F^4}(\vec{\phi}\cdot\vec{\phi})^2\right\}$$

$$+ 2M^2(h_1+l_3) + l_4\frac{1}{F^2}\partial_\mu\vec{\phi}\cdot\partial^\mu\vec{\phi} - 2(l_3+l_4)\frac{M^2}{F^2}\vec{\phi}\cdot\vec{\phi} \tag{7.1.3}$$

Here the first and second line in each of (7.1.2) and (7.1.3) arise respectively from $\mathscr{L}_{\text{eff}}^{(2)}$ and $\mathscr{L}_{\text{eff}}^{(4)}$. Again equating the thermal generating functionals of QCD and effective theory as in (6.1.1) and expanding both sides to first order in $\delta M^2(x)$, we get

$$-\frac{1}{2B}\langle\bar{q}q\rangle = \langle T\mathscr{N}(x)\rangle, \tag{7.1.4}$$

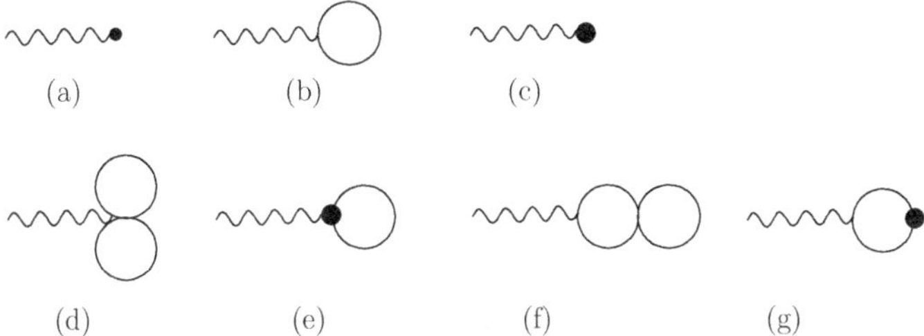

Figure 7.1 Graphs for the quark condensate up to two loops. Graphs (a,b,c) already appear in Figure 3.3. Also, the graphs (b,d,f) are the same as in Figure 5.1. As before, dots and filled circles represent vertices from $\mathscr{L}_{\text{eff}}^{(2)}$ and $\mathscr{L}_{\text{eff}}^{(4)}$ respectively.

in the Heisenberg representation. To calculate it perturbatively we have to go over to the interaction representation. Up to two loops we have the graphs of Figure 7.1 and it suffices to consider perturbation in $\mathscr{L}_{int}$ to first order. Omitting notational details, the quark condensate (7.1.4) can be expanded as

$$\langle \bar{q}q \rangle = -2B\langle T\mathscr{N}(x)\rangle + B\langle T\vec{\phi}(x)\cdot\vec{\phi}(x)i\int d^4y\{\mathscr{L}_{int}(\phi_1) - \mathscr{L}_{int}(\phi_2)\}\rangle.$$

As in the one-loop calculation, the basic quantity is the propagator in coordinate space at the origin,

$$\langle T\phi(x)\phi(x)\rangle = M^2\overline{G} = 2\lambda M^2 + N(\beta, M) \tag{7.1.5}$$

where we take the renormalisation scale μ in (D.12) to be M itself. Omitting the factor $-2BF^2$, the sum of graphs (a), (b) and (c) give

$$(a+b+c): \quad 1 - 2\eta\left(\frac{3}{2}\lambda - h_1 - l_3\right) - \frac{3N}{2F^2} \tag{7.1.6}$$

which we already evaluated in Section 6.1. Among the higher-order graphs, let us indicate how to evaluate (f), as an example. The structure of this graph was already found in (5.1.13), ignoring derivatives and isospin in the interaction and combinatorics arising from Wick contractions. Introducing these numerical factors, we get

$$\langle \bar{q}q \rangle_{(f)} = -2BF^2M^2\overline{G}\left(\frac{-i}{F^4}\right)\int\frac{d^4k}{(2\pi)^4}\left(k^2 - \frac{M^2}{4}\right)(\boldsymbol{D\tau D})_{11} \tag{7.1.7}$$

where $\boldsymbol{D}$ is the matrix propagator (4.2.17). It can be simplified by using the mass derivative formula (5.1.36), which for the present case gives

$$\int\frac{d^4k}{(2\pi)^4}(1, k^2)\boldsymbol{D\tau D} = -\frac{\partial}{\partial M^2}\int\frac{d^4k}{(2\pi)^4}(1, k^2)\boldsymbol{D}. \tag{7.1.8}$$

In the dimensional regularisation that we are using here, the factor k^2 in the integrand can be replaced by M^2. Thus these integrals reduce essentially to

λ, N and $\partial N/\partial M^2$. Let us now list the results for the higher-order graphs. Again leaving out the factor $-2BF^2$, these are

$$(d) \ : \ \frac{5}{8F^4}(2\lambda M^2 + N)^2$$

$$(e) \ : \ -\frac{3M^2}{F^4}(2\lambda M^2 + N)(l_4 + 2l_3)$$

$$(f) \ : \ -\frac{1}{4F^4}(2\lambda M^2 + N)\left(14M^2\lambda + 4N + 3M^2\frac{\partial N}{\partial M^2}\right)$$

$$(g) \ : \ \frac{3M^2}{F^4}\left\{2(l_4 - l_3)\lambda M^2 + l_4 N - l_3 M^2\frac{\partial N}{\partial M^2}\right\}. \tag{7.1.9}$$

Finally, we sum the contributions from all the graphs to get

$$\langle \bar{q}q \rangle = -2BF^2\left\{1 - \frac{3N}{2F^2} - \frac{3N^2}{8F^4}\right.$$
$$+2\eta(h_1^r + l_3^r) - 3\eta^2 l_3^r\frac{\partial N}{\partial M^2} - 3\eta\frac{N}{F^2}\left(2l_3^r + \frac{1}{4}\frac{\partial N}{\partial M^2}\right)$$
$$\left.+\eta^2\left(\frac{3}{2}\lambda^2 + 18l_3^r\lambda\right)\right\}. \tag{7.1.10}$$

Observe that T-dependent divergent terms cancel out among themselves. It reflects a general feature of thermal field theory: if the coupling constants are renormalised in such a way that the perturbative series for vacuum amplitudes are finite, then the corresponding series for thermal amplitudes are also guaranteed to be finite. The divergent terms in the last line in (7.1.10) would also arise in the calculation of the condensate in vacuum and so appropriate terms in $\mathscr{L}_{\text{eff}}^{(6)}$ are required to cancel them, which will now be understood to be present in (7.1.10). We next isolate the β-independent terms (those without N and its derivative) from (7.1.10) to get the vacuum condensate

$$\langle 0|\bar{q}q|0 \rangle = -2BF^2\left\{1 + 2\eta(h_1^r + l_3^r) + \eta^2\left(\frac{3}{2}\lambda^2 + 18l_3^r\lambda\right)\right\} \tag{7.1.11}$$

again with the presence of terms from $\mathscr{L}_{\text{eff}}^{(6)}$ being understood. Dividing (7.1.10) by (7.1.11) we get

$$\langle \bar{q}q \rangle = \langle 0|\bar{q}q|0 \rangle\left\{1 - \frac{3N}{2F^2} - \frac{3N^2}{8F^4}\right.$$
$$\left.+\frac{3N\eta}{F^2}(h_1^r - l_3^r) - \frac{3\eta}{4F^2}N\frac{\partial N}{\partial M^2} - 3\eta^2 l_3^r\frac{\partial N}{\partial M^2}\right\}. \tag{7.1.12}$$

Let us now take the limit $M \to 0$ in (7.1.12). Then only the terms in the first line survive. Noting the evaluation (D.16), we get [1]

$$\langle \bar{q}q \rangle = \langle 0|\bar{q}q|0 \rangle\left(1 - \frac{T^2}{8F^2} - \frac{T^4}{384F^4}\right). \tag{7.1.13}$$

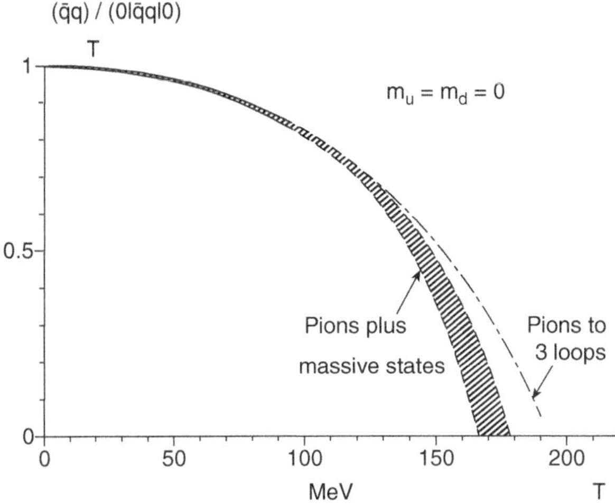

Figure 7.2 Temperature dependence of the quark condensate in the chiral limit. The shaded region reflects the uncertainty in the three-loop formula and contribution from massive states. (From P. Gerber and H. Leutwyler, Nucl. Phys. **B 321**, 387 (1989).)

The condensate is also calculated to three loops [2]

$$\langle \bar{q}q \rangle = \langle 0|\bar{q}q|0\rangle \left(1 - \frac{T^2}{8F^2} - \frac{T^4}{384F^4} - \frac{T^6}{288F^6} \ln \frac{\Lambda_q}{T} \right). \tag{7.1.14}$$

Here the scale Λ_q is fixed by the renormalised coupling constants l_1^r and l_2^r, which appear in $\mathscr{L}_{\text{eff}}^{(4)}$, and has the value $\Lambda_q = (470 \pm 110)$ MeV. A comprehensive discussion of the behaviour of the condensate is given in [2, 3]. Below we briefly describe the salient features.

As mentioned earlier, the quark condensate in *vacuum* is the order parameter of the spontaneous symmetry breaking, which is determined, in principle, by the scale of QCD theory. At finite temperature, (7.1.14) shows that it gradually 'melts' as temperature rises, indicating the possibility of restoring the symmetry at higher temperature. One may even think of using this formula to calculate the critical temperature. Taking $F = 88$ MeV in the chiral limit[1], it leads to $\langle \bar{q}q \rangle = 0$ at a temperature of $T \simeq 190$ MeV, where however the formula cannot be trusted. The reason is that the tree graph gives $\langle 0|\bar{q}q|0\rangle$, while the temperature dependence of $\langle \bar{q}q \rangle$ arises from loop graphs, representing corrections to the tree graph result. So if the 'corrections' cancel the leading term at temperature T, the perturbative expansion will cease to be valid well before this temperature.

To extend this calculation to higher temperatures, we look for contributions from massive states. The ρ-meson, for example, behaves $\sim \exp(-m_\rho/T)$, which,

---

[1] The more accurate estimate is $F = 85.9 \pm 1.0$ MeV [4].

though exponentially small at low T, increases rapidly as T rises. Further, the increasing density of massive states with rise in mass makes this contribution sizeable. We are dealing here with an asymptotic series for $\langle \bar{q}q \rangle$, where perturbative expansion will not show up this behaviour to any finite order. So we need to include the massive states separately in the calculation.

An estimate of the condensate from the massive states can easily be made [2]. Applying the formula (I.7) of Appendix I to the operator $\bar{q}q$, we can get it as

$$\Delta \langle \bar{q}q \rangle = \sum_i \int \frac{d^3 k_i\, n(\omega_i)}{(2\pi)^3 2\omega_i} \langle k_i | \bar{q}q | k_i \rangle \tag{7.1.15}$$

where i runs over massive states, taking the spin and isospin multiplicities into account. A rough estimate of the matrix element in (7.1.15) follows from the non-relativistic quark model. Here the matrix elements of $\bar{u}u$ and $\bar{d}d$ are approximately equal to those of $u^\dagger u$ and $d^\dagger d$, which count the number N_i of valence quarks of type u and d in the state i,

$$\frac{1}{2\omega_i} \langle k_i | \bar{q}q | k_i \rangle = N_i. \tag{7.1.16}$$

Denoting the particle number density by

$$\int \frac{d^3 k_i}{(2\pi)^3} n(\omega_i) = \overline{n}_i$$

formula (7.1.15) reduces to

$$\Delta \langle \bar{q}q \rangle = \sum_i \overline{n}_i N_i. \tag{7.1.17}$$

It is possible to check the accuracy of the approximation (7.1.16) by calculating the matrix elements for the K-meson and nucleon, using the Feynman–Hellmann theorem (Problem 7.1) and the mass formulae for these particles. Comparing with these calculations, the naive formula appears to underestimate the matrix elements by a factor of about 1.5 [2].

The contribution of pions and massive states are shown in Figure 7.2. It is seen that pions alone reduce the condensate by a factor of about 2, when T reaches 150 MeV, beyond which the perturbation expansion cannot be trusted. Also the dilute gas approximation for massive states can be shown to be valid only for T less than 150 MeV [2]. So the critical temperature remains beyond the range of validity of this calculation. Indeed, phase transition of quarks and gluons cannot take place from a configuration of isolated hadrons in their ground states. Rather, such a transition can only occur at a temperature T, when the gas is dense and a substantial fraction of particles is in excited states. Nevertheless, as the order parameter falls rapidly at the upper end of validity of the calculation, it appears meaningful to make an estimate of the critical temperature. Figure 7.2 indicates that in the chiral limit, the phase transition occurs around $T_c \simeq 170$ MeV. For massive quarks the estimate turns out to be somewhat higher [2].

7.2 Pion Decay Rate

We calculate the decay rate for pion in medium in two ways. One is to use the reaction rate formulae derived in Appendix E. The other is to find the imaginary part of the pion self-energy (to two loops), which is related to the decay rate by (5.3.19).

(A) From Rate Formulae

Let a pion of isospin index a and momentum q decay through the scattering process

$$\pi_a(q) + \pi_b(k_1) \to \pi_c(k_2) + \pi_d(k_3)$$

and regenerate by the reverse process. Then the net decay rate obtained from (E.20) and (E.21) is

$$\Gamma(\omega) = \Gamma_d(\omega) - \Gamma_i(\omega)$$
$$= \frac{1}{2\omega}\frac{1}{2}\int \prod_{i=1}^{3} \frac{d^3k_i}{(2\pi)^3 2\omega_i}(2\pi)^4\delta(q + k_1 - k_2 - k_3)|T|^2$$
$$\times \{n_1(1 + n_2)(1 + n_3) - (1 + n_1)n_2 n_3\} \tag{7.2.1}$$

where the factor $\frac{1}{2}$ compensates for the double counting of states in the phase space of two identical pions, labelled 2 and 3. Here $|T|^2$ is the isospin summed, squared $\pi\pi$ scattering amplitude. Using the amplitude (3.7.17) to leading order, we get

$$|T|^2\delta_{aa'} = \sum_{b,c,d} M_{cd,ab}M_{cd,a'b}$$
$$= \delta_{aa'}\{2(s^2 + t^2 + u^2) - 9M^4\}/F^4. \tag{7.2.2}$$

Noting (E.23) the factor in (7.2.1) involving distribution functions can be written in the form

$$n_1(1 + n_2)(1 + n_3) - (1 + n_1)n_2 n_3 = \frac{2\sinh(\beta\omega/2)}{\prod_{i=1}^{3} 2\sinh(\beta\omega_i/2)}. \tag{7.2.3}$$

Inserting (7.2.3) in (7.2.1) we get the required net pion decay rate [5]

$$\Gamma(\omega) = \frac{\sinh(\beta\omega/2)}{2\omega}\int d\nu_1 d\nu_2 d\nu_3 (2\pi)^4\delta^4(q + k_1 - k_2 - k_3)|T|^2 \tag{7.2.4}$$

where

$$d\nu_i = \frac{d^3k_i}{(2\pi)^3 4\omega_i \sinh(\beta\omega_i/2)}.$$

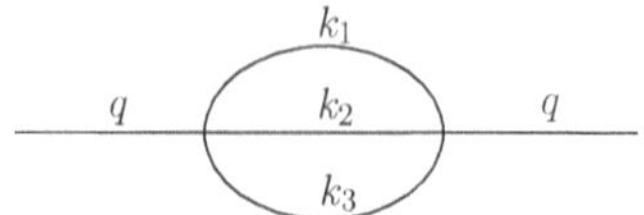

Figure 7.3 Two-loop self-energy graph for pion propagator.

(B) From Pion Self-Energy

We now calculate the imaginary part of the self-energy correcting the free pion propagator in medium. Calculated to one loop in Section 6.2, it was real. We now find it to two loops (Figure 7.3). As usual, we first write the expression in vacuum.

Vacuum Self-Energy

The self-energy correction to pion propagator $\langle 0|T\phi^i(x)\phi^j(x')|0\rangle$ is obtained from the second-order perturbation formula

$$\frac{i^2}{2}\int d^4z\, d^4z'\, \langle 0|T\phi^i(x)\phi^j(x')\mathscr{L}_{\text{int}}^{(2)}(z)\mathscr{L}_{\text{int}}^{(2)}(z')|0\rangle \tag{7.2.5}$$

in the interaction (3.7.15), where the fields are in the interaction representation. To discuss the structure of different terms resulting from contractions corresponding to Figure 7.3, we rewrite the interaction Lagrangian as

$$\mathscr{L}_{\text{int}}^{(2)} = \frac{1}{6F^2}\left(L_1 - L_2 + \frac{M^2}{4}L_3\right), \tag{7.2.6}$$

with $L_1 = \vec{\phi}\cdot\partial_\mu\vec{\phi}\,\vec{\phi}\cdot\partial^\mu\vec{\phi}$, $L_2 = \vec{\phi}\cdot\vec{\phi}\,\partial_\mu\vec{\phi}\cdot\partial^\mu\vec{\phi}$, and $L_3 = (\vec{\phi}\cdot\vec{\phi})^2$. Then (7.2.5) becomes (omitting integrals over z and z')

$$\frac{i^2}{2}\left(\frac{1}{6F^2}\right)^2\langle 0|T\phi^i(x)\phi^j(x')\{L_1(z)L_1(z') + L_2(z)L_2(z') - 2L_1(z)L_2(z')$$

$$+ (M^2/2)(L_1(z) - L_2(z))L_3(z') + (M^2/4)^2L_3(z)L_3(z')\}|0\rangle. \tag{7.2.7}$$

It turns out that the contractions from the two pieces $L_1(z)L_3(z')$ and $L_2(z)L_3(z')$ result in the same expression, so that the fourth term in (7.2.7) vanishes, showing the absence of any term proportional to M^2. Further, all of the first three pieces have terms of the same structure, differing only in their coefficients arising from the isospin index of the pion field. Taking Fourier transform of (7.2.7) after contractions, we write the complete propagator with self-energy correction

$$\Delta'(q) = \Delta_F(q) + \Delta_F(q)\Sigma(q)\Delta_F(q) + \cdots \tag{7.2.8}$$

where Σ is the self-energy function

$$\Sigma(q) = -\frac{1}{6}\int\frac{d^4k_1}{(2\pi)^4}\frac{d^4k_2}{(2\pi)^4}\frac{d^4k_3}{(2\pi)^4}(2\pi)^4\delta^4(q - k_1 - k_2 - k_3)$$

$$\times |T|^2\Delta_F(k_1)\Delta_F(k_2)\Delta_F(k_3), \tag{7.2.9}$$

with

$$|T|^2 = (5M^4 + 2f)/(3F^4). \tag{7.2.10}$$

Here f is a polynomial in momenta, written symmetrically as

$$f = \{(q+k_1)\cdot(k_2-k_3)\}^2 + \{(q+k_2)\cdot(k_3-k_1)\}^2 + \{(q+k_3)\cdot(k_1-k_2)\}^2. \tag{7.2.11}$$

We shall presently calculate $|T|^2$ in the imaginary part of the above self-energy. As the notation indicates, it will agree with the evaluation (7.2.2).

Thermal Self-Energy

As in earlier cases, it suffices to calculate the 11-component of the self-energy matrix, which may be obtained from (7.2.9) after recalling (5.1.33)

$$\Sigma_{11}(q) = -\frac{1}{6} \int \frac{d^4k_1}{(2\pi)^4} \frac{d^4k_2}{(2\pi)^4} |T|^2 D_{11}(k_1) D_{11}(k_2) D_{11}(q - k_1 - k_2). \tag{7.2.12}$$

Its imaginary part is easily obtained in the same way as we did for the one-loop graph in Section 5.3. We write all the three free propagators in their spectral representations (5.3.2) and rewrite Σ so as to isolate the integrals over the time components of k_1 and k_2

$$\Sigma_{11}(q_0, \boldsymbol{q}) = -\frac{1}{6} \int \frac{d^3k_1}{(2\pi)^3} \frac{d^3k_2}{(2\pi)^3} \int \frac{dk_0'}{2\pi} \frac{dk_0''}{2\pi} \frac{dk_0'''}{2\pi} |T|^2$$
$$\times \rho_0(k_0', \boldsymbol{k}_1) \rho_0(k_0'', \boldsymbol{k}_2) \rho_0(k_0''', \boldsymbol{q} - \boldsymbol{k}_1 - \boldsymbol{k}_2) K \tag{7.2.13}$$

where K is the double integral over the energy denominators of the propagators

$$K = \int \frac{dk_{10}\, dk_{20}}{2\pi\ 2\pi} \left(\frac{1+f'}{k_0' - k_{10} - i\eta} - \frac{f'}{k_0' - k_{10} + i\eta} \right) \left(\frac{1+f''}{k_0'' - k_{20} - i\eta} - \frac{f''}{k_0'' - k_{20} + i\eta} \right)$$
$$\left(\frac{1+f'''}{k_0''' - (q_0 - k_{10} - k_{20}) - i\eta} - \frac{f'''}{k_0''' - (q_0 - k_{10} - k_{20}) + i\eta} \right) \tag{7.2.14}$$

which we now evaluate. Notice that we keep the momentum dependent polynomial $|T|^2$ outside the k_{10} and k_{20} integrals with the same justification as in Section 5.4. Let us first integrate over the variable k_{10}, which occurs in the first and third factors in (7.2.14) to get

$$i\left\{ \frac{(1+f')(1+f''')}{k_{20} + k_0' + k_0''' - q_0 + i\eta} - \frac{f'f'''}{k_{20} + k_0' + k_0''' - q_0 - i\eta} \right\}. \tag{7.2.15}$$

Next carry out the k_{20} integration over the second factor in (7.2.14) and the one just obtained, getting finally

$$K = -\left(\frac{(1+f')(1+f'')(1+f''')}{k_0' + k_0'' + k_0''' - q_0 - i\eta} - \frac{f'f''f'''}{k_0' + k_0'' + k_0''' - q_0 + i\eta} \right) \tag{7.2.16}$$

whose imaginary part is immediately obtained as

$$\mathrm{Im}K = -\pi\{(1+f')(1+f'')(1+f'') + f'f''f'''\}\delta(q_0 - k_0' - k_0'' - k_0''')$$

$$= -\pi\coth(\beta q_0/2)\{(1+f')(1+f'')(1+f'') - f'f''f'''\}\delta(q_0 - k_0' - k_0'' - k_0''')$$

$$(7.2.17)$$

Having obtained $\mathrm{Im}K$ and noting (5.2.8), we get from (7.2.13) the imaginary part of the diagonalised self-energy

$$\mathrm{Im}\overline{\Sigma}(q_0, \boldsymbol{q}) = \frac{\pi}{6}\epsilon(q_0)\int\frac{d^3k_1}{(2\pi)^3}\frac{d^3k_2}{(2\pi)^3}\int\frac{dk_0'}{2\pi}\frac{dk_0''}{2\pi}\frac{dk_0'''}{2\pi}\rho_0(k_0')\rho_0(k_0'')\rho_0(k_0''')$$

$$\times |T|^2\{(1+f')(1+f'')(1+f'') - f'f''f'''\}\delta(q_0 - k_0' - k_0'' - k_0''') \qquad (7.2.18)$$

where we suppress the three-momentum dependence of the three spectral functions for brevity. With all the intermediate pions on mass shells ($k_i^2 = M^2$) in $\mathrm{Im}\overline{\Sigma}$, we can evaluate (7.2.11) for f. As in Section 3.7 we use the Mandelstam variables, $s = (q - k_1)^2$, $t = (q - k_2)^2$, $u = (q - k_3)^2$ with $s + t + u = 4M^2$. Then, for example, the first term in f becomes

$$(q + k_1) \cdot (k_2 - k_3) = \{2q - (k_2 + k_3)\} \cdot (k_2 - k_3) = -t + u.$$

We thus get from (7.2.10) the same expression as (7.2.2).

In the example considered in Section 5.3, we had two particles in the intermediate state giving rise to four different cuts in the energy variable. Now with three particles in the intermediate state, we shall have eight cuts in the energy plane, corresponding to $\delta(q_0 \pm \omega_1 \pm \omega_2 \pm \omega_3)$, where $\omega_i = \sqrt{k_i + M^2}$, $i = 1, 2, 3$. Of these possibilities, we consider only those corresponding to scattering of pions with momenta q and k_i. Owing to the complete symmetry of the expression with respect to the three intermediate pions, we simply multiply by 3 the contribution from one scattering process, say, $q + k_1 \rightarrow k_2 + k_3$ corresponding to $\delta(q_0 + \omega_1 - \omega_2 - \omega_3)$. We then integrate k_0', k_0'' and k_0''' over the respective delta functions and convert the fs to ns as in Section 5.3. Restoring the k_3 integral and noting (5.3.24), we get the net pion decay rate

$$\Gamma(q) = \frac{1}{4\omega}\int\prod_{i=1}^{3}\frac{d^3k_i}{(2\pi)^32\omega_i}(2\pi)^4\delta^4(q + k_1 - k_2 - k_3)$$

$$\{n_1(1+n_2)(1+n_3) - (1+n_1)n_2n_3\}|T|^2 \qquad (7.2.19)$$

which is the same as obtained in (A) from the rate formulae. Thus the probability amplitude of a pion of momentum q is attenuated in the medium by $1/e$ in time $1/\Gamma(q)$. It has speed $|\boldsymbol{q}|/\omega$ and so travels during this time a distance $\lambda = |\boldsymbol{q}|/(\omega\Gamma(q))$, called the mean free path. It is plotted in Figure 7.4 as a function of pion momentum at temperatures 120 MeV and 150 MeV.

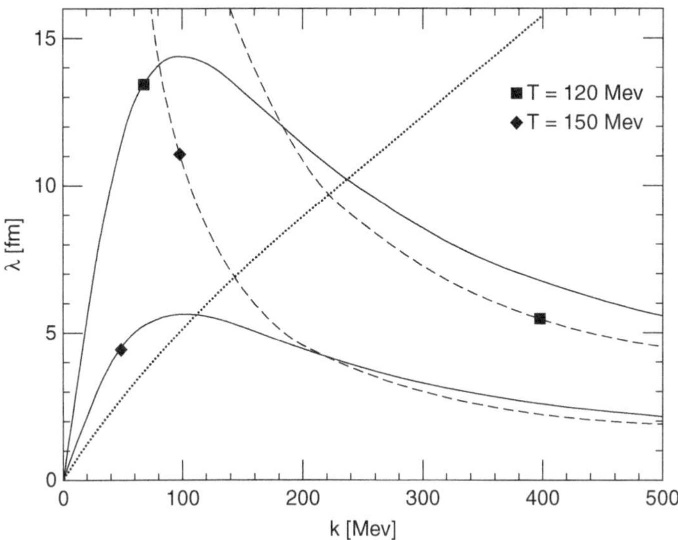

Figure 7.4 Mean free path of pion as a function of its momentum at two different temperatures. The solid lines correspond to the physical value of pion mass, while the dashed lines correspond to chiral limit. The dotted line gives the qualitative behaviour in $\lambda\phi^4$ theory. (From J.L. Goity and H. Leutwyler, Phys. Lett., **228**, 517 (1989).)

7.3 Pion Mass and Coupling

Here we extend the previous one-loop calculation of pion pole parameters to two loops in the axial-vector two-point function [6–8]. We shall first evaluate the Feynman graphs for the vacuum two-point function and then convert it to finite temperature. After cancellation of temperature-dependent divergent parts, we shall finally get the form of the pion pole, which gives the desired corrections.

Vacuum Amplitude

It is convenient to divide all the graphs into four sets shown in Figures 7.5 to 7.8. In the first three sets the loop integrals, if present, factorise. The first set given by Figure 7.5 contains graphs (a to e), which are the same as in Figure 3.2 (excluding counterparts of (b) and (c)), that were evaluated in Section 3.10. Let us rewrite that result as

$$T^{(1)}_{\mu\nu}(q)_{(a\cdots e)} = q_\mu q_\nu F^2 \Gamma \{\Delta_F + \Delta_F \Sigma \Delta_F + O(\eta^2)\} \qquad (7.3.1)$$

where $\Gamma = 1 + \gamma\eta$, $\Sigma = \eta(M^2\sigma + \sigma'/\Delta)$ with γ, σ and σ' given by (3.10.15). The remaining graphs (f to l) in this set need not be evaluated separately, if we note that they originate from the one-particle *reducible* self-energy. So they can be obtained from an iteration of the Dyson equation for the complete propagator. Thus we can sum the contribution of all graphs of Figure 7.5 as

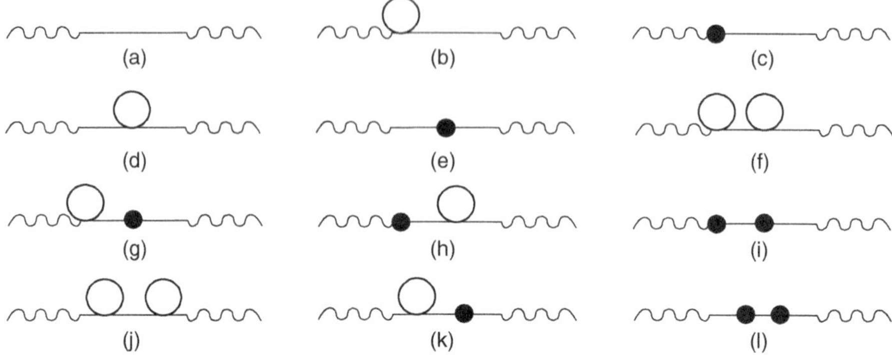

Figure 7.5 The free amplitude with corrections from one-loop graphs along with counterterm graphs and those two-loop graphs that are iterations of the former ones. Vertices of $\mathscr{L}^{(2)}$ and $\mathscr{L}^{(4)}$ are shown as points and filled circles respectively. Wavy and straight lines denote axial current and pion respectively.

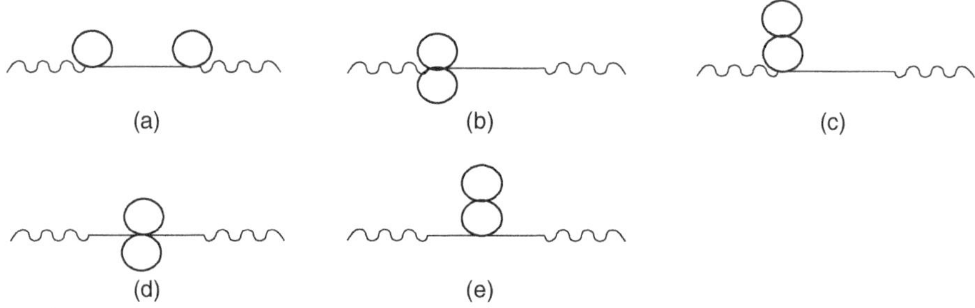

Figure 7.6 Remaining factorisable two-loop graphs with vertices from $\mathscr{L}^{(2)}$ only.

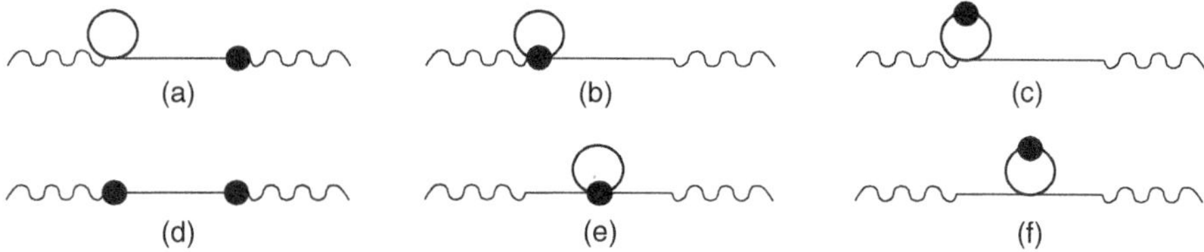

Figure 7.7 Further counterterm graphs.

Figure 7.8 Non-factorisable two-loop graphs.

$$T^{(1)}_{\mu\nu}(q) = q_\mu q_\nu F^2 \Gamma \{\Delta_F + \Delta_F \Sigma \Delta_F + \Delta_F \Sigma \Delta_F \Sigma \Delta_F + O(\eta^3)\}. \tag{7.3.2}$$

Collecting the coefficients of Δ_F and Δ_F^2 it may be written as

$$T^{(1)}_{\mu\nu}(q) = q_\mu q_\nu F^2 \{\gamma_1 \Delta_F(q) + \sigma_1 M^2 \Delta_F^2(q)\} \tag{7.3.3}$$

where

$$\gamma_1 = 1 + 2\eta(l_4 - G) - 4\eta^2(l_4 - G)(3l_4 - G)/3,$$
$$\sigma_1 = -\eta(4l_3 + G)/2 + 2\eta^2 G(4l_3 + G)/3. \tag{7.3.4}$$

Here we have cancelled factors as $(q^2 - m^2)\Delta_F^2(q) = -\Delta_F(q)$, which is also justified at finite temperature.

The graphs of Figure 7.6 contain only vertices of $\mathscr{L}^{(2)}$ and their sum is again of the form

$$T^{(2)}_{\mu\nu}(q) = q_\mu q_\nu F^2 \{\gamma_2 \Delta_F(q) + \sigma_2 M^2 \Delta_F^2(q)\} \tag{7.3.5}$$

with

$$\gamma_2 = \eta^2 G(8G + 3G')/3, \qquad \sigma_2 = \eta^2 G(3G + 2G')/8 \tag{7.3.6}$$

where in addition to G, defined by (5.1.4), we encounter a new divergent integral,

$$G'(M) = -i \int \frac{d^4 k}{(2\pi)^4} \Delta_F^2(k) = -\frac{\partial}{\partial M^2}(M^2 G) = -2\lambda - \frac{1}{16\pi^2} \tag{7.3.7}$$

with the renormalisation scale set at the pion mass.

The third set of graphs from Figure 7.7 with vertices from $\mathscr{L}^{(2)}$ and $\mathscr{L}^{(4)}$ gives

$$T^{(3)}_{\mu\nu}(q) = q_\mu q_\nu F^2 \{\gamma_3 \Delta_F(q) + \sigma_3 M^2 \Delta_F^2(q)\}$$
$$+ 8F^2 \eta^2 (l_1 + 2l_2)(q_\mu G_{\nu\lambda} + q_\nu G_{\mu\lambda})q^\lambda \Delta_F(q) \tag{7.3.8}$$

where

$$\gamma_3 = \eta^2\{(36l_1 + 12l_2 - 25l_4)G + 12l_3 G' + 12l_4^2\}/3,$$
$$\sigma_3 = \eta^2\{(36l_1 + 12l_2 + 16l_3 - 3l_4)G + 3l_3 G' + 24(l_1 + 2l_2)q^\lambda q^\sigma G_{\lambda\sigma}/M^2\}/3. \tag{7.3.9}$$

Here we have still another divergent integral,

$$G_{\mu\nu}(M) = \frac{-i}{M^4} \int \frac{d^4 k}{(2\pi)^4} k_\mu k_\nu \Delta_F(k). \tag{7.3.10}$$

Its divergence can be easily found, but we do not need it in the present calculation.

Finally we have the amplitude from the non-factorisable two-loop graphs of Figure 7.8,

$$T^{(4)}_{\mu\nu}(q) = \frac{2}{9F^2} \{q_\mu \Delta_F(q)\Gamma_\nu(q) + \Gamma_\mu(q)\Delta_F(q)q_\nu\}$$
$$- \frac{1}{18F^2} q_\mu q_\nu \Delta_F(q)\Sigma(q)\Delta_F(q). \tag{7.3.11}$$

Here the vertex function $\Gamma_\mu(q)$ of graph (a) is

$$\Gamma_\mu(q) = \int \frac{d^4k_1}{(2\pi)^4} \frac{d^4k_2}{(2\pi)^4} G_\mu(q, k_1, k_2)\Delta(k_1)\Delta(k_2)\Delta_F(q - k_1 - k_2), \qquad (7.3.12)$$

where

$$G_\mu = (2q_\mu - 3k_{1\mu} - 3k_{2\mu})(k_1^2 + k_2^2 + 4k_1 \cdot k_2 + 2q \cdot (k_1 + k_2) - 2q^2 + M^2) \quad (7.3.13)$$

The self-energy function $\Sigma(q)$ of graph (b) is already evaluated in (7.2.9). Integrating out k_3 it is

$$\Sigma(q) = \int \frac{d^4k_1}{(2\pi)^4} \frac{d^4k_2}{(2\pi)^4} (5M^4 + 2f)\Delta_F(k_1)\Delta_F(k_2)\Delta_F(q - k_1 - k_2) \qquad (7.3.14)$$

where f is given by (7.2.11). In the previous section we found its imaginary part; here we shall be concerned with the real part.

The vertex and self-energy integrals can be simplified by using the symmetries of integrands under the interchange of the integration variables [7]. Defining

$$K(q) = -\frac{1}{M^2} \int \frac{d^4k_1}{(2\pi)^4} \frac{d^4k_2}{(2\pi)^4} \Delta_F(k_1)\Delta_F(k_2)\Delta_F(q - k_1 - k_2) \qquad (7.3.15)$$

$$K_{\mu\nu}(q) = -\frac{1}{M^4} \int \frac{d^4k_1}{(2\pi)^4} \frac{d^4k_2}{(2\pi)^4} k_{1\mu}k_{1\nu}\Delta_F(k_1)\Delta_F(k_2)\Delta_F(q - k_1 - k_2), \qquad (7.3.16)$$

the pole part of (7.3.11) may be written as

$$T_{\mu\nu}^{(4)} = q_\mu q_\nu F^2\eta^2 \left\{ \frac{2}{3}G^2\Delta_F(q) + \frac{1}{6}(8G^2 - K + 24q^\rho q^\sigma K_{\rho\sigma}/M^2)M^2\Delta_F^2(q) \right\}$$
$$+ 4F^2\eta^2(q_\mu K_{\nu\lambda} + q_\nu K_{\mu\lambda})q^\lambda \Delta_F(q). \qquad (7.3.17)$$

The sum of the above amplitudes along with an insertion from $\mathscr{L}^{(6)}$ (not calculated above) would give the complete vacuum amplitude. After renormalisation we can extend the one-loop results (3.10.21) and (3.10.22) to two loops for the pion pole parameters. Instead, we turn to the corresponding thermal amplitudes to find the temperature dependence of these parameters.

Thermal Amplitude

It is by now familiar how to convert the vacuum amplitudes to the corresponding thermal diagonalised amplitudes, denoted by a bar. Accordingly we replace

$$G \to \overline{G} = \frac{-i}{M^2} \int \frac{d^4k}{(2\pi)^4} D_{11}(k)$$

$$G' \to \overline{G}' = i\frac{\partial}{\partial M^2} \int \frac{d^4k}{(2\pi)^4} D_{11}(k)$$

$$G_{\mu\nu} \to \overline{G}_{\mu\nu} = \frac{-i}{M^4} \int \frac{d^4k}{(2\pi)^4} k_\mu k_\nu D_{11}(k)$$

$$K(q) \to \overline{K}(q) = \frac{-1}{M^2} \int \frac{d^4k_1}{(2\pi)^4} \frac{d^4k_2}{(2\pi)^4} D_{11}(k_1) D_{11}(k_2) D_{11}(q - k_1 - k_2)$$

$$K_{\mu\nu}(q) \to \overline{K}_{\mu\nu}(q) = \frac{-1}{M^4} \int \frac{d^4k_1}{(2\pi)^4} \frac{d^4k_2}{(2\pi)^4} k_{1\mu} k_{1\nu} D_{11}(k_1) D_{11}(k_2) D_{11}(q - k_1 - k_2).$$

$$(7.3.18)$$

(As we are interested in *real* parts, the diagonalised amplitudes $\overline{K}$ and $\overline{K}_{\mu\nu}$ are given simply by their 11-components.) As an example, the vacuum amplitude $T^{(1)}_{\mu\nu}(q)$ goes over to $\overline{T}^{(1)}_{\mu\nu}(q)$,

$$\overline{T}^{(1)}_{\mu\nu}(q) = q_\mu q_\nu F^2 \{\overline{\gamma}_1 \Delta_F(q) + \overline{\sigma}_1 M^2 \Delta_F^2(q)\} \tag{7.3.19}$$

with

$$\overline{\gamma}_1 = 1 + 2\eta(l_4 - \overline{G}) - 4\eta^2(l_4 - \overline{G})(3l_4 - \overline{G})/3,$$

$$\overline{\sigma}_1 = -\eta(4l_3 + \overline{G})/2 + 2\eta^2 \overline{G}(4l_3 + \overline{G})/3. \tag{7.3.20}$$

Each term in $\overline{T}^{(i)}_{\mu\nu}(q)$ can be separated into three parts. One is the vacuum part, which remains divergent, as we already pointed out. Another is the β-dependent divergent part, which should cancel out in the complete amplitude. The last one is the β-dependent finite piece we are looking for, which will shift the physical pole parameters at finite temperature.

The $\overline{G}$s containing a single propagator can be separated into its vacuum and thermal part immediately,

$$\overline{G} = G + N, \qquad \overline{G}' = G' + N', \qquad \overline{G}_{\mu\nu} = G_{\mu\nu} + N_{\mu\nu} \tag{7.3.21}$$

where

$$N = \int \frac{d^4k}{(2\pi)^3} n(k)\delta(k^2 - M^2)$$

$$N' = -\frac{\partial}{\partial M^2} N$$

$$N_{\mu\nu} = \int \frac{d^4k}{(2\pi)^3} k_\mu k_\nu n(k)\delta(k^2 - M^2). \tag{7.3.22}$$

The $\overline{K}$s contain three propagators. Along with the vacuum divergence, they have β dependent divergence, arising from terms linear in $n(k)$,

$$\overline{K} \to K + \frac{3}{M^2} \int \frac{d^4k}{(2\pi)^3} n(k)\delta(k^2 - M^2) L(q - k)$$

$$\overline{K}_{\mu\nu} \to K_{\mu\nu} + \frac{1}{M^4} \int \frac{d^4k}{(2\pi)^3} k_\mu k_\nu n(k)\delta(k^2 - M^2) L(q - k)$$

$$+ \frac{2}{M^2} \int \frac{d^4k}{(2\pi)^3} n(k)\delta(k^2 - M^2) L_{\mu\nu}(q - k) \tag{7.3.23}$$

where

$$L(p) = -i \int \frac{d^4k}{(2\pi)^4} \Delta(k)\Delta(p-k)$$

$$L_{\mu\nu}(p) = \frac{-i}{M^2} \int \frac{d^4k}{(2\pi)^4} k_\mu k_\nu \Delta(k)\Delta(p-k). \tag{7.3.24}$$

The integrals L and $L_{\mu\nu}$ can be worked out in dimensional regularisation to give the divergences

$$L(p) \to -2\lambda$$

$$L_{\mu\nu}(p) \to \frac{\lambda}{M^2} \left\{ \left(\frac{p^2}{6} - M^2 \right) g_{\mu\nu} - \frac{2}{3} p_\mu p_\nu \right\}. \tag{7.3.25}$$

Inserting (7.3.25) in (7.3.23) we can write

$$\overline{K} = K - 6\lambda N - \frac{3}{16\pi^2} N + \widetilde{K}$$

$$\overline{K}_{\mu\nu} = K_{\mu\nu} + \frac{\lambda}{3} \left\{ 10 N_{\mu\nu} + 4N \left(g_{\mu\nu} + \frac{q_\mu q_\nu}{M^2} \right) - \frac{N}{M^2}(q^2 - M^2) g_{\mu\nu} \right\}$$

$$+ \frac{1}{288\pi^2 M^2} [-28 M^2 N_{\mu\nu} + \{ (q^2 + M^2) g_{\mu\nu} - 10 q_\mu q_\nu \} N] + \widetilde{K}_{\mu\nu}$$

$$\tag{7.3.26}$$

separating $\overline{K}$ and $\overline{K}_{\mu\nu}$ into three parts, as mentioned above: in each of these expressions the first term is the vacuum part, the second term proportional to λ is the β-dependent divergent part, and the third and fourth terms are β-dependent finite parts. We do not write the fourth terms (denoted by a tilde) explicitly; they involve integrals, to be evaluated numerically.

It is simple to find the β-dependent divergent pieces in the amplitudes $\overline{T}^{(i)}$. For $i = 1, 2, 3$ the divergences arise from the low-energy constants (3.10.3), G and G', giving

$$\overline{T}_{\mu\nu}^{(1+2+3)} \to \frac{\lambda}{3} F^2 \eta^2 \left[q_\mu q_\nu \{ 12N\Delta_F + (13NM^2 + 40 q^\lambda q^\sigma N_{\lambda\sigma}) \Delta_F^2 \} \right.$$

$$\left. + 40(q_\mu N_{\nu\lambda} q^\lambda + q_\nu N_{\mu\lambda} q^\lambda) \Delta_F \right]. \tag{7.3.27}$$

For $\overline{T}^{(4)}$ such divergences arise from those of $G, \overline{K}$ and $\overline{K}_{\mu\nu}$, which is exactly equal and opposite to (7.3.27), thereby cancelling the β-dependent divergence in the complete pion pole amplitude, as it should, for the success of the renormalisation programme. We shall also omit the vacuum parts, replacing M and F by their physical values M_π and F_π.

Let us now collect the β-dependent finite parts in the amplitude. To write these in a compact way, we define the following combinations of low-energy constants,

$$l = \frac{1}{12\pi^2} \left(\bar{l}_1 + \bar{l}_2 - \frac{14}{3} \right)$$

$$l' = \frac{1}{48\pi^2}\left(6\bar{l}_1 + 4\bar{l}_2 - 9\bar{l}_4 - \frac{7}{3}\right)$$

$$l'' = \frac{1}{48\pi^2}\left(6\bar{l}_1 + 4\bar{l}_2 - 6\bar{l}_3 - 3\bar{l}_4 - \frac{55}{12}\right) \tag{7.3.28}$$

and the β-dependent but momentum-independent parameters

$$A = N(2N + l') + N'\left(N - \frac{\bar{l}_3}{16\pi^2}\right)$$

$$B = N\left(\frac{19}{8}N + l''\right) + \frac{N'}{4}\left(N - \frac{\bar{l}_3}{16\pi^2}\right). \tag{7.3.29}$$

After some algebra we get the terms in Δ_F and Δ_F^2 as

$$\overline{T}_{\mu\nu} = F_\pi^2\{q_\mu q_\nu(1 - 2\eta N + \eta^2 A) + \eta^2(q_\mu S_{\nu\lambda} + q_\nu S_{\mu\lambda})q^\lambda\}\Delta_F(q)$$

$$- q_\mu q_\nu F_\pi^2\left\{\frac{1}{2}\eta N - \eta^2\left(B - \frac{\widetilde{K}(q)}{6} + \frac{q^\lambda q^\sigma}{M_\pi^2}S_{\lambda\sigma}(q)\right)\right\}M_\pi^2\Delta^2(q) \tag{7.3.30}$$

where we introduce the momentum-dependent quantity

$$S_{\mu\nu}(q) = lN_{\mu\nu} + 4\widetilde{K}_{\mu\nu}(q). \tag{7.3.31}$$

To write the pion pole in q_0^2, we expand the q_0 dependence of $\widetilde{K}$ and $S_{\mu\nu}$ in the neighbourhood of $q_0^2 = \boldsymbol{q}^2 + M^2 \equiv \omega^2$ at fixed $\boldsymbol{q}$,

$$\widetilde{K}(q) = \widetilde{K}^{(0)}(\omega) + (q_0^2 - \omega^2)\widetilde{K}^{(1)}(\omega) + \cdots, \quad \widetilde{K}^{(1)}(\omega) = \frac{1}{2\omega}\frac{\partial}{\partial q_0}\widetilde{K}(q)|_{q_0=\omega}, \tag{7.3.32}$$

and similarly for $S_{\mu\nu}(q)$. We can now combine the terms in (7.3.30) to get the real part of pion pole term correct to order η^2 [7, 8]

$$\overline{T}_{\mu\nu}(q) = -\frac{f_\mu(q,\beta)f_\nu(q,\beta)}{q_0^2 - \Omega^2(\omega) + i\epsilon}, \tag{7.3.33}$$

where

$$\Omega^2(\omega) = \boldsymbol{q}^2 + M_\pi^2\left\{1 + \frac{1}{2}N\eta - \left(B - \frac{\widetilde{K}^{(0)}}{6} + \frac{1}{M_\pi^2}(q^\lambda q^\sigma S_{\lambda\sigma})^{(0)} - N^2\right)\eta^2\right\}$$

$$f_\mu(q) = F_\pi\left[q_\mu\left\{1 - N\eta + \frac{1}{2}\left(A + \frac{M_\pi^2}{6}\widetilde{K}^{(1)} - (q^\lambda q^\sigma S_{\lambda\sigma})^{(1)} - N^2\right)\eta^2\right\}\right.$$

$$\left. + (S_{\mu\lambda}q^\lambda)^{(0)}\eta^2\right]. \tag{7.3.34}$$

Following our discussion in the previous chapter, we have here two pion decay constants, the temporal and the spatial,

$$f_0(q) = q_0 F_{\pi,t}(q), \qquad f_i(q) = q_i F_{\pi,s}(q). \tag{7.3.35}$$

Also, the effective pion mass in pion medium is given by $\Omega(\omega)|_{q=0}$. For their quantitative values, we have to evaluate the different integrals representing the quantities in (7.3.34), for which we refer the reader to [7, 8].

Problems

Problem 7.1: Find how the mass of a physical particle changes with quark mass using the Feynman–Hellman theorem.

Solution: The Feynman–Hellman theorem is a general result. (See, for example, [9].) We use it in the context of QCD. The total Hamiltonian of QCD may be written as

$$H = H_0 + H_I \qquad H_I = \hat{m} \int d^3 x\, \bar{q}q, \qquad \hat{m} = (m_u + m_d)/2 \qquad (\text{P7.1})$$

where H_0 is the Hamiltonian for massless quarks and H_I the mass term acting as perturbation with (real) mass $\hat{m}$. Consider the eigenvalue equation

$$H|\boldsymbol{k}\rangle = \omega|\boldsymbol{k}\rangle \qquad (\text{P7.2})$$

where $|\boldsymbol{k}\rangle$ is a physical single particle state of mass m and energy ω, normalised according to (1.1.1).

We enclose the system in a finite volume V. Then the continuum normalisation (1.1.1) goes over to

$$\langle \boldsymbol{k'}|\boldsymbol{k}\rangle = 2\omega V \delta_{k'k}.$$

So the eigenvalue equation (P7.2) gives

$$\omega = \frac{\langle \boldsymbol{k}|H|\boldsymbol{k}\rangle}{2\omega V}. \qquad (\text{P7.3})$$

Differentiating with respect to $\hat{m}$ gives

$$\frac{d\omega}{d\hat{m}} = \frac{1}{2\omega V}\langle \boldsymbol{k}|\frac{dH_I}{d\hat{m}}|\boldsymbol{k}\rangle$$
$$+ \frac{d}{d\hat{m}}\left(\frac{1}{2\omega V}\right)\langle \boldsymbol{k}|H|\boldsymbol{k}\rangle + \frac{1}{2\omega V}\left\{\frac{d}{d\hat{m}}\langle \boldsymbol{k}|\right\}H|\boldsymbol{k}\rangle + \frac{1}{2\omega V}\langle \boldsymbol{k}|H\left\{\frac{d}{d\hat{m}}|\boldsymbol{k}\rangle\right\}.$$
$$(\text{P7.4})$$

As H is Hermitian, we can use (P7.2) in the last two terms, when the sum of the last three terms reduces to

$$\omega\frac{d}{d\hat{m}}\left(\frac{\langle \boldsymbol{k}|\boldsymbol{k}\rangle}{2\omega V}\right) = 0$$

and (P7.4) becomes

$$\frac{d\omega^2}{d\hat{m}} = \langle \boldsymbol{k}|\bar{q}q|\boldsymbol{k}\rangle. \qquad (\text{P7.5})$$

We can now take the infinite volume limit. With the energy-momentum relation $\omega^2 = m^2 + \boldsymbol{k}^2$, we get finally

$$\frac{dm^2}{d\hat{m}} = \langle \boldsymbol{k}|\bar{q}q|\boldsymbol{k}\rangle. \qquad (\text{P7.6})$$

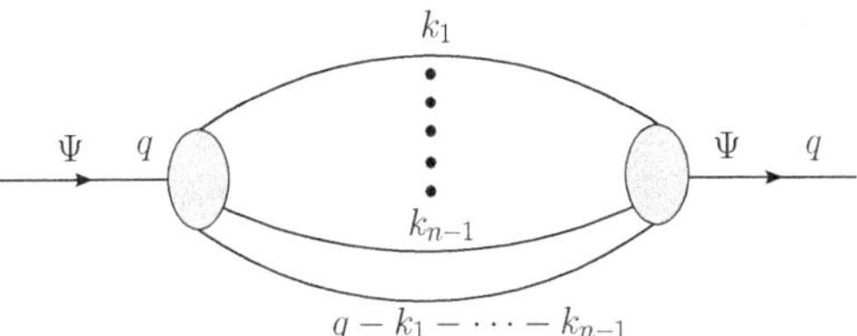

Figure 7.9 Self-energy graph with multiparticle intermediate state.

Thus the response of $(mass)^2$ of a physical particle to a change in the quark mass is given by the expectation value of $\bar{q}q$ in the physical eigenstate of the particle.

Problem 7.2: Find the imaginary part of the self-energy graph of Figure 7.9 with n particles in the intermediate state.

Solution: We shall find the result taking all the n particles to be bosons. To begin with, let the particles be also spinless. We express the self-energy in terms of independent loop momenta, using the overall delta function in momentum. Then the 11-component of the matrix is

$$\Sigma_{11}^{(n)}(q) = (-i)^{n-1} \int \prod_{i=1}^{n-1} \frac{d^4 k_i}{(2\pi)^4} D_{11}(k_1) \cdots D_{11}(q - k_1 - \cdots - k_{n-1})|M|^2 \quad \text{(P7.7)}$$

where M is the vertex amplitude. Following Section 5.3 we use the spectral representation (5.3.2) for the propagators and write it as

$$\Sigma_{11}^{(n)}(q) = \int \prod_{i=1}^{n-1} \frac{d^3 k_i}{(2\pi)^3} \int \prod_{j=1}^{n} \frac{dk'_{j,0}}{2\pi} \rho_0(k'_{1,0}, \boldsymbol{k}_i) \cdots \rho_0(k'_{n,0}, \boldsymbol{q} - \boldsymbol{k}_1 \cdots \boldsymbol{k}_{n-1})|M|^2 Q$$

$$\text{(P7.8)}$$

where

$$Q = (-i)^{n-1} \int \frac{dk_{1,0}}{2\pi} \cdots \int \frac{dk_{n-1,0}}{2\pi} \left(\frac{1 + f_1}{k'_{1,0} - k_{1,0} - i\eta} - \frac{f_1}{k'_{1,0} - k_{1,0} + i\eta} \right) \cdots$$

$$\left(\frac{1 + f_n}{k'_{n,0} - (q_0 - k_{1,0} \cdots - k_{n-1,0}) - i\eta} - \frac{f_n}{k'_{n,0} - (q_0 - k_{1,0} \cdots - k_{n-1,0}) + i\eta} \right).$$

$$\text{(P7.9)}$$

We now carry out the $k_{i,0}$ integrations in Q one by one. The result is fairly clear from our manipulations in Section 7.2. Beginning with $k_{1,0}$, the first and last factors in (P7.9) depend on it. Integrating this variable over these two factors gives an expression like (7.2.15). Next, integrate $k_{2,0}$ over the resulting factor and the second one in (P7.9), and so on. We finally get

$$Q = \frac{(1 + f_1) \cdots (1 + f_n)}{k'_{n,0} - (q - k'_{1,0} \cdots - k'_{n-1,0}) - i\eta} - \frac{f_1 \cdots f_n}{k'_{n,0} - (q - k'_{1,0} \cdots - k'_{n-1,0}) + i\eta} \tag{P7.10}$$

whose imaginary part is

$$\mathrm{Im}Q = \pi\{(1 + f_1) \cdots (1 + f_n) + f_1 \cdots f_n\}\delta(q_0 - k'_{1,0} \cdots - k'_{n,0}). \tag{P7.11}$$

Using the delta function we can extract the trigonometric factor from $\mathrm{Im}Q$ by changing the plus sign to minus sign between the two terms in (P7.11). Then we get the imaginary part of the diagonalised self-energy as

$$\mathrm{Im}\overline{\Sigma} = \pi\epsilon(q_0) \int \prod_{i=1}^{n-1} \frac{d^3 k_i}{(2\pi)^3} \int \prod_{j=1}^{n} \frac{dk'_{j,0}}{2\pi} \rho_0(k'_{1,0}, \boldsymbol{k}_1) \cdots \rho(k'_{n,0}, \boldsymbol{q} - \boldsymbol{k}_1 \cdots \boldsymbol{k}_{n-1})$$

$$\times |M|^2\{(1 + f_1) \cdots (1 + f_n) - f_1 \cdots f_n\}\delta(q_0 - k'_{1,0} \cdots - k'_{n,0}) \tag{P7.12}$$

generalising (5.3.8) to n-particle intermediate state. We then collect the decay and the inverse decay amplitudes from this formula to validate (5.3.19) for this case.

If the intermediate particles have spin, their spectral functions will contain spin sums like (1.6.1). As in the two-particle case of Section 5.4, we can bring these polarizations at the two vertices to define the matrix element M. If the external particle has also spin with spin function $U(q, \sigma)$, we get in this general case

$$U^*\mathrm{Im}\overline{\Sigma}U = \omega(\Gamma_d - \Gamma_i) \tag{P7.13}$$

where $\Gamma_{d,i}$ are given by (E.20) and (E.21) of Appendix E.

References

[1] J. Gasser and H. Leutwyler, Phys. Lett. **B 184**, 83 (1984).
[2] P. Gerber and H. Leutwyler, Nucl. Phys. **B321**, 387 (1987).
[3] H. Leutwyler in *QCD 20 Years Later*, edited by P.M. Zerwas and H.A. Kastrup, World Scientific (1993).
[4] G. Colangelo and S. Durr, Eur. Phys. J. **C 33**, 543 (2004).
[5] J.L. Goity and H. Leutwyler, Phys. Lett. **228**, 517 (1989).
[6] R.D. Pisarski and M. Tytgat, Phys. Rev. **D 54**, 2989 (1996).
[7] D. Toublan, Phys. Rev. **56**, 5629 (1997).
[8] S. Mallik and S. Sarkar, Eur. Phys. J. **C 49**, 755 (2007).
[9] C. Cohen-Tannoudji, B. Diu and F. Laloe, *Quantum Mechanics*, vol. 2, John Wiley and Sons (1977).

8

Heavy Ion Collisions

The subject of heavy ion collisions has many aspects. In this chapter we shall deal with a particular one, namely the dilepton (and photon) production, which is directly related to the spectral function at finite temperature. Though introduced in Chapter 4, the complete spectral functions have not been applied to any problem so far. We shall now use them to calculate the strong interaction part of this process.

Collision of heavy nuclei at ultra-relativistic energies creates an excited system at high temperature and density. The asymptotic freedom of QCD theory predicts the interaction between quarks and gluons to be weak in this condition, so that colour is liberated. We studied in Chapter 3 the spontaneous breakdown of chiral (flavour) symmetry in the hadronic phase. With colour liberated, this symmetry is expected to be restored, leading to a transition from the hadronic phase to the quark–gluon plasma (QGP) phase. Detection of dileptons (and photons) from such a system prove a powerful probe to study this transition: having no final state interaction, they reach the detector from the collision region with the information of the region at the time of their creation.

The lepton–antilepton pairs are, however, produced at all stages of the collision [1, 2]. In the initial pre-equilibrium stage, they are produced mostly in hard processes with invariant mass $\gtrsim 3$ GeV. A rapid thermalisation follows in the hot central region producing QGP, when they are produced mainly from quark–antiquark annihilation. As the system expands and cools into a hadron gas, pion and kaon annihilation constitute their major source. Finally in the freeze-out stage, the dominant sources are hadrons and three-body (Dalitz) decays of mesons. All these processes produce dileptons in the region of invariant mass $\lesssim 1$ GeV.

The dilepton emission rate formula that we shall derive below is quite general, valid in both phases in equilibrium. In the text we evaluate it in the hot hadronic phase, when the energy density is stored mostly in the pion, kaon and resonances like $\rho, \omega, \cdots$. Later in Problem 9.1 we shall take up the QGP phase. Comparison of calculations with the experimental data [3] could suggest the mechanism by

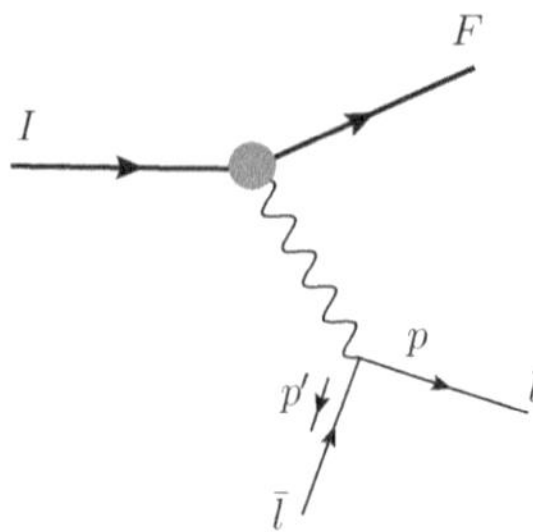

Figure 8.1 Dilepton production amplitude. An initial hadron state I decays into a final hadron state F and a dilepton. The wavy line is the photon propagator.

which the hadronic phase approaches the QGP phase and hence the restoration of chiral symmetry.

8.1 Dilepton Production

Several works are addressed to this problem [4–6]. We shall follow the work of Weldon [5] to find the general rate formula involving the two-point function of electromagnetic current and then evaluate it in the hadronic phase with the effective theory.

Rate Formula

Consider the S-matrix element for a transition, where an initial hadronic state I goes over to a final hadronic state F along with the emission of a lepton–antilepton pair, with momenta and spin z-components p, σ and p', σ' (Figure 8.1)

$$\langle F, l(p, \sigma), \bar{l}(p', \sigma') | S | I \rangle , \tag{8.1.1}$$

where the scattering matrix operator S is

$$S = T \exp\left(i \int d^4 x \mathscr{L}_{int}(x)\right) . \tag{8.1.2}$$

With the lepton current

$$j_\mu(x) = \overline{\psi}(x)\gamma_\mu\psi(x)$$

and the hadron current

$$J_\mu(x) = \frac{2}{3}\overline{u}(x)\gamma_\mu u(x) - \frac{1}{3}\overline{d}(x)\gamma_\mu d(x) \tag{8.1.3}$$

in the fundamental theory, the interaction Lagrangian is given by the (minimal) coupling of the electromagnetic field $A_\mu(x)$ to these currents,

$$\mathscr{L}_{\text{int}} = -e(j^\mu(x) + J^\mu(x))A_\mu(x). \tag{8.1.4}$$

To second order in this interaction, the above matrix element is

$$\langle F, l, \bar{l}|S|I\rangle = -e^2 \int d^4x\, d^4y \langle F|J^\mu(x)|I\rangle\, \langle l, \bar{l}|j^\nu(y)|0\rangle\, \langle 0|TA_\mu(x)A_\nu(y)|0\rangle.$$

$$(8.1.5)$$

The strong interaction is contained entirely in the first matrix element. The other two can immediately be worked out. The second one is leptonic

$$\langle l(p,\sigma), \bar{l}(p',\sigma')|j^\nu(y)|0\rangle = \bar{u}(p,\sigma)\gamma^\nu v(p',\sigma')e^{i(p+p')\cdot y}$$

while the last one is the photon propagator

$$\langle 0|TA_\mu(x)A_\nu(y)|0\rangle = \int \frac{d^4k}{(2\pi)^4} e^{-ik\cdot(x-y)} \frac{-ig_{\mu\nu}}{k^2 + i\epsilon}$$

Integration over y in (8.1.5) gives a delta function $(2\pi)^4\delta(p + p' + k)$ which removes the k-integral to give (8.1.5) as

$$\langle F, l, \bar{l}|S|I\rangle = \frac{ie^2}{q^2}\bar{u}(p,\sigma)\gamma_\mu v(p',\sigma') \int d^4x\, e^{iq\cdot x}\langle F|J^\mu(x)|I\rangle \qquad (8.1.6)$$

where $q = p + p'$, the dilepton four-momentum. The overall energy-momentum conserving delta function resides in the integral: translating the current operator to the origin, it gives

$$\int d^4x\, e^{iq\cdot x}\langle F|J^\mu(x)|I\rangle = (2\pi)^4\delta(P_I - P_F - q)\langle F|J^\mu(0)|I\rangle. \qquad (8.1.7)$$

In Appendix E we calculate transition pobability by putting the system in a large spatial box of volume V and also in a 'time box' of extension T during which the interaction is turned on. Then we can interpret the square of the delta function in the usual way. If we do not observe lepton polarisations, we have to sum over the spin projections σ and σ'. Then the squared matrix element may be written as

$$|\langle F, l, \bar{l}|S|I\rangle|^2 = VT(2\pi)^4\delta(P_I - P_F - q)\frac{e^4}{q^4}l^{\mu\nu}\langle F|J_\mu(0)|I\rangle\,\langle I|J_\nu(0)|F\rangle \quad (8.1.8)$$

where $l_{\mu\nu}$ is the lepton tensor

$$\begin{aligned}
l^{\mu\nu} &= \sum_{\sigma,\sigma'} \bar{u}(p,\sigma)\gamma^\mu v(p',\sigma')\bar{v}(p',\sigma')\gamma^\nu u(p,\sigma) \\
&= \mathrm{tr}\,(\slashed{p} + m_l)\gamma^\mu(\slashed{p}' - m_l)\gamma^\nu \\
&= 4\{p^\mu p'^\nu + p'^\mu p^\nu - (p\cdot p' + m_l^2)g^{\mu\nu}\}
\end{aligned} \qquad (8.1.9)$$

where m_l is the lepton mass and we use (1.3.11) for the spin sums.

So far we have taken the initial state of the system to be a specific hadronic state. Now we assume the system to be thermal, so that we take an ensemble average of (8.1.8), that is, sum over all initial states I, each weighted by the corresponding Boltzmann factor $\exp(-\beta E_I)/Z$, where Z is the partition function of the system. The delta function in (8.1.8) allows us to write $E_I = E_F + q_0$, where q_0 is the dilepton energy. Then we use (8.1.7) in the reverse direction to absorb the delta function. The sum over initial states I can then be performed by completeness of states, $\sum_I |I\rangle\langle I| = 1$. In this way we get from (8.1.8)

$$\sum_I e^{-\beta E_I} |\langle F, l, \bar{l}|S|I\rangle|^2 / Z$$

$$= VT \frac{e^4}{q^4} e^{-\beta q^0} l^{\mu\nu} \int d^4 x \, e^{iq\cdot x} e^{-\beta E_F} \langle F|J_\mu(x)J_\nu(0)|F\rangle / Z. \tag{8.1.10}$$

Finally we calculate the inclusive probability, where the hadrons in the final state are not observed,

$$\sum_F \sum_I e^{-\beta E_I} |\langle F, l, \bar{l}|S|I\rangle|^2 / Z \equiv R. \tag{8.1.11}$$

Earlier we defined the ensemble average with the (complete set of) initial states; now it turns out to be defined equivalently with the final states.

$$R = VT \frac{e^4}{q^4} e^{-\beta q^0} l^{\mu\nu} M^+_{\mu\nu}(q) \tag{8.1.12}$$

where we write in our notation the tensor M^+ as

$$M^+_{\mu\nu}(q) = \int d^4 x \, e^{iq\cdot x} \langle J_\mu(x)J_\nu(0)\rangle. \tag{8.1.13}$$

The single particle lepton states $|\boldsymbol{p}\rangle$ are normalised in the continuum according to (1.1.1). We now use states $|\boldsymbol{p}\rangle^B$, which are normalised to unity in the box. The relation between the two sets is already obtained in (E.4),

$$|\boldsymbol{p}\rangle \to \sqrt{2\omega V} |\boldsymbol{p}\rangle^B.$$

So the matrix element squared (8.1.11) may be related to that with box states as

$$R = V^2 2\omega 2\omega' R^B.$$

It is the quantity R^B, which gives the required probability $\mathscr{P}$ for the system to emit a dilepton pair,

$$\mathscr{P} \equiv R^B = \frac{R}{V^2 2\omega 2\omega'}. \tag{8.1.14}$$

This is the probability for dilepton emission into one specific box state $|\boldsymbol{p}\rangle^B \times |\boldsymbol{p}'\rangle^B$ of momenta p_1 and p_2. The total number dN_p of single particle box states between momenta $\boldsymbol{p}$ and $\boldsymbol{p}+d\boldsymbol{p}$ is given by (E.1). Hence the total probability of emission of lepton–antilepton pair with momenta in the intervals $d^3\boldsymbol{p}$ and $d^3\boldsymbol{p}'$ is

$$d\mathscr{P} = \mathscr{P} dN_p dN_{p'} = R \frac{d^3\boldsymbol{p}}{(2\pi)^3 2\omega} \frac{d^3\boldsymbol{p}'}{(2\pi)^3 2\omega'}. \tag{8.1.15}$$

Putting for R from (8.1.12) and inserting $1 = \int d^4q\,\delta(q - p - p')$ in (8.1.15), we get the dilepton emission *rate per unit volume* as

$$N = \frac{1}{4} \int d^4q\,\delta^{(4)}(q - p - p')\frac{e^4}{q^4}e^{-\beta q_0}\,l^{\mu\nu}M^+_{\mu\nu}\frac{d^3p}{(2\pi)^3\omega}\frac{d^3p'}{(2\pi)^3\omega'}. \qquad (8.1.16)$$

Finally, the total differential multiplicity is

$$\frac{dN}{d^4q} = \frac{e^4}{q^4}e^{-\beta q_0}L^{\mu\nu}M^+_{\mu\nu} \qquad (8.1.17)$$

where

$$L^{\mu\nu} = \frac{1}{4} \int \frac{d^3p}{(2\pi)^3\omega}\frac{d^3p'}{(2\pi)^3\omega'}\delta^{(4)}(q - p - p')l^{\mu\nu}. \qquad (8.1.18)$$

To evaluate $L_{\mu\nu}$ note that it is conserved,

$$q^\mu L_{\mu\nu} = q^\nu L_{\mu\nu} = 0\,,$$

so its Lorentz structure must be of the form

$$L^{\mu\nu} = A(q^2)(q^\mu q^\nu - q^2 g^{\mu\nu}). \qquad (8.1.19)$$

The scalar function $A(q^2)$ can be evaluated in any frame, say the centre-of-mass of the dilepton, $\boldsymbol{p} + \boldsymbol{p}' = 0$. We thus get

$$A(q^2) = \frac{1}{(2\pi)^6}\frac{2\pi}{3}\left(1 + \frac{2m_l^2}{q^2}\right)\sqrt{1 - \frac{4m_l^2}{q^2}}. \qquad (8.1.20)$$

Noting that $M^+_{\mu\nu}$ is conserved and if $m_l^2 << q^2$, we can write (8.1.17) as

$$\frac{dN}{d^4q} = \frac{\alpha^2}{6\pi^3 q^2}e^{-\beta q_0}(-g^{\mu\nu}M^+_{\mu\nu}) \qquad (8.1.21)$$

where $\alpha \equiv e^2/(4\pi)$ is the fine structure constant.

Strong Interaction

In Section 4.4 we derived the spectral representation for the time-ordered two-point function of local operators $\Phi_l(x)$ (elementary or composite). It applies to the present case by identifying $\Phi_l(x)$ with $J_\mu(x)$, when (4.4.14) is replaced by

$$M_{\mu\nu}(x, x') = \theta_c(\tau - \tau')i\langle J_\mu(x)J_\nu(x')\rangle + \theta_c(\tau' - \tau)i\langle J_\nu(x')J_\mu(x)\rangle. \qquad (8.1.22)$$

Its Fourier transform can be put in the form of a matrix like (4.4.16), which can be diagonalised with the diagonal element

$$\overline{M}_{\mu\nu}(q) = \int_{-\infty}^{\infty}\frac{dq_0'}{2\pi}\frac{\rho_{\mu\nu}(q_0', \boldsymbol{q})}{q_0' - q_0 - i\eta\epsilon(q_0)}. \qquad (8.1.23)$$

Here the spectral function $\rho_{\mu\nu}$ is defined as the Fourier transform of the commutator

$$\rho_{\mu\nu}(q_0, \boldsymbol{q}) = \int d^4y\, e^{iq\cdot(y-y')} \langle [J_\mu(y), J_\nu(y')] \rangle \tag{8.1.24}$$

which, by (8.1.23) is directly related to $\overline{M}_{\mu\nu}(q)$,

$$\rho_{\mu\nu}(q) = 2\,\mathrm{Im}\overline{M}_{\mu\nu}(q), \qquad q_0 > 0. \tag{8.1.25}$$

The amplitude $M_{\mu\nu}^+(q)$ in the formula (8.1.21) is then also related to $\overline{M}_{\mu\nu}(q)$: from (4.4.12) we get

$$M_{\mu\nu}^+(q) = \frac{e^{\beta q_0}}{e^{\beta q_0} - 1}\rho_{\mu\nu}(q) = \frac{2e^{\beta q_0}}{e^{\beta q_0} - 1}\mathrm{Im}\overline{M}_{\mu\nu}(q). \tag{8.1.26}$$

on using (8.1.25). Putting (8.1.26) in (8.1.21) the dilepton production rate becomes

$$\frac{dN}{d^4q} = \frac{\alpha^2}{3\pi^3 q^2}\frac{W}{e^{\beta q_0} - 1}, \qquad W = -g^{\mu\nu}\mathrm{Im}\overline{M}_{\mu\nu}. \tag{8.1.27}$$

In the quark–gluon phase, we could directly work with the electromagnetic current (8.1.3) carried by the quark fields, as we show in Problem 8.1. However, in the hadronic phase, we have to work in the effective theory of QCD, to which we now turn.

Effective Theory

The electromagnetic current may be split into

$$\begin{aligned}
J_\mu(x) &= \frac{1}{2}(\bar{u}\gamma_\mu u - \bar{d}\gamma_\mu d) + \frac{1}{6}(\bar{u}\gamma_\mu u + \bar{d}\gamma_\mu d) \\
&= \bar{q}\gamma_\mu \frac{\tau^3}{2}q + \frac{1}{6}\bar{q}\gamma_\mu q \\
&\equiv V_\mu^{(3)}(x) + \frac{1}{6}V_\mu^{(0)}
\end{aligned} \tag{8.1.28}$$

the (third component of) isovector and isoscalar currents. So the current is dominated by the isovector part, which is known to couple strongly to the ρ meson field. On this basis we shall now outline a treatment to one loop of

$$M_{\mu\nu}^+ = \int d^4x\, e^{iq\cdot x}\langle V_\mu^{(3)}(x)V_\nu^{(3)}(0)\rangle \tag{8.1.29}$$

in the effective theory developed in Sections 3.13 and 6.4. There we obtained only the shift of the mass and coupling of the ρ meson. Now we are interested in the entire spectral function.

The diagonalised amplitude relevant for the process may be written from (6.4.2) as

$$\overline{M}_{\mu\nu}(q) = \left(\frac{\widetilde{F}_\rho^T}{m_\rho}\right)^2 q^4 D_{\mu\nu} \tag{8.1.30}$$

where $\overline{D}_{\mu\nu}$ is the complete ρ propagator satisfying

$$\overline{D}_{\mu\nu} = \widetilde{\Delta}_{\mu\nu} + \widetilde{\Delta}_{\mu\rho}\overline{\Pi}^{\rho\sigma}\overline{D}_{\sigma\nu}. \tag{8.1.31}$$

Here $\overline{\Pi}_{\mu\nu}$ denotes the self-energy from all the loops in the ρ propagator. The current-ρ meson vertex is

$$\widetilde{F}^T_\rho = F_\rho\left(1 - \frac{T^2}{12F^2}\right). \tag{8.1.32}$$

Compared to (6.4.21), we omit the g_1^2 term in (8.1.32), as the graph giving this correction will be included in the calculation of $\overline{\Pi}_{\mu\nu}$.

In Sections 6.4 we decomposed all tensors in (8.1.31) in terms of two kinematic tensors (6.4.5), when $\overline{D}_{\mu\nu}$ can be solved with scalar functions $\overline{D}_{t,l}$ given by (6.4.15). Noting $P^\mu_\mu = -2$ and $Q^\mu_\mu = -q^2$, we can express W in terms of $\overline{D}_{t,l}$. Inserting the latter in (8.1.27) we finally get

$$\frac{dN}{d^4q} = \frac{\alpha^2}{3\pi^3}\left(\frac{\widetilde{F}^T_\rho}{m_\rho}\right)^2 q^2\frac{2\,\mathrm{Im}D_t + \mathrm{Im}D_l}{e^{\beta q_0} - 1} \tag{8.1.33}$$

where

$$\mathrm{Im}D_{t,l} = \frac{(1, q^2)\mathrm{Im}\Pi_{t,l}}{\{q^2 - m_\rho^2 + (1, q^2)\mathrm{Re}\Pi_{t,l}\}^2 + \{(1, q^2)\mathrm{Im}\Pi_{t,l}\}^2}. \tag{8.1.34}$$

In the next two sections we describe the analytic structure of $\overline{\Pi}$ and its evaluation.

8.2 Analytic Structure

In Section 5.3 and 5.4 we discussed physical properties of the imaginary part of (one-loop) self-energies. We shall now consider the self-energy function for a particle like the ρ meson and find its analytic structure. As we did in Section 6.4, a tensor amplitude can be decomposed into scalar ones. Taking the particles in the loop to be bosons, the 11-component of such a scalar amplitude has the form

$$\Pi_{11}(q) = -i\int\frac{d^4k}{(2\pi)^4}N(q,k)D_{11}(k,m)D_{11}(q-k,m'). \tag{8.2.1}$$

We shall evaluate it using spectral representation for the propagators, following steps similar to those already worked out in Section 5.3. The generalisation now concerns the polynomial $N(q,k)$ replacing the coupling constant in (5.3.1). It arises from two sources: one is the possible derivative couplings in the interaction at the vertices of graphs. The other is the form of the general spectral function (1.6.5), which is a product of the scalar spectral function multiplied with a polynomial in momentum. In the following we shall suppress for brevity the spatial components of momenta in N and display only the time components as $N(k_0, q_0 - k_0)$. We thus write (8.2.1) as

$$\Pi_{11}(q) = \int \frac{d^3k}{(2\pi)^3} \int \frac{dk_0'}{2\pi} \rho_0(k_0', \boldsymbol{k}) \int \frac{dk_0''}{2\pi} \rho_0(k_0'', \boldsymbol{q} - \boldsymbol{k}) \cdot J \tag{8.2.2}$$

where

$$J = -i \int \frac{dk_0}{(2\pi)} N(k_0, q_0 - k_0) \left(\frac{1+f'}{k_0' - k_0 - i\eta} - \frac{f'}{k_0' - k_0 + i\eta} \right)$$
$$\times \left(\frac{1+f''}{k_0'' - (q_0 - k_0) - i\eta} - \frac{f''}{k_0'' - (q_0 - k_0) + i\eta} \right)$$

with f', f'' defined as in Section 5.3.

To begin with, we put the variables of N at the poles of the respective spectral representation, i.e. set $k_0 = k_0', q_0 - k_0 = k_0''$. While this is no approximation for the imaginary part of Π_{11}, it is a mild one for the real part, affecting it only away from the poles. With N now appearing outside the k_0 integral, the latter can be evaluated as before to get (5.3.5). Extracting a trigonometric function from the imaginary part as we did there, we can represent the diagonal, analytic amplitude $\overline{\Pi}$ with both real and imaginary parts as

$$\overline{\Pi}(q_0, \boldsymbol{q}) = - \int \frac{d^3k}{(2\pi)^3} \int \frac{dk_0'}{2\pi} \rho_0(k_0', \boldsymbol{k}, m) \int \frac{dk_0''}{2\pi} \rho_0(k_0'', \boldsymbol{q} - \boldsymbol{k}, m')$$
$$\times N(k_0', k_0'') \frac{(1+f')(1+f'') - f'f''}{q_0 - (k_0' + k_0'') + i\eta\epsilon(q_0)}. \tag{8.2.3}$$

It is now simple to do the k_0' and k_0'' integrals with the spectral function (5.3.9). With $q_0 \equiv E$ and $\epsilon(q_0) \equiv \epsilon$, we get

$$\overline{\Pi}(E, \boldsymbol{q}) = - \int \frac{d^3k}{(2\pi)^3 4\omega\omega'} \left(\frac{N(\omega, \omega')(1 + n + n')}{E - \omega - \omega' + i\eta\epsilon} - \frac{N(\omega, -\omega')(n - n')}{E - \omega + \omega' + i\eta\epsilon} \right.$$
$$\left. + \frac{N(-\omega, \omega')(n - n')}{E + \omega - \omega + i\eta\epsilon} - \frac{N(-\omega, -\omega')(1 + n + n')}{E + \omega + \omega' + i\eta\epsilon} \right) \tag{8.2.4}$$

where $\omega = \sqrt{\boldsymbol{k}^2 + m^2}$, $\omega' = \sqrt{(\boldsymbol{q} - \boldsymbol{k})^2 + m'^2}$, $n = (e^{\beta\omega} - 1)^{-1}$ and $n' = (e^{\beta\omega'} - 1)^{-1}$. Its real part is given by the principal value integrals, while the imaginary part is

$$\mathrm{Im}\,\overline{\Pi}(E, \boldsymbol{q}) = \pi\epsilon(E) \int \frac{d^3k}{(2\pi)^3 4\omega\omega'} \times$$
$$\{ N(\omega, \omega')(1 + n + n')\delta(E - \omega - \omega') - N(\omega, -\omega')(n - n')\delta(E - \omega + \omega')$$
$$+ N(-\omega, \omega')(n - n')\delta(E + \omega - \omega') - N(-\omega, -\omega')(1 + n + n')\delta(E + \omega + \omega') \}. \tag{8.2.5}$$

To find the analytic structure of $\overline{\Pi}$, we first recall the comment at the end of Section 4.1 that in a medium we have a Lorentz scalar, namely $\omega \equiv u \cdot q$, where u^μ is the four-velocity of the medium, in addition to q^2. We choose the independent scalar variables as ω and $\bar{q} = \sqrt{\omega^2 - q^2}$, which reduce to $q_0 \equiv E$ and $|\boldsymbol{q}|$ in the medium at rest. So the variables of $\overline{\overline{\Pi}}$ shown above really represent two independent Lorentz scalars in a particular frame. Compared to vacuum, we thus have a richer structure in medium to analyse $\Pi(E, \boldsymbol{q})$ as a function of E at fixed $\boldsymbol{q}$ at different temperatures (and densities). Its imaginary parts and hence the cuts are situated in regions in which the delta functions in (8.2.5) are non-vanishing. For the first and fourth terms this condition is

$$E = \pm(\omega + \omega') \tag{8.2.6}$$

while for the second and third terms, it is

$$E = \pm(\omega - \omega'). \tag{8.2.7}$$

In the analysis below, we need a Schwartz inequality [7]. Considering the Euclidean four-vectors

$$(\boldsymbol{k}, m) \qquad \text{and} \qquad (\boldsymbol{q} - \boldsymbol{k}, m')$$

we get this inequality as

$$(\boldsymbol{k}^2 + m^2)\{(\boldsymbol{q} - \boldsymbol{k})^2 + m'^2\} \geq \{\boldsymbol{k} \cdot (\boldsymbol{q} - \boldsymbol{k}) + mm'\}^2 \tag{8.2.8}$$

$$\text{or} \qquad \omega\omega' \geq |\boldsymbol{k} \cdot (\boldsymbol{q} - \boldsymbol{k}) + mm'|. \tag{8.2.9}$$

We now square (8.2.6) to get

$$\begin{aligned} E^2 &= m^2 + m'^2 + \boldsymbol{k}^2 + (\boldsymbol{q} - \boldsymbol{k})^2 + 2\omega\omega' \\ &= (m + m')^2 + \boldsymbol{q}^2 + 2\{\omega\omega' - \boldsymbol{k} \cdot (\boldsymbol{q} - \boldsymbol{k}) - mm'\} \\ &\geq (m + m')^2 + \boldsymbol{q}^2 \end{aligned} \tag{8.2.10}$$

on using the above inequality. It may be written equivalently as $q^2 \geq (m + m')^2$, as indeed one does in the vacuum theory. We define $E_1 = \sqrt{(m + m')^2 + \boldsymbol{q}^2}$ to write (8.2.10) as

$$(E - E_1)(E + E_1) \geq 0$$

giving the two branch cuts in the E plane

$$E_1 \leq E \leq \infty \qquad \text{and} \qquad -\infty \leq E \leq -E_1 \tag{8.2.11}$$

corresponding to the first and fourth terms in (8.2.5) respectively. As in the vacuum theory we call these unitary cuts.

Similarly squaring (8.2.7) we get

$$\begin{aligned} E^2 &= m^2 + m'^2 + \boldsymbol{k}^2 + (\boldsymbol{q} - \boldsymbol{k})^2 - 2\omega\omega' \\ &= (m - m')^2 + \boldsymbol{q}^2 - 2\{\omega\omega' - \boldsymbol{k} \cdot (\boldsymbol{k} - \boldsymbol{q}) - mm'\} \\ &\leq (m - m')^2 + \boldsymbol{q}^2 \end{aligned} \tag{8.2.12}$$

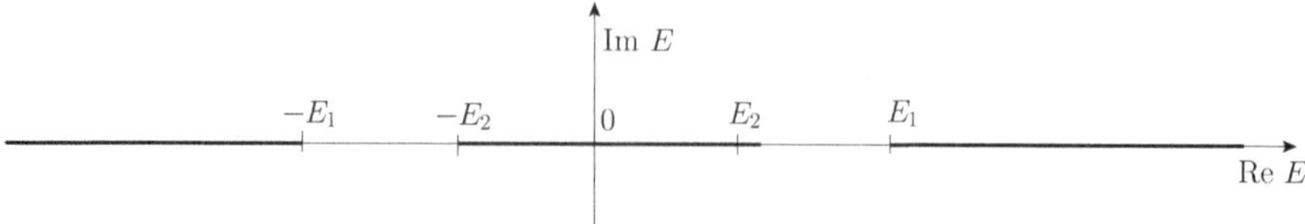

Figure 8.2 Branch cuts of self-energy function in E plane at fixed $\boldsymbol{q}$. Here $E_1 = \sqrt{(m+m')^2 + \boldsymbol{q}^2}$ and $E_2 = \sqrt{(m'-m)^2 + \boldsymbol{q}^2}$. The cut in the middle is the Landau cut, while the other two correspond to the unitary cut.

which again may be written equivalently as $q^2 \leq (m - m')^2$, but it has no counterpart in the vacuum theory. Define $E_2 = \sqrt{(m-m')^2 + \boldsymbol{q}^2}$ to write (8.2.12) as

$$(E - E_2)(E + E_2) \leq 0$$

when one gets the branch cut

$$-E_2 \leq E \leq E_2 \tag{8.2.13}$$

corresponding to the second and third terms in (8.2.5). This is called the Landau cut. The position of the unitary and Landau cuts are shown in Figure 8.2.

The physical origin of these cuts is clear from the discussion in Section 5.3. The unitary cuts arise from the states, which can communicate with the external particle. These states are, of course, the same as in vacuum, but, as we see above, the probabilities of their occurrence in medium are modified by the distribution functions. On the other hand, the Landau cut appears only in medium and arises from scattering of the external particle with particles in medium.

8.3 Evaluation

We shall now put the imaginary part of the self-energy (8.2.5) in a form that is convenient for numerical evaluation. Writing $d^3k = 2\pi|\boldsymbol{k}|\,\omega d\omega\, d(\cos\theta)$, where θ is the angle between $\boldsymbol{q}$ and $\boldsymbol{k}$, we can readily integrate over $\cos\theta$ using the δ-functions. But we have to ensure that θ remains physical as we integrate over ω. As we shall see presently, it reduces the a priori range (from m to ∞) of ω integration.

Label consecutively the four terms in (8.2.5) by the indices $1, \cdots, 4$. Consider the first term

$$\mathrm{Im}\,\overline{\Pi}_1(E, \boldsymbol{q}) = \pi \int \frac{d^3k}{(2\pi)^3 4\omega\omega'} N(\omega, \omega')(1 + n + n')\delta(E - \omega - \omega'). \tag{8.3.1}$$

In the previous section we used the delta function to find its non-vanishing region in E, namely $E \geq E_1$. Now we find the particular value of $\cos\theta$, call it $\overline{\cos\theta}$, for which it is satisfied. Squaring $E - \omega = \omega'$, we get

$$\overline{\cos\theta} = \frac{2E\omega - R^2}{2|\boldsymbol{q}||\boldsymbol{k}|}, \qquad R^2 = q^2 + m^2 - m'^2, \qquad q^2 = E^2 - |\boldsymbol{q}|^2 \qquad (8.3.2)$$

Then

$$\delta(E - \omega - \omega') = \frac{\omega'}{|\boldsymbol{q}||\boldsymbol{k}|}\delta(\cos\theta - \overline{\cos\theta})$$

so that (8.3.1) becomes

$$\operatorname{Im}\overline{\overline{\Pi}}_1(E,\boldsymbol{q}) = \frac{1}{16\pi|\boldsymbol{q}|}\int d\omega\theta((1 - |\overline{\cos\theta}|)(1 + n + n')N(\omega). \qquad (8.3.3)$$

The theta function constraint

$$\left(\frac{2E\omega - R^2}{2|\boldsymbol{q}||\boldsymbol{k}|}\right)^2 \leq 1$$

gives rise to a quadratic expression in ω

$$4q^2\omega^2 - 4R^2E\omega + R^4 + 4|\boldsymbol{q}|^2m^2 \leq 0$$

which factorises to

$$q^2(\omega - \omega_+)(\omega - \omega_-) \leq 0 \qquad (8.3.4)$$

with the roots $\omega_\pm$ given by

$$\omega_\pm = \frac{1}{2q^2}(ER^2 \pm |\boldsymbol{q}|\sqrt{R^4 - 4q^2m^2}). \qquad (8.3.5)$$

(We note here that $(R^4 - 4q^2m^2)/4q^2$ is the three-momentum squared in the centre of mass of the two particles in the intermediate state: Expanded out it shows all the four thresholds.) We can now find the range of ω integration in (8.3.3) constrained by (8.3.4). With

$$q^2 > 0, \qquad E > |\boldsymbol{q}|, \qquad R^2 > \sqrt{R^4 - 4q^2m^2},$$

we see that both ω_+ and ω_- are positive with $\omega_+ > \omega_-$, restricting the range of ω to $\omega_- < \omega < \omega_+$,

$$\operatorname{Im}\overline{\overline{\Pi}}_{1(a)}(E,\boldsymbol{q}) = \frac{1}{16\pi|\boldsymbol{q}|}\int_{\omega_-}^{\omega_+} d\omega(1 + n + n')\,N(\omega), \qquad E \geq E_1. \qquad (8.3.6)$$

Consider next the second term in (8.2.5)

$$\operatorname{Im}\overline{\overline{\Pi}}_2(E,\boldsymbol{q}) = -\pi\epsilon(E)\int \frac{d^3k}{(2\pi)^3 4\omega\omega'}N(\omega, -\omega')(n - n')\delta(E - \omega + \omega'). \qquad (8.3.7)$$

It exists on the cut $-E_2 < E < E_2$, which can conveniently be split into three pieces as shown in Figure 8.3. Again $\overline{\cos\theta}$ is given by (8.3.2) with the restriction (8.3.4) on the ω integration. But the variables take different values. On the pieces (a) and (b) of the cut we have

$$q^2 > 0, \qquad |E| > |\boldsymbol{q}|, \qquad |R|^2 > \sqrt{R^4 - 4q^2m^2}$$

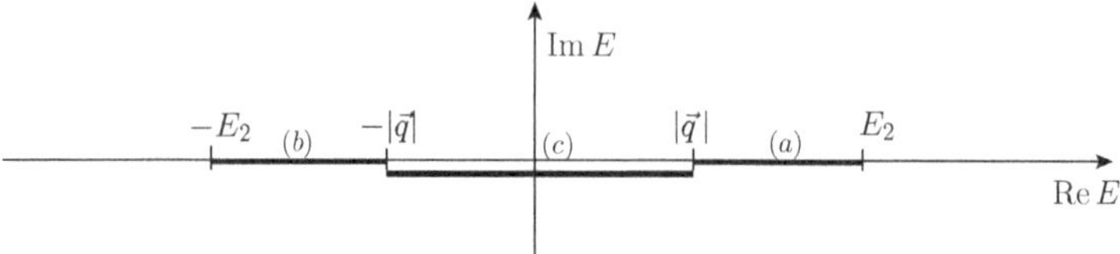

Figure 8.3 The splitting of the Landau cut into three pieces. The middle piece is slightly lowered for clarity.

so that the signs of $\omega_\pm$ are determined by the first term in (8.3.5). Now on (a) $R^2 < 0$, $E > 0$, making both $\omega_\pm$ negative, while on (b) $R^2 < 0$, $E < 0$, making both $\omega_\pm$ positive with $\omega_+ > \omega_-$. We thus get

$$\operatorname{Im}\overline{\Pi}_{2(a)}(E, \boldsymbol{q}) = 0, \qquad\qquad |\boldsymbol{q}| < E < E_2$$

$$\operatorname{Im}\overline{\Pi}_{2(b)}(E, \boldsymbol{q}) = \frac{\epsilon(E)}{16\pi|\boldsymbol{q}|} \int_{\omega_-}^{\omega_+} d\omega(n - n')N(\omega), \qquad -E_2 < E < -|\boldsymbol{q}|. \quad (8.3.8)$$

On the piece (c) of the cut we have

$$q^2 < 0, \qquad |E| < |\boldsymbol{q}|, \qquad |R|^2 < \sqrt{R^4 - 4q^2m^2}$$

so that the signs of $\omega_\pm$ are determined by the second term in (8.3.5), giving $\omega_+ < 0, \omega_- > 0$. Thus we have

$$\operatorname{Im}\overline{\Pi}_{2(c)}(E, \boldsymbol{q}) = \frac{-\epsilon(E)}{16\pi|\boldsymbol{q}|} \int_{\omega_-}^{\infty} d\omega(n - n')N(\omega), \qquad -|\boldsymbol{q}| < E < |\boldsymbol{q}|. \quad (8.3.9)$$

For the third and fourth terms in (8.2.5), $\overline{\cos\theta}$ will take a different expression, but they can be analysed in the same way as above. (In the present calculation we need the imaginary parts for positive values of E only. However, to construct dispersion relations for the amplitudes, we have to include cut contributions from all values of E.)

*   *   *

The numerical evaluation of the imaginary parts of each self-energy loop can now be carried out. For the real parts, we have to calculate principal value integrals. Here we expect divergence in the vacuum contribution, which should cancel with counterterms, leaving physical values of mass and coupling constant and a finite contribution from vacuum, whose real part (and possibly imaginary part also) is small and can be ignored. Retaining the thermal parts, one thus gets the self-energy from different loop graphs [8, 9]. Loop graphs involving baryons can be treated in a similar way [10]. For somewhat different approaches, see [11, 12].

It is still a long way to compare the calculated dilepton rate with experiment. One must include the rates from all stages of heavy ion collision, particularly the QGP phase. Also recall that the rate is calculated in the rest-frame of matter.

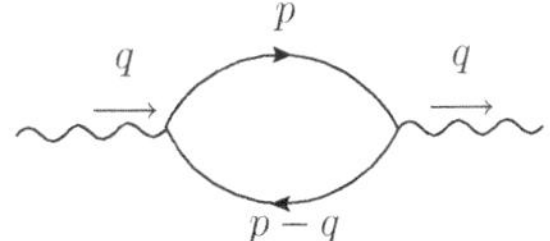

Figure 8.4 Dilepton production in QGP phase.

To compare with experiment, this rate must be transformed to the lab frame and integrated over the space–time evolution of the so-called fireball, whose description is beyond the scope of this book.

8.4 Comments

The primary motivation for measuring dilepton (and photon) yield is to find how the hadronic phase transforms itself into the QGP phase, as the temperature and/or density rise to their critical values. This deconfinement is expected to be accompanied by restoration of chiral symmetry. It would then imply degeneracy of isospin multiplets of opposite parity. Thus if the observed hadrons were to continue in the QGP phase, then, for example, ρ and a_1 as well as $\frac{1}{2}^+[N(940)]$ and $\frac{1}{2}^-[N(1535)]$ would be degenerate in mass. The fate of the pion triplet would be most interesting, as, besides being a bound state of u and d quarks like all other light hadrons, they are also the Goldstone bosons resulting from spontaneous symmetry breaking.

In the context of dilepton production, we can study the behaviour of vector mesons, particularly the ρ, as the temperature and/or density increase. As we observe from (8.1.33) and (8.1.34), there are three quantities which may be affected, namely its *mass*, *width* and *coupling* to the electromagnetic current. At low temperatures we have seen in Section 6.3 that its mass remains constant, the coupling becomes weak and the width is expected to increase. It remains to be seen how they really behave at higher temperature and density.

There are other signatures of QGP phase, for example, suppression of charmonium production and enhancement of strange particle formation. These effects follow basically from the weakness of colour gauge interaction at high temperature and density. But quantitative estimates of these effects are rather difficult.

Problems

Problem 8.1: Work out the dilepton emission rate in QGP phase to lowest order.

Solution: The rate in the QGP phase is still given by (8.1.27), but we do not need any effective theory to find $\overline{M}_{\mu\nu}$. In the QGP phase we can work directly with the quark current (8.1.3). As usual, we begin with the 11-component of the two-point function

$$M_{\mu\nu}(q)_{11} = i \int d^4x\, e^{iq\cdot x} \langle T J_\mu(x) J_\nu(0)\rangle_{11} \qquad \text{(P8.1)}$$

which to lowest order is given by (Figure 8.4)

$$M_{\mu\nu}(q)_{11} = \frac{5i}{3} \int \frac{d^4p}{(2\pi)^4}\, \mathrm{tr}\,\{S_{11}(p)\gamma_\nu S_{11}(p-q)\gamma_\mu\}. \qquad \text{(P8.2)}$$

Inserting (4.4.42) for the propagators and recalling the free spectral function (1.3.13) for the Dirac field

$$\sigma(p) = (\slashed{p} + m_q)\rho(p) \qquad \text{(P8.3)}$$

we can write it as

$$M_{\mu\nu}(q)_{11} = \int \frac{d^3p}{(2\pi)^3} \int \frac{dp_0'}{2\pi}\rho_0(p_0', \boldsymbol{p}) \int \frac{dp_0''}{2\pi}\rho_0(p_0'', \boldsymbol{p} - \boldsymbol{q}) \cdot K_{\mu\nu} \qquad \text{(P8.4)}$$

where

$$K_{\mu\nu} = i \int \frac{dp_0}{2\pi}\bar{l}_{\mu\nu} \left(\frac{1-\widetilde{f}'}{p_0' - p_0 - i\eta} + \frac{\widetilde{f}'}{p_0' - p_0 + i\eta} \right) \times$$

$$\left(\frac{1-\widetilde{f}''}{p_0'' - (p_0 - q_0) - i\eta} + \frac{\widetilde{f}''}{p_0'' - (p_0 - q_0) + i\eta} \right) \qquad \text{(P8.5)}$$

with $\bar{l}_{\mu\nu} = (5/3)\,\mathrm{tr}\,\{(\slashed{p}+m_q)\gamma_\nu(\slashed{p}-\slashed{q}+m_q)\gamma_\mu\}$ and we denote for short $\widetilde{f}' = \widetilde{f}(p_0')$, $\widetilde{f}'' = \widetilde{f}(p_0'')$.

We now have to evaluate $W \equiv -g^{\mu\nu}\mathrm{Im}\overline{M}_{\mu\nu}$ appearing in the dilepton rate formula (8.1.27). For the imaginary part, we can fix the energy variables in N at the poles of the spectral representations as we did in Section 8.2, when N can be brought out of the integral. Then it can be evaluated as in Section 5.3. Using (5.2.8) we find the imaginary part of the diagonalised amplitude, giving

$$W(q) = -\pi\epsilon(q_0) \int \frac{d^3p}{(2\pi)^3} \int\frac{dp_0'}{2\pi}\rho_0(p_0', \boldsymbol{p}) \int \frac{dp_0''}{2\pi}\rho_0(p_0'', \boldsymbol{p} - \boldsymbol{q})$$

$$\times \bar{l}_\mu^\mu(p_0', p_0'')(\widetilde{f}'' - \widetilde{f}')\delta(p_0'' - p_0' + q_0). \qquad \text{(P8.6)}$$

For the problem at hand we need $W(q)$ in the physical region only, $q_0 \geq \omega + \omega'$. So (P8.6) reduces to

$$W(q) = \pi \int \frac{d^3p}{(2\pi)^3\, 4\omega\omega'}\bar{l}_\mu^\mu(\omega, -\omega')\{\widetilde{f}(-\omega') - \widetilde{f}(\omega)\}\delta(q_0 - \omega - \omega'). \qquad \text{(P8.7)}$$

On the mass shell $\bar{l}_\mu^\mu = 20(q^2 + 2m_q^2)/3 \simeq 20q^2/3$, when (P8.7) becomes

$$W = \frac{20\pi}{3}q^2 \int \frac{d^3p}{(2\pi)^3\, 4\omega\omega'}\{1 - \widetilde{n}(\omega) - \widetilde{n}(\omega')\}\delta(q_0 - \omega - \omega'). \qquad \text{(P8.8)}$$

The kinematics for unequal mass case is worked out in Setion 8.3, which simplifies for the equal mass case giving

$$W = \frac{5q^2}{12\pi|\boldsymbol{q}|}\frac{1}{2}\int_{\omega_-}^{\omega_+} d\omega \left\{1 - \frac{2}{e^{\beta\omega}+1} + 1 - \frac{2}{e^{\beta(q_0-\omega)}+1}\right\}$$

$$= \frac{5q^2}{6\pi\beta|\boldsymbol{q}|}\ln\left(\frac{\cosh(\beta\omega_+/2)}{\cosh(\beta\omega_-/2)}\right) \tag{P8.9}$$

where the limits are $\omega_\pm = \frac{1}{2}(q_0 \pm |\boldsymbol{q}|v)$, $v = \sqrt{1 - 4m_q^2/q^2} \simeq 1$. Inserting (P8.9) in (8.1.27) we shall get the dilepton emission rate in QGP phase.

Problem 8.2: Find the rate of emission of photons from the system formed in heavy ion collisions.

Solution: Here we consider an initial hadronic state I going to a final hadronic state F, accompanied by a real photon (unlike virtual photon in the dilepton case) of momentum k and spin projection σ. This transition amplitude to leading order is

$$\langle F, \gamma(k,\sigma)|S|I\rangle = -ie\int d^4x \langle F|J^\mu(x)|I\rangle\langle\gamma(k,\sigma)|A_\mu(x)|0\rangle$$

$$= -ie\int d^4x\, e^{ik\cdot x}\langle F|J_\mu(x)|I\rangle\epsilon^\mu(k,\sigma). \tag{P8.10}$$

The squared matrix element with sum over photon spin gives

$$|\langle F,\gamma|S|I\rangle|^2 = e^2 VT\delta^4(P_I - P_F - k)\langle F|J_\mu(0)|I\rangle\langle I|J_\nu(0)|F\rangle \sum_\sigma \epsilon^\mu(k,\sigma)\epsilon^{\nu*}(k,\sigma) \tag{P8.11}$$

where the spin sum is given by (1.4.12).

We now follow the dilepton case. Assume the initial state to be thermal, getting the probability

$$\sum_I e^{-\beta E_I}|\langle F,\gamma(k)|S|I\rangle|^2 = VTe^2 e^{-\beta k_0}\int d^4x\, e^{ik\cdot x}e^{-\beta E_F}\langle F|J_\mu(x)J_\nu(0)|F\rangle(-g_{\mu\nu}). \tag{P8.12}$$

We then sum over probabilities for all final hadronic states to get

$$\sum_F \sum_I e^{-\beta E_I}|\langle F,\gamma(k)|S|I\rangle|^2 = VTe^2 e^{-\beta k_0}M^+_{\mu\nu}(k)(-g^{\mu\nu}) \tag{P8.13}$$

where $M^+_{\mu\nu}$ is given by the same expression (8.1.13) appearing in the dilepton case. The total photon emission rate per unit volume is

$$N = e^2 \int \frac{d^3k}{(2\pi)^3 2\omega} e^{-\beta q_0} M^+_{\mu\nu}(-g^{\mu\nu}) \tag{P8.14}$$

giving the differential rate as

$$\frac{dN}{d^3k} = \frac{\alpha}{2\pi^2\omega}\frac{W}{e^{\beta q_0}-1} \qquad W = -g^{\mu\nu}M^+_{\mu\nu}. \tag{P8.15}$$

Thus the strong interaction part is the same as in the dilepton case and so can be evaluated in the same way.

References

[1] R. Rapp and J. Wambach, Adv. Nucl. Phys. **25**, 1 (2000).

[2] J. Alam, S. Sarkar, P. Roy, T. Hatsuda and B. Sinha, Annals Phys. **286**, 159 (2001).

[3] I. Tserruya, Acta Phys. Pol., Proc. Suppl. B**3**, 679 (2010).

[4] L.D. McLerran and T. Toimela, Phys. Rev. **D 31**, 545 (1985).

[5] H.A. Weldon, Phys. Rev. **D 42**, 2384 (1990).

[6] C. Gale and J.I. Kapusta, Nucl. Phys. **B 357**, 65 (1991).

[7] A. Das, *Finite Temperature Field Theory*, World Scientific, Singapore (1998).

[8] R. Rapp and C. Gale, Phys. Rev. **C 60**, 024903 (1999).

[9] S. Ghosh, S. Mallik and S. Sarkar, Eur. Phys. J. **C 70**, 251 (2010).

[10] S. Ghosh, S. Sarkar and S. Mallik, Phys. Rev. **C 83**. 018201 (2011).

[11] Z. Huang, Phys. Lett. B **361** 132 (1995).

[12] J.V. Steele, H. Yamagishi and I. Zahed, Phys. Lett. B **384**, 255 (1996).

9

Non-Equilibrium Processes

So far in this book we have studied properties of systems in thermal equilibrium. We now consider transport phenomena in fluids out of equilibrium. There are two distinct methods to calculate the transport coefficients. One is based on the *Boltzmann equation* in relativistic kinetic theory [1]. The other finds the *response* of the system to the energy–momentum tensor [2]. Here we take up the latter procedure, leaving the former to a brief review in Problem 9.1.

Assuming the system to be not far from equilibrium, we consider only responses linear in the perturbation causing non-equilibrium in the fluid medium. The phenomenological form of its energy–momentum tensor involves gradients of fluid velocity and temperature. The terms in this tensor without these gradients describe perfect fluid, while terms containing them represent dissipative effects in real fluids. The latter terms have (transport) coefficients to be determined. The method of linear response calculates these coefficients from thermal and quantum fluctuations arising in loop graphs of quantum field theory.

The first five sections are reviews leading to field theoretical expressions for the transport coefficients. Following Weinberg [3] we first describe elements of relativistic hydrodynamics to get the phenomenological expression for the energy–momentum tensor of imperfect fluids. Besides defining transport coefficients, it offers physical insight into the problem. We then present Zubarev's formulation of non-equilibrium statistical theory [2] and find the thermal expectation value of the components of the corresponding tensor operator. The transport coefficients so obtained are written as standard equilibrium correlation functions of the retarded type as in Hosoya et al. [4]. In the last section we evaluate these coefficients for the pion gas using the real time method of thermal field theory developed in this book. As already mentioned in Section 5.4, this method has the advantage over the imaginary time method in that it does not require any analytic continuation from discrete frequencies.

9.1 Relativistic Hydrodynamics

Consider a fluid having at each point a velocity three-vector $\boldsymbol{v}$ in a lab frame of reference (at rest in the lab). In a comoving frame (moving with the fluid) at some space–time point, the fluid will be at rest at that point. A *perfect* or *ideal* fluid is one which appears isotropic around this point from the comoving (denoted by tilde) frame. So in this frame the components of the energy–momentum tensor of such a fluid must have the form

$$\widetilde{T}^{00}_{(o)} = \varepsilon, \qquad \widetilde{T}^{i0}_{(o)} = \widetilde{T}^{0i}_{(o)} = 0, \qquad \widetilde{T}^{ij}_{(o)} = p\,\delta^{ij}, \tag{9.1.1}$$

the subscript (o) denoting perfect fluid. (Non-zero values of $\widetilde{T}^{i0}_{(o)}$ and any term other than δ^{ij} in $\widetilde{T}^{ij}_{(o)}$ would select out special direction in space, violating the assumption of isotropy.) The quantities ε and p are the proper energy density and pressure respectively.

As the comoving frame moves with velocity $\boldsymbol{v}$ relative to the lab frame, the lab coordinates x^μ and the comoving coordinates $\widetilde{x}^\nu$ are related by a Lorentz transformation

$$x^\mu = \Lambda^\mu{}_\nu(\boldsymbol{v})\widetilde{x}^\nu$$

where $\Lambda^\mu{}_\nu$ is a boost,

$$\Lambda^0{}_0 = \gamma, \quad \Lambda^i{}_0 = \Lambda^0{}_i = \gamma v^i, \quad \Lambda^i{}_j = \delta^{ij} + \frac{v^i v^j}{\boldsymbol{v}^2}(\gamma - 1) \tag{9.1.2}$$

with $\gamma = 1/\sqrt{1 - \boldsymbol{v}^2}$. The four-velocity u^μ of a fluid particle[1] is defined as $u^\mu = dx^\mu/d\tau$, where $d\tau$ is the proper time, $d\tau^2 = dt^2 - d\boldsymbol{x}^2$. So in the comoving frame the fluid particle at rest has the four-velocity

$$\widetilde{u}^\mu = (1, 0), \tag{9.1.3}$$

which transforms in the lab frame to

$$u^\mu = \Lambda^\mu{}_\nu(\boldsymbol{v})\widetilde{u}^\nu$$

giving

$$u^i = \gamma v^i, \quad u^0 = \gamma \; ; \quad u^\mu u_\mu = 1. \tag{9.1.4}$$

The second rank tensor $T^{\mu\nu}_{(o)}$ transforms as

$$T^{\mu\nu}_{(o)} = \Lambda^\mu{}_\rho(\boldsymbol{v})\Lambda^\nu{}_\sigma(\boldsymbol{v})\widetilde{T}^{\rho\sigma}_{(o)}$$

giving the components

$$T^{00}_{(o)} = \gamma^2(\varepsilon + p\boldsymbol{v}^2), \quad T^{i0}_{(o)} = \gamma^2(p + \varepsilon)v^i, \quad T^{ij}_{(o)} = p\delta^{ij} + \gamma^2(p + \varepsilon)v^i v^j \tag{9.1.5}$$

---

[1] Note the distinction between u^i and v^i : $u^i = dx^i/d\tau$, $v^i = dx^i/dt$.

which may be assembled into the tensor[2]

$$T^{\mu\nu}_{(o)} = -pg^{\mu\nu} + (p + \varepsilon)u^\mu u^\nu. \tag{9.1.6}$$

Besides $T^{\mu\nu}_{(o)}$ we also consider the particle number. If n is the particle number density in the comoving frame at a given space–time point, the particle current four-vector $\widetilde{N}^\mu$ at that point has components

$$\widetilde{N}^0_{(o)} = n, \qquad \widetilde{N}^i_{(o)} = 0. \tag{9.1.7}$$

Then, as before, we find in the lab frame

$$N^\mu_{(o)} = \Lambda^\mu{}_\nu(\boldsymbol{v})\widetilde{N}^\nu_{(o)}$$

giving

$$N^0_{(o)} = \gamma n, \qquad N^i_{(o)} = \gamma n v^i \tag{9.1.8}$$

or putting together

$$N^\mu_{(o)} = n u^\mu. \tag{9.1.9}$$

Real fluids are generally somewhat *imperfect*; their motion dissipates fluid kinetic energy into heat. Let it modify (9.1.6) and (9.1.9) by terms $\Delta T^{\mu\nu}$ and ΔN^μ respectively, when we have for such fluids

$$\begin{aligned} T^{\mu\nu} &= T^{\mu\nu}_{(o)} + \Delta T^{\mu\nu} \\ N^\mu &= N^\mu_{(o)} + \Delta N^\mu. \end{aligned} \tag{9.1.10}$$

We now have to state how ε, n, p and u^μ are affected by these correction terms. Generally, ε and n are taken as the *total* energy density and particle number density in the comoving frame

$$\widetilde{T}^{00} \equiv \varepsilon, \qquad \widetilde{N}^0 \equiv n \tag{9.1.11}$$

and p is defined by the same function of ε and n as in the perfect fluid case. Finally u^μ can be interpreted as velocity of energy transport [5] or particle transport [6]. We follow the latter interpretation, so that $\widetilde{N}^i$ also remains unaltered. So in the comoving frame the dissipative terms in $T^{\mu\nu}$ and N^μ are restricted by

$$\Delta \widetilde{T}^{00} = 0, \qquad \Delta \widetilde{N}^0 = \Delta \widetilde{N}^i = 0 \tag{9.1.12}$$

and therefore in a general Lorentz frame by

$$u_\mu u_\nu \Delta T^{\mu\nu} = 0, \qquad \Delta N^\mu = 0. \tag{9.1.13}$$

The motion of the fluid is governed by the conservation laws of energy–momentum and of particle number

$$0 = \partial_\nu T^{\mu\nu}, \qquad 0 = \partial_\mu N^\mu = \partial_\mu(n u^\mu). \tag{9.1.14}$$

---

[2] We could have obtained the tensor (9.1.6) by guessing its form and checking that it equals (9.1.1) in the comoving frame, when it must be valid in all Lorentz frames – an argument we shall use later.

For a perfect fluid, $T_{\mu\nu}$ is known and these laws immediately give the fundamental equations of relativistic hydrodynamics. For an imperfect fluid, $\Delta T_{\mu\nu}$ is unknown at this stage. But it is possible to turn the problem around and determine the most general form of $\Delta T_{\mu\nu}$ allowed by these conservation laws and the second law of thermodynamics [5, 6]. We describe this phenomenology in the next section.

9.2 Phenomenological Energy-Momentum Tensor

Let us multiply the first equation of (9.1.14) by u_μ to get

$$0 = u_\mu \partial_\nu T^{\mu\nu}_{(o)} + u_\mu \partial_\nu \Delta T^{\mu\nu}. \tag{9.2.1}$$

As $T^{\mu\nu}_{(0)}$ is given by (9.1.6), we can evaluate the first term

$$
\begin{aligned}
u_\mu \partial_\nu T^{\mu\nu}_{(o)} &\equiv -u^\nu \partial_\nu p + u_\mu \partial_\nu [u^\mu (p+\varepsilon) u^\nu] \\
&= -u^\nu \partial_\nu p + \partial_\nu \left(\frac{p+\varepsilon}{n} n u^\nu \right) \\
&= -u^\nu \partial_\nu p + n u^\nu \partial_\nu \left(\frac{p+\varepsilon}{n} \right) \\
&= n u^\nu \left[p \partial_\nu \left(\frac{1}{n} \right) + \partial_\nu \left(\frac{\varepsilon}{n} \right) \right]
\end{aligned}
\tag{9.2.2}
$$

where we use the kinematic relation, $1 = u_\mu u^\mu$ and so $0 = \partial_\nu (u_\mu u^\mu) = 2 u_\mu \partial_\nu u^\mu$ as well as the second equation of (9.1.14).

To interpret (9.2.2) we recall the second law of thermodynamics in the form

$$kT dS = p dV + dE \tag{9.2.3}$$

specifying the change in entropy S of a system arising from independent changes in volume V and internal energy E. (With Boltzmann's constant k multiplying T in (9.2.3), S is dimensionless.) To recognise the identity of the right-hand sides of (9.2.2) and (9.2.3), we divide the latter relation by Vn to bring the extensive quantities as measured per particle. Calling σ the entropy per particle, we can then write (9.2.2) as

$$u_\mu \partial_\nu T^{\mu\nu}_{(0)} = n u^\nu kT \partial_\nu \sigma = T \partial_\nu (n k \sigma u^\nu) \tag{9.2.4}$$

on noting again the conservation of particle current. Inserting this result in (9.2.1) we get

$$\partial_\mu (n k \sigma u^\mu) = -\frac{1}{T} u_\nu \partial_\mu \Delta T^{\mu\nu}. \tag{9.2.5}$$

For a proper identification of $\Delta T^{\mu\nu}$, we replace the derivative on $\Delta T^{\mu\nu}$ with a total derivative and compensating terms. Shifting the total derivative on the other side, we get

$$\partial_\mu S^\mu = -\frac{1}{T^2} \partial_\mu T u_\nu \Delta T^{\mu\nu} + \frac{1}{T} \partial_\mu u_\nu \Delta T^{\mu\nu} \tag{9.2.6}$$

where S^μ may be interpreted as the entropy current four-vector

$$S^\mu = nk\sigma u^\mu + \frac{1}{T}u_\nu \Delta T^{\mu\nu}. \tag{9.2.7}$$

Integrating (9.2.6) over space, the spatial derivative on the left side goes to zero as usual, and we see that the *rate* of change of entropy density S^0 is given by the right side of this equation. In the absence of $\Delta T^{\mu\nu}$ (perfect fluid), this rate is zero, S^0 remaining constant. If we include non-zero $\Delta T^{\mu\nu}$, it gives the rate, which, by the second law of thermodynamics, must be positive for all fluid configurations. Requiring this positivity, one can obtain the form of $\Delta T^{\mu\nu}$ itself [5]. To anticipate this form, we note that this rate depends linearly on the gradients of temperature and velocity. Hence different components of $\Delta T^{\mu\nu}$ must be given by such linear combinations of these gradients, which turn the rate into a sum of quadratic terms with positive coefficients. The gradients of other variables like ρ, p and n cannot be present in these expressions, as it would not ensure this positivity.

It is convenient to impose this positivity in the comoving frame. *Omitting tilde from now on*, we recall that in this frame $u^\mu = (1,0)$, $\partial_\mu u^0 = 0$, and $\Delta T^{00} = 0$. Then (9.2.6) becomes

$$\partial_\mu S^\mu = \frac{1}{T}\left(-\frac{1}{T}\partial_i T + \partial_0 u_i\right)\Delta T^{i0} + \frac{1}{T}\partial_i u_j \Delta T^{ij}. \tag{9.2.8}$$

We split ΔT^{ij} into traceless and trace parts

$$\Delta T^{ij} = \overline{\Delta T}^{\,ij} + \frac{1}{3}\delta^{ij}\Delta T_{kk} \tag{9.2.9}$$

where $\overline{\Delta T}$ is the traceless part

$$\overline{\Delta T}^{\,ij} = (\delta^{ik}\delta^{jl} - \frac{1}{3}\delta^{ij}\delta^{kl})\Delta T_{kl}. \tag{9.2.10}$$

(As usual, a sum over a repeated index is implied.) Inserting (9.2.9) in the second term of (9.2.8) and noting that only the symmetric and traceless part of $\partial_i u_j$ can couple to $\overline{\Delta T}^{\,ij}$, it becomes

$$\partial_i u_j \Delta T^{ij} = \frac{1}{2}(\partial_i u_j + \partial_j u_i - \frac{2}{3}\delta_{ij}\partial_k u_k)\overline{\Delta T}^{\,ij} + \partial_k u_k \frac{1}{3}\Delta T^{ii}. \tag{9.2.11}$$

The right side of (9.2.8) will remain always positive, if the components of $\Delta T^{\mu\nu}$ are taken as

$$\Delta T^{i0} = -\kappa\left(-\frac{1}{T}\partial^i T + \partial^0 u^i\right)$$

$$\overline{\Delta T}^{\,ij} = \eta\left(\partial^i u^j + \partial^j u^i - \frac{2}{3}\delta^{ij}\partial_k u_k\right)$$

$$\frac{1}{3}\Delta T^{ii} = \zeta \partial_k u_k \tag{9.2.12}$$

with coefficients κ, η and ζ all positive. Looking back at (9.2.5), we find that we could not have argued for positivity of $\partial_\mu S^\mu$, unless we took the total derivative on the other side. Also these expressions coincide with those for the non-relativistic theory [5], except for the $\partial^0 u^i$ term in ΔT^{i0}, which implies that in the relativistic case the heat flow is due to not only temperature gradient but also acceleration. So the constants κ, η and ζ can be identified as the coefficients of *heat conduction, shear viscosity* and *bulk viscosity.*

To write the tensor components (9.2.12) in an arbitrary frame, we introduce the projection tensor $\Delta_{\mu\nu}$

$$\Delta_{\mu\nu} = g_{\mu\nu} - u_\mu u_\nu, \qquad \Delta_{\mu\nu}u^\mu = \Delta_{\mu\nu}u^\nu = 0 \qquad (9.2.13)$$

and note the forms of u_μ and $\Delta_{\mu\nu}$ in the comoving frame

$$u_\mu = (1,0,0,0), \qquad \Delta_{\mu\nu} = \text{diag}\,(0,-1,-1,-1). \qquad (9.2.14)$$

We already used u^μ in (9.1.13) to convert the time index in the comoving frame to a general Lorentz index. In the same way we can use $\Delta_{\mu\nu}$ to deal with space indices. Then the equations (9.2.12) become

$$\Delta_{\alpha\mu}u_\nu\Delta T^{\mu\nu} = -\kappa\Delta_{\alpha\mu}Q^\mu$$

$$(\Delta_{\alpha\mu}\Delta_{\beta\nu} - \frac{1}{3}\Delta_{\alpha\beta}\Delta_{\mu\nu})\Delta T^{\mu\nu} = \eta\Delta_{\alpha\mu}\Delta_{\beta\nu}W^{\mu\nu}$$

$$-\frac{1}{3}\Delta_{\mu\nu}\Delta T^{\mu\nu} = -\zeta\partial_\lambda u^\lambda \qquad (9.2.15)$$

where we introduce

$$Q^\mu = -\frac{1}{T}\partial^\mu T + Du^\mu, \qquad D = u_\nu\partial^\nu$$

$$W^{\mu\nu} = \partial^\mu u^\nu + \partial^\nu u^\mu - \frac{2}{3}g^{\mu\nu}\partial_\alpha u^\alpha \qquad (9.2.16)$$

as the covariant form of the heat flow vector and shear tensor.

We thus arrive phenomenologically at the *form* of the energy–momentum tensor at the macroscopic level using relativistic hydrodynamics, leaving the transport coefficients undetermined. We shall now *calculate* this tensor statistically at the microscopic level using relativistic quantum field theory. As the entropy production points out, the statistical framework must incorporate situations away from equilibrium. The next section describes such a non-equilibrium framework.

9.3 Non-Equilibrium Statistical Theory

In previous chapters we treated systems in global thermal equilibrium, which are represented by the density operator

$$\rho_0 = \frac{\exp(-\beta H)}{\text{Tr}\,\exp(-\beta H)}, \qquad H = \int d^3x\,\mathcal{H}(\boldsymbol{x},t). \qquad (9.3.1)$$

But, as we already remarked in the last section, transport processes involve non-equilibrium thermal states. Accordingly we describe, following Zubarev [2], a generalisation of the density operator for such states[3]

$$\rho = \frac{\exp(-\int d^3x\, C(\boldsymbol{x}, t))}{\mathrm{Tr}\, \exp(-\int d^3x\, C(\boldsymbol{x}, t))} \tag{9.3.2}$$

where, by usual statistical arguments, $C(\boldsymbol{x}, t)$ can be written in the form

$$C(\boldsymbol{x}, t) = F^{\nu}(\boldsymbol{x}, t)\mathscr{T}_{0\nu}(\boldsymbol{x}, t), \qquad (?) \tag{9.3.3}$$

with

$$F^{\nu}(\boldsymbol{x}, t) = \beta(\boldsymbol{x}, t)u^{\nu}(\boldsymbol{x}, t). \tag{9.3.4}$$

Observe that β and u^{ν} are treated here as thermodynamic *fields*, which become constants in equilibrium situation. Also we are dealing here with the energy–momentum tensor *operator* with components $\mathscr{T}_{\mu\nu}$.

The expression (9.3.3) for $C(\boldsymbol{x}, t)$ is unphysical in that it implies non-equilibrium condition setting in suddenly at time t. It would be realistic, if this condition develops gradually over an interval of time prior to t. Mathematically it is convenient to take this interval as infinite [8]. So, noting $\epsilon \int_{-\infty}^{t} dt_1 e^{\epsilon(t_1 - t)} = 1$, we may properly redefine it as[4]

$$C(\boldsymbol{x}, t) = \epsilon \int_{-\infty}^{t} dt_1 e^{\epsilon(t_1 - t)} F^{\nu}(\boldsymbol{x}, t_1)\mathscr{T}_{0\nu}(\boldsymbol{x}, t_1), \qquad \epsilon \to 0+. \tag{9.3.5}$$

This definition eliminates oscillations and has a built-in retarded structure. As we always work in the thermodynamic limit ($V = L^3 \to \infty$), the limit $\epsilon \to 0$ is to be taken as usual at the end of the calculation.

As a check on the density operator (9.3.2), let us show that it satisfies the Liouville equation

$$\frac{d\rho}{dt} = 0, \tag{9.3.6}$$

in the Heisenberg picture, where the differentiation is carried out for both the time-dependent thermodynamic fields and field operators entering in the definition of ρ. All we have to show is that C is independent of time. Differentiating with respect to time, it gives

$$\frac{dC(\boldsymbol{x}, t)}{dt} = \epsilon F^{\nu}(\boldsymbol{x}, t)\mathscr{T}_{0\nu}(\boldsymbol{x}, t) - \epsilon^2 \int_{-\infty}^{t} dt_1 e^{\epsilon(t_1 - t)} F^{\nu}(\boldsymbol{x}, t_1)\mathscr{T}_{0\nu}(\boldsymbol{x}, t_1). \tag{9.3.7}$$

Here $\mathscr{T}_{0\nu}(\boldsymbol{x}, t)$ and $F^{\nu}(\boldsymbol{x}, t)$ are clearly finite, so that the right side of (9.3.7) goes to zero as $\epsilon \to 0^+$.

---

[3] For the leading contribution that we are interested in, it suffices to consider only the real line of the time contour.

[4] Zubarev compares this averaging with the one used by Gell-Mann and Goldberger in their formulation of quantum theory of scattering [7, 8].

To see the physical content of ρ and introduce an approximation scheme accordingly, we integrate $C(\boldsymbol{x}, t)$ by parts to get

$$C(\boldsymbol{x}, t) = F^{\nu}(\boldsymbol{x}, t) \mathscr{T}_{0\nu}(\boldsymbol{x}, t) - \int_{-\infty}^{t} dt_1 e^{\epsilon(t_1 - t)} (F^{\nu} \partial^0 \mathscr{T}_{0\nu} + \partial^0 F^{\nu} \mathscr{T}_{0\nu}). \quad (9.3.8)$$

Using the conservation of the energy–momentum operator, $\partial_{\mu} \mathscr{T}^{\mu\nu} = 0$, we can replace the term with time derivative of its components by space derivative. Next consider the space integral of $C(\boldsymbol{x}, t)$ and integrate this term with space derivative by parts. After dropping the total derivative, it can be put as

$$\int d^3 x\, C(\boldsymbol{x}, t) = \int d^3 x\, F^{\nu}(\boldsymbol{x}, t) \mathscr{T}_{0\nu}(\boldsymbol{x}, t) - \int d^3 x \int_{-\infty}^{t} dt_1 e^{\epsilon(t_1 - t)} \mathscr{T}_{\mu\nu}(\boldsymbol{x}, t_1) \partial^{\mu} F^{\nu}(\boldsymbol{x}, t_1)$$

$$\equiv A - B \qquad\qquad\qquad (9.3.9)$$

where A and B depend respectively on the thermodynamic fields and their gradients. The non-equilibrium density operator (9.3.2) is now given by

$$\rho = \frac{e^{-A+B}}{\text{Tr } e^{-A+B}}. \qquad (9.3.10)$$

We also define a *local* equilibrium density operator

$$\rho_l = \frac{e^{-A}}{\text{Tr } e^{-A}}. \qquad (9.3.11)$$

The notion of local equilibrium may be understood intuitively as follows. We think of dividing the macroscopic extension of the system into mesoscopic regions. Suppose the collision time of particles in the system is short compared to the relaxation time of the whole system. Then each mesoscopic region by itself will attain (quasi-)thermal equilibrium with local average values of thermodynamic fields, well before the whole system is in global thermal equilibrium. The density operator ρ_l describes this local equilibrium of the system. During the time between the local and global equilibrium, gradients of thermodynamic fields operate among the mesoscopic regions.

Extending our earlier notation for global equilibium with ρ_0, we denote the ensemble averages of an operator $\mathcal{O}(\boldsymbol{x}, t)$ with the three density operators as

$$\langle \mathcal{O} \rangle \equiv \text{Tr} \left(\rho_0 \mathcal{O} \right) \qquad \langle \mathcal{O} \rangle_l \equiv \text{Tr} \left(\rho_l \mathcal{O} \right) \qquad \text{and} \qquad \langle\!\langle \mathcal{O} \rangle\!\rangle \equiv \text{Tr} \left(\rho\, \mathcal{O} \right).$$

If the effects of thermodynamic fields are small, we can expand ρ given by (9.3.10) in powers of B. To first power[5] it is given by (C.16) of Appendix C

$$\rho = \left[1 + \int_0^1 d\tau \left(e^{-A\tau} B e^{A\tau} - \langle B \rangle_l \right) \right] \rho_l. \qquad (9.3.12)$$

---

[5] The density operator ρ and the viscosity coefficient are calculated to order B^2 in [9].

Using this expression for ρ, we can evaluate the non-equilibrium average of the operator $\mathcal{O}$

$$\langle\!\langle \mathcal{O}(x,t)\rangle\!\rangle = \langle \mathcal{O}(x,t)\rangle_l + \int_0^1 d\tau \langle \mathcal{O}\left(e^{-A\tau}Be^{A\tau} - \langle B\rangle_l\right)\rangle_l \tag{9.3.13}$$

giving the linear response of $\mathcal{O}$ to the thermodynamic forces. Inserting B from (9.3.9), it is convenient to write it as

$$\langle\!\langle \mathcal{O}(x,t)\rangle\!\rangle = \langle \mathcal{O}(x,t)\rangle_l + \int d^3x' \int_{-\infty}^t dt' e^{\epsilon(t'-t)}(\mathcal{O}(x,t), \mathcal{T}_{\rho\sigma}(x',t'))\partial^\rho F^\sigma((x',t')) \tag{9.3.14}$$

where the correlation function is defined as

$$(\mathcal{O}(x,t), \mathcal{T}_{\rho\sigma}(x',t')) = \int_0^1 d\tau \langle \mathcal{O}(x,t)(\overline{\mathcal{T}}_{\rho\sigma}(x',t') - \langle \mathcal{T}_{\rho\sigma}(x',t')\rangle_l))\rangle_l \tag{9.3.15}$$

with $\overline{\mathcal{T}}_{\mu\nu}(x',t') = e^{-A\tau}\mathcal{T}_{\mu\nu}(x',t')e^{A\tau}$. Note that the upper limit of t' integral is given by the time argument of the operator $\mathcal{O}$. Formula (9.3.14) relates a non-equilibrium average to one of local equilibrium.

9.4 Statistical Energy-Momentum Tensor

We can now use (9.3.14) to find statistically the linear response to the energy–momentum operator from the non-equilibrium system. For this purpose we first decompose the tensor operator into suitable components, using the vector and tensor available in the problem, namely u^μ and $\Delta^{\mu\nu}$ defined by (9.1.4) and (9.2.13). As an example, let us take a four-vector V_μ, which can be decomposed along u^μ and perpendicular to it

$$V^\mu = (V^\alpha u_\alpha)u^\mu + V_\alpha \Delta^{\alpha\mu}.$$

In the same way we can decompose any second-rank tensor, in particular, $\mathcal{T}^{\mu\nu}$ into the following components[6]

$$\varepsilon = u_\mu u_\nu \mathcal{T}^{\mu\nu}$$

$$p = -\frac{1}{3}\Delta_{\mu\nu}\mathcal{T}^{\mu\nu}$$

$$P_\alpha = \Delta_{\alpha\mu}u_\nu \mathcal{T}^{\mu\nu}$$

$$\pi_{\alpha\beta} = \left(\Delta_{\alpha\mu}\Delta_{\beta\nu} - \frac{1}{3}\Delta_{\alpha\beta}\Delta_{\mu\nu}\right)\mathcal{T}^{\mu\nu}. \tag{9.4.1}$$

---

[6] It is simple to check that this decomposition reproduces all the 10 independent components of $\mathcal{T}^{\mu\nu}$. The operators ε and p give two components. Then the four-vector P^μ has $4 - 1 = 3$ independent components, as $P_\alpha u^\alpha = 0$ gives one relation among them. Finally $\pi_{\alpha\beta}$ has $10 - 4 - 1 = 5$ independent components, as $\pi_{\alpha\beta}u^\alpha = 0$ and $\pi_\alpha^\alpha = 0$ give 4 and 1 relations among the components.

Here ε and p are now operators for energy and pressure, as can be seen in the comoving frame, where

$$\varepsilon = \mathscr{T}^{00}, \qquad p = \frac{1}{3}\mathscr{T}^{ii}.$$

By construction P_α and $\pi_{\alpha\beta}$ satisfy

$$P_\alpha u^\alpha = 0, \quad \pi_{\alpha\beta}u^\beta = 0, \quad \pi_{\alpha\beta}\Delta^{\beta\gamma} = 0, \quad \pi_\alpha^\alpha = 0. \tag{9.4.2}$$

The operator $\mathscr{T}^{\mu\nu}$ itself may be written in terms of the components (9.4.1) as

$$\mathscr{T}^{\mu\nu} = \varepsilon u^\mu u^\nu - p\Delta^{\mu\nu} + P^\mu u^\nu + P^\nu u^\mu + \pi^{\mu\nu} \tag{9.4.3}$$

which may be verified by inserting it in (9.4.1) and reproducing them. It is satisfying to see that the decomposition we find here for $\mathscr{T}^{\mu\nu}$ agrees with (9.2.15) for $\Delta T_{\mu\nu}$ obtained in the hydrodynamical analysis. We shall actually identify the two descriptions in the next section.

We are now in a position to express the non-equilibrium part B in the density operator in terms of the scalar, vector and tensor operators. Multiplying (9.4.3) with the derivative of (9.3.4), we get

$$\mathscr{T}_{\mu\nu}\partial^\mu F^\nu = \beta\pi_{\mu\nu}\partial^\mu u^\nu + \beta P_\mu(\beta^{-1}\partial^\mu\beta + Du^\mu) + \varepsilon D\beta - \beta p\partial_\mu u^\mu \tag{9.4.4}$$

with the differential operator $D = u^\mu\partial_\mu$, being the time derivative in the comoving frame. It is possible to relate $D\beta$ to $\partial_\mu u^\mu$. Since it is already of first order in the gradient, we can use (9.2.1) for the ideal fluid in this context

$$u_\mu\partial_\nu\langle\mathscr{T}^{\mu\nu}_{(o)}\rangle_l = 0 \tag{9.4.5}$$

giving

$$h\partial_\mu u^\mu + D\langle\varepsilon\rangle_l = 0 \tag{9.4.6}$$

where h is the enthalpy density, $h = \langle\varepsilon\rangle_l + \langle p\rangle_l$. As $\langle\varepsilon\rangle_l$ is a function of temperature only, we may write

$$D\langle\varepsilon\rangle_l = u^\mu\frac{\partial\langle\varepsilon\rangle_l}{\partial\beta}\,\partial_\mu\beta = D\beta\,\frac{\partial\langle\varepsilon\rangle_l}{\partial\beta}. \tag{9.4.7}$$

Inserting (9.4.7) into (9.4.6) we get

$$D\beta = -\frac{\partial\beta}{\partial\langle\varepsilon\rangle_l}\,h\partial_\mu u^\mu. \tag{9.4.8}$$

Further, using the Gibbs–Duhem relation of thermodynamics in the entropy representation [10]

$$h = -\beta\frac{\partial\langle p\rangle_l}{\partial\beta} \tag{9.4.9}$$

we get the required relation

$$D\beta = \beta\frac{\partial\langle p\rangle_l}{\partial\langle\varepsilon\rangle_l}\,\partial_\mu u^\mu. \tag{9.4.10}$$

Eliminating $D\beta$ from (9.4.4) with this relation, we get finally

$$\mathscr{T}_{\mu\nu}\partial^\mu F^\nu = \beta\pi_{\mu\nu}\partial^\mu u^\nu + \beta P_\mu(\beta^{-1}\partial^\mu\beta + Du^\mu) - \beta p'\partial_\mu u^\mu \qquad (9.4.11)$$

where

$$p'(\boldsymbol{x},t) = p(\boldsymbol{x},t) - \frac{\partial\langle p(\boldsymbol{x},t)\rangle_l}{\partial\langle\varepsilon(\boldsymbol{x},t)\rangle_l}\,\epsilon(\boldsymbol{x},t).$$

The quantities $\partial_\mu u^\nu$, $\beta^{-1}\partial^\mu\beta + Du^\mu$ and $\partial_\mu u^\mu$ may now be interpreted as thermodynamic forces conjugate to components $\pi_{\mu\nu}$, P_μ and p' of the energy-momentum tensor respectively. Note that the energy operator ε alone is absent from the decomposition (9.4.11), resulting in no linear response to it.

We can now write the linear responses of the operators $\pi_{\mu\nu}$, P_μ and p' in terms of correlation functions of the form (9.3.15). The latter may be simplified by invoking Curie's principle [1], which states that in an isotropic medium correlation functions between operators of different tensor structures vanish. We thus get

$$\langle\!\langle\pi_{\mu\nu}(\boldsymbol{x},t)\rangle\!\rangle = \int d^3x' \int_{-\infty}^t dt'\, e^{\epsilon(t'-t)}(\pi_{\mu\nu}(\boldsymbol{x},t),\pi_{\rho\sigma}(\boldsymbol{x}',t'))\beta(\boldsymbol{x}',t')\partial^\rho u^\sigma$$

$$\langle\!\langle P_\mu(\boldsymbol{x},t)\rangle\!\rangle = \int d^3x' \int_{-\infty}^t dt'\, e^{\epsilon(t'-t)}(P_\mu(\boldsymbol{x},t),P_\rho(\boldsymbol{x}',t'))\beta(\boldsymbol{x}',t')(\beta^{-1}\partial^\rho\beta + Du^\rho)$$

$$\langle\!\langle p(\boldsymbol{x},t)\rangle\!\rangle = \langle p(\boldsymbol{x},t)\rangle_l + \int d^3x' \int_{-\infty}^t dt'\, e^{\epsilon(t'-t)}(p'(\boldsymbol{x},t),p'(\boldsymbol{x}',t'))\beta(\boldsymbol{x}',t')\partial_\rho u^\rho.$$

$$(9.4.12)$$

In the first two equations above we use

$$\langle P_\mu\rangle_l = \langle\pi_{\mu\nu}\rangle_l = 0$$

which follow from parity conservation in the comoving system and hence in all frames. The last equation in (9.4.12) is actually for the operator p', where we use

$$\langle\!\langle p'\rangle\!\rangle - \langle p'\rangle_l = \langle\!\langle p\rangle\!\rangle - \langle p\rangle_l - \frac{\partial\langle p\rangle_l}{\partial\langle\varepsilon\rangle_l}(\langle\!\langle\varepsilon\rangle\!\rangle - \langle\varepsilon\rangle_l) = \langle\!\langle p\rangle\!\rangle - \langle p\rangle_l$$

as we already noted the absence of linear response to ε.

In the isotropic medium the tensor structure of the correlation functions can be further simplified. Going over to the comoving frame, we have

$$(P_k,P_l) = L_P\delta_{kl}$$

$$(\pi_{kl},\pi_{mn}) = L_\pi\frac{1}{2}(\delta_{km}\delta_{ln} + \delta_{kn}\delta_{lm} - \frac{2}{3}\delta_{kl}\delta_{mn}) \qquad (9.4.13)$$

where L_P and L_π are scalar functions and so frame independent. So in an arbitrary frame, these become

$$(P_\mu,P_\rho) = -L_P\Delta_{\mu\rho}$$

$$(\pi_{\mu\nu},\pi_{\rho\sigma}) = L_\pi\frac{1}{2}(\Delta_{\mu\rho}\Delta_{\nu\sigma} + \Delta_{\mu\sigma}\Delta_{\nu\rho} - \frac{2}{3}\Delta_{\mu\nu}\Delta_{\rho\sigma}). \qquad (9.4.14)$$

Contracting the indices we immediately find the scalars L_P and L_π

$$(P_\mu, P^\mu) = -3L_P, \qquad (\pi_{\mu\nu}, \pi^{\mu\nu}) = 5L_\pi \qquad (9.4.15)$$

If the thermodynamic forces vary slowly over the correlation lengths of the two-point functions, we may take these outside the integral. Also we note

$$(\Delta_{\mu\rho}\Delta_{\nu\sigma} + \Delta_{\mu\sigma}\Delta_{\nu\rho} - \frac{2}{3}\Delta_{\mu\nu}\Delta_{\rho\sigma})\partial^\rho u^\sigma$$

$$= \Delta_{\mu\rho}\Delta_{\nu\sigma}(\partial^\rho u^\sigma + \partial^\sigma u^\rho - \frac{2}{3}g^{\rho\sigma}\partial_\alpha u^\alpha) \equiv \Delta_{\mu\rho}\Delta_{\nu\sigma}W^{\rho\sigma},$$

$$\Delta_{\mu\rho}(\beta^{-1}\partial^\rho\beta + Du^\rho) \equiv \Delta_{\mu\rho}Q^\rho \qquad (9.4.16)$$

in the notation of (9.2.16). So the responses (9.4.12) simplify to

$$\langle\!\langle \pi_{\mu\nu}(\boldsymbol{x},t) \rangle\!\rangle = \beta(\boldsymbol{x},t)\int d^3x' \int_{-\infty}^{t} dt' e^{\epsilon(t'-t)} L_\pi(\boldsymbol{x},t;\boldsymbol{x}',t')\Delta_{\mu\rho}\Delta_{\nu\sigma}W^{\rho\sigma}$$

$$\langle\!\langle P_\mu(\boldsymbol{x},t) \rangle\!\rangle = -\beta(\boldsymbol{x},t)\int d^3x' \int_{-\infty}^{t} dt' e^{\epsilon(t'-t)} L_P(\boldsymbol{x}.t;\boldsymbol{x}',t')\Delta_{\mu\rho}Q^\rho$$

$$\langle\!\langle p(\boldsymbol{x},t) \rangle\!\rangle = \langle p(\boldsymbol{x},t)\rangle_l + \beta(\boldsymbol{x},t)\int d^3x' \int_{-\infty}^{t} dt' e^{\epsilon(t'-t)} (p'(\boldsymbol{x},t),p'(\boldsymbol{x}',t'))\partial_\rho u^\rho.$$

$$(9.4.17)$$

9.5 Transport Coefficients

In Section 9.2 we defined phenomenologically the transport coefficients for a real fluid in terms of the components of $\Delta T^{\mu\nu}$. Let us recall that $\Delta T^{\mu\nu}$ stands for the *deviation* in the value of the energy–momentum tensor from that for the ideal fluid. In the previous section we worked out the corrections to the same components (9.4.1) from the quantum and thermal fluctuations in the ideal fluid. We now identify the corrections (9.4.17) as the source of deviations (9.2.15) from ideal fluid, obtaining the transport coefficient 'fields' as[7]

$$\eta(\boldsymbol{x},t) = \frac{\beta(\boldsymbol{x},t)}{10}\int d^3x' \int_{-\infty}^{t} dt_1 e^{\epsilon(t_1-t)}(\pi_{\mu\nu}(\boldsymbol{x},t),\pi^{\mu\nu}(\boldsymbol{x}',t_1))$$

$$\kappa(\boldsymbol{x},t) = -\frac{\beta(\boldsymbol{x},t)}{3}\int d^3x' \int_{-\infty}^{t} dt_1 e^{\epsilon(t_1-t)}(P_\mu(\boldsymbol{x},t),P^\mu(\boldsymbol{x}',t_1))$$

$$\zeta(\boldsymbol{x},t) = \beta(\boldsymbol{x},t)\int d^3x' \int_{-\infty}^{t} dt_1 e^{\epsilon(t_1-t)}(p'(\boldsymbol{x},t),p'(\boldsymbol{x}',t_1)). \qquad (9.5.1)$$

Having obtained the expressions for the coefficient fields, let us work these out to lowest order, when all thermodynamic fields take constant values, given by their spatial averages, so that ρ_l goes over to ρ_0 [4]. With T representing any

---

[7] In this identification we extend the phenomenological results to mesoscopic regions.

of the operators $\pi_{\mu\nu}$, P_μ and p', the correlation functions of the form (9.3.15) appearing in (9.5.1) take the form

$$
\begin{aligned}
& (T(\boldsymbol{x},t), T(\boldsymbol{x}',t_1)) \\
& = \int_0^1 d\tau \{\langle T(\boldsymbol{x},t)e^{-\beta H\tau}T(\boldsymbol{x}',t_1)e^{\beta H\tau}\rangle - \langle T(\boldsymbol{x},t)\rangle\langle T(\boldsymbol{x}',t_1)\rangle\} \\
& = \frac{1}{\beta}\int_0^\beta d\tau\{\langle T(\boldsymbol{x},t)T(\boldsymbol{x}',t_1+i\tau)\rangle - \langle T(\boldsymbol{x},t)\rangle\langle T(\boldsymbol{x}',t_1)\rangle\} \qquad (9.5.2)
\end{aligned}
$$

on replacing $\beta\tau$ by τ and using the time displacement formula (4.1.4). To bring out its retarded structure, we note that the correlation vanishes as $t_1 \to -\infty$,

$$
\lim_{t_1\to-\infty}\langle T(\boldsymbol{x},t)T(\boldsymbol{x}',t_1+i\tau)\rangle \to \langle T(\boldsymbol{x},t)\rangle\langle T(\boldsymbol{x}',t_1+i\tau)\rangle. \qquad (9.5.3)
$$

Also recall that a one-point function is a constant, independent of its argument. We can then write (9.5.2) as

$$
(T(\boldsymbol{x},t), T(\boldsymbol{x}',t_1)) = \frac{1}{\beta}\int_0^\beta d\tau \int_{-\infty}^{t_1} dt' \frac{d}{dt'}\langle T(\boldsymbol{x},t)T(\boldsymbol{x}',t'+i\tau)\rangle. \qquad (9.5.4)
$$

As t' and τ occur in the combination $t'+i\tau$, we can replace the derivative with respect to t' by one with respect to $i\tau$, when the τ integral becomes trivial. We get

$$
(T(\boldsymbol{x},t), T(\boldsymbol{x}',t_1)) = -\frac{i}{\beta}\int_{-\infty}^{t_1} dt' \langle T(\boldsymbol{x},t)T(\boldsymbol{x}',t'+i\beta) - T(\boldsymbol{x},t)T(\boldsymbol{x}',t')\rangle. \qquad (9.5.5)
$$

Again using the time translation formula (in the reverse direction) and the cyclicity of the trace, the first term inside the integral becomes

$$
\begin{aligned}
\langle T(\boldsymbol{x},t)T(\boldsymbol{x}',t'+i\beta)\rangle &= \mathrm{Tr}\left(\rho_0 T(\boldsymbol{x},t)e^{-\beta H}T(\boldsymbol{x}',t')e^{\beta H}\right) \\
&= \langle T(\boldsymbol{x}',t')T(\boldsymbol{x},t)\rangle. \qquad (9.5.6)
\end{aligned}
$$

Further, the limits of t' and t_1 integrals restrict the times to $t' < t_1 < t$. We may then write the correlation function (9.5.5)

$$
(T(\boldsymbol{x},t), T(\boldsymbol{x}',t_1)) = \frac{1}{\beta}\int_{-\infty}^{t_1} dt' \langle T(\boldsymbol{x},t), T(\boldsymbol{x}',t')\rangle^R \qquad (9.5.7)
$$

in terms of the conventional retarded, thermal two-point function

$$
\langle T(\boldsymbol{x},t), T(\boldsymbol{x}',t')\rangle^R = i\theta(t-t')\langle[T(\boldsymbol{x},t), T(\boldsymbol{x}',t')]\rangle. \qquad (9.5.8)
$$

As it actually depends on the coordinate difference, we can set conveniently $\boldsymbol{x} = t = 0$. Then inserting (9.5.8) in each of equations (9.5.1), we finally get the coefficients as

$$
\eta = \frac{1}{10}\int_{-\infty}^0 dt_1 e^{\epsilon t_1}\int_{-\infty}^{t_1} dt' \int d^3x' \langle \pi_{\mu\nu}(0,0), \pi^{\mu\nu}(\boldsymbol{x}',t')\rangle^R
$$

$$\kappa = -\frac{1}{3} \int_{-\infty}^{0} dt_1 e^{\epsilon t_1} \int_{-\infty}^{t_1} dt' \int d^3x' \langle P_\mu(0,0), P^\mu(\boldsymbol{x}',t') \rangle^R$$

$$\zeta = \int_{-\infty}^{0} dt_1 e^{\epsilon t_1} \int_{-\infty}^{t_1} dt' \int d^3x' \langle p'(0,0), p'(\boldsymbol{x}',t') \rangle^R. \tag{9.5.9}$$

The three transport coefficients (9.5.9) can be simplified in the same way. Take the example of η, whose expression we rewrite as

$$\eta = \frac{1}{10} \int_{-\infty}^{0} dt_1 e^{\epsilon t_1} \int_{-\infty}^{t_1} dt' \int d^3x' \Pi_{11}^R(0,x'), \qquad x' = (\boldsymbol{x}',t') \tag{9.5.10}$$

where $\Pi_{11}^R(0,x')$ stands for

$$\Pi_{11}^R(x,x') = i\theta(t-t') \langle [\pi_{\mu\nu}(x), \pi^{\mu\nu}(x')] \rangle \tag{9.5.11}$$

with the first argument set equal to zero. The subscripts 11 remind us that the time arguments of the two-point function are on the real time axis. We can extract the t' dependence of $\Pi_{11}^R(0,x')$ in (9.5.10) by taking the Fourier transform

$$\Pi_{11}^R(x,x') = \int \frac{d^4q}{(2\pi)^4} e^{-iq\cdot(x-x')} \Pi_{11}^R(q). \tag{9.5.12}$$

Setting $x = 0$, the Fourier inversion gives

$$\Pi_{11}^R(q) = \int d^4x' \, e^{-iq\cdot x'} \Pi_{11}^R(0,x'). \tag{9.5.13}$$

To get the spatial integral appearing in (9.5.10), we set $\boldsymbol{q} = 0$ in (9.5.13)

$$\Pi_{11}^R(q_0) = \int_{-\infty}^{0} dt' e^{-iq_0 t'} \int d^3x' \Pi_{11}^R(0,x') \tag{9.5.14}$$

where the range of t' integration is given by the theta function in (9.5.11). Its temporal inversion gives

$$\int d^3x' \Pi_{11}^R(0,x') = \int_{-\infty}^{+\infty} \frac{dq_0}{2\pi} e^{iq_0 t'} \Pi_{11}^R(q_0). \tag{9.5.15}$$

We insert this result in (9.5.10), interchange integrals and regularise the t' integral by replacing q_0 in the exponential with $q_0 - i\epsilon$ to get

$$\eta = \frac{1}{10} \int_{-\infty}^{+\infty} \frac{dq_0}{2\pi} \Pi_{11}^R(q_0) \int_{-\infty}^{0} dt_1 e^{\epsilon t_1} \int_{-\infty}^{t_1} dt' e^{i(q_0 - i\epsilon)t'}. \tag{9.5.16}$$

The two time integrals can be done immediately, giving

$$\eta = -\frac{1}{10} \int_{-\infty}^{+\infty} \frac{dq_0}{2\pi} \Pi_{11}^R(q_0) \frac{1}{(q_0 - i\epsilon)^2}. \tag{9.5.17}$$

As seen from (9.5.14), $\Pi_{11}^R(q_0)$ is an analytic function in the upper-half q_0 plane. So we can close the integration line in (9.5.17) with a semicircle and evaluate the integral by the residue of the pole at $q_0 = i\epsilon$ to get the Kubo-type formula [11]

$$\eta = \frac{-i}{10} \frac{d}{dq_0} \Pi_{11}^R(q_0) \Big|_{q_0=0}. \tag{9.5.18}$$

Thus the coefficient depends only on the long wavelength and low frequency fluctuations taking place in the fluid in equilibrium.

9.6 Evaluation

We now evaluate (9.5.18) to find the coefficient of viscosity of the hadronic medium at low temperature ($T < 150\,\text{MeV}$) with all chemical potentials set equal to zero. Then, as we discussed in Chapter 6, pions will dominate the medium, the presence of more massive states like kaon being exponentially small. The pion system can be described in the framework of chiral perturbation theory, with the effective Lagrangian

$$\mathcal{L}_{\text{eff}} = \mathcal{L}_0 + \mathcal{L}_{\text{int}}, \quad \mathcal{L}_0 = \frac{1}{2} \partial_\mu \vec{\phi} \cdot \partial^\mu \vec{\phi} - \frac{M^2}{2} \vec{\phi} \cdot \vec{\phi} \tag{9.6.1}$$

and $\mathcal{L}_{\text{int}}$ consists of terms with (increasingly higher) derivatives of the pion field. So, unlike the case considered in Appendix B, these terms will produce additional terms in the the energy–momentum tensor, besides contributing to the $g^{\mu\nu}$ term. But to lowest order, in which we work here, we can ignore these interaction terms and write the tensor with the free Lagrangian,

$$\mathcal{T}^{\mu\nu} = \partial^\mu \vec{\phi} \cdot \partial^\nu \vec{\phi} - g^{\mu\nu} \mathcal{L}_0. \tag{9.6.2}$$

(The interaction terms are, of course, needed to find the complete propagator. In the calculation below, they appear in the decay width of pion in the medium.) Inserting (9.6.2) in (9.4.1), the $g^{\mu\nu}$ term drops out, giving

$$\pi_{\alpha\beta} = (\Delta_{\alpha\mu}\Delta_{\beta\nu} - \frac{1}{3}\Delta_{\alpha\beta}\Delta_{\mu\nu})\partial^\mu \vec{\phi} \cdot \partial^\nu \vec{\phi}. \tag{9.6.3}$$

In calculating two-point functions perturbatively for this problem, we must use complete propagators replacing the free ones used in the conventional perturbation expansion. This is the so-called skeleton expansion [4], where all self-energy insertions, which are already included in the propagator, must be removed from Feynman graphs to avoid double counting. Mathematically, it removes pinch singularities as we shall see below. Physically, it indicates the necessity of interactions to make the transport coefficients finite.

As already discussed in Section 4.5, we cannot calculate directly a retarded correlation function, like the one appearing in (9.5.18), in conventional perturbation theory. In the literature this difficulty is dealt with by introducing the corresponding time-ordered correlation function in imaginary time, which can be calculated perturbatively. The result is then continued analytically back to real time, getting the retarded function. In the real time method that we use here, such continuation from imaginary time is not needed. We simply calculate the

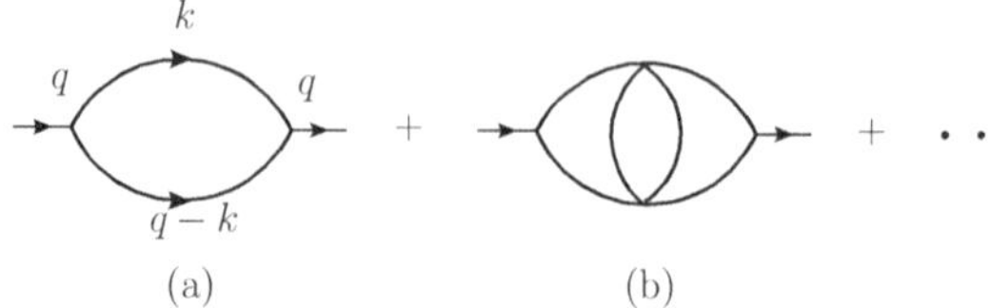

Figure 9.1 Skeleton expansion of the two-point function. We evaluate the lowest order graph (a); it is the same graph as Figure 5.3, now with complete propagators drawn in thick lines.

time-ordered function in real time and get the retarded one from it by using the relation (4.5.6).

Following the above discussion, we begin with the time-ordered two-point function

$$\Pi_{11}(q) = i \int d^4x\, e^{iq\cdot(x-y)} \langle T\pi_{\alpha\beta}(x)\pi^{\alpha\beta}(y)\rangle. \tag{9.6.4}$$

To lowest order in the skeleton expansion it gives (Figure 9.1)

$$\Pi_{11}(q) = -i \int \frac{d^4k}{(2\pi)^4} N(q,k) D'_{11}(k) D'_{11}(q-k) \tag{9.6.5}$$

where D'_{11} is the 11 component (4.4.18) of the complete, thermal propagator matrix. The factor N arises from derivatives on pion fields in (9.6.3). We now work in the comoving frame, where $u^\mu = (1,0)$. Then only space derivatives contribute to N, giving $N = 4|\mathbf{k}|^4$. The integral in (9.6.5) is essentially the same as (5.3.1) we treated earlier. In particular, the k_0 integral is the same as before with the result (5.3.5). The matrix, of which $\Pi_{11}(q)$ is the 11-component, can again be diagonalised with the diagonal elements given by an analytic function $\overline{\Pi}(k)$, whose real and imaginary parts are

$$\mathrm{Re}\overline{\Pi}(q) = -\int \frac{d^3k}{(2\pi)^3} N(\mathbf{k}) \int \frac{dk'_0}{2\pi} \rho(k'_0,\mathbf{k}) \int \frac{dk''_0}{2\pi} \rho(k''_0, \mathbf{q}-\mathbf{k})$$
$$\times \{(1+f')(1+f'') - f'f''\} \mathscr{P}\frac{1}{q_0 - k'_0 - k''_0} \tag{9.6.6}$$

where $\mathscr{P}$ denotes the principal value, and

$$\mathrm{Im}\overline{\Pi}(q) = \pi\epsilon(q_0) \int \frac{d^3k}{(2\pi)^3} N(\mathbf{k}) \int \frac{dk'_0}{2\pi} \rho(k'_0,\mathbf{k}) \int \frac{dk''_0}{2\pi} \rho(k''_0, \mathbf{q}-\mathbf{k})$$
$$\times \{(1+f')(1+f'') - f'f''\}\delta(q_0 - k'_0 - k''_0). \tag{9.6.7}$$

We can now build the analytic function $\overline{\Pi}(q)$ by adding the real and imaginary parts

$$\overline{\Pi}(q) = -\int \frac{d^3k}{(2\pi)^3} N(\mathbf{k}) \int \frac{dk'_0}{2\pi} \rho(k'_0,\mathbf{k}) \int \frac{dk''_0}{2\pi} \rho(k''_0, \mathbf{q}-\mathbf{k})$$
$$\times \frac{(1+f')(1+f'') - f'f''}{q_0 - k'_0 - k''_0 + i\epsilon(q_0)\eta}. \tag{9.6.8}$$

Having obtained this time-ordered amplitude, we use relation (4.5.6) to change the cut prescription and get the 11-component of the corresponding retarded amplitude

$$\Pi_{11}^R(q) = -\int \frac{d^3k}{(2\pi)^3} N(\boldsymbol{k}) \int \frac{dk_0'}{2\pi} \rho(k_0', \boldsymbol{k}) \int \frac{dk_0''}{2\pi} \rho(k_0'', \boldsymbol{q} - \boldsymbol{k})$$
$$\times \frac{(1 + f')(1 + f'') - f'f''}{q_0 - k_0' - k_0'' + i\eta}. \tag{9.6.9}$$

It is convenient to rewrite $(1 + f')(1 + f'')$ also as $f'f''$ by changing the signs of the integration variables k_0' and k_0'' (see footnote on page 127). Then setting $\boldsymbol{q} = 0$, we get the quantity appearing in (9.5.18)

$$\Pi_{11}^R(q_0) = \int \frac{d^3k}{(2\pi)^3} N(\boldsymbol{k}) \int \frac{dk_0'}{2\pi} \rho(k_0', \boldsymbol{k}) f(k_0') \int \frac{dk_0''}{2\pi} \rho(k_0'', \boldsymbol{k}) f(k_0'') W(q_0, k_0', k_0'')$$
$$\tag{9.6.10}$$

where

$$W(q_0) = \frac{1}{q_0 - k_0' - k_0'' + i\eta} - \frac{1}{q_0 + k_0' + k_0'' + i\eta}. \tag{9.6.11}$$

We see that the entire q_0 dependence of Π_{11}^R is contained in W given by (9.6.11). So we can carry out explicitly the derivative in (9.5.18). Before we do so, it is convenient to change the integration variables k_0' and k_0'' in (9.6.10) to $\overline{\omega}$ and ω defined by

$$\overline{\omega} = k_0' + k_0'', \qquad \omega = \frac{1}{2}(k_0' - k_0'')$$

the Jacobian of the transformation being unity. Then (9.5.18) gives

$$\eta = \frac{-i}{5} \int \frac{d^3k}{(2\pi)^3} N(\boldsymbol{k}) \int_{-\infty}^{+\infty} \frac{d\omega}{2\pi} \int_{-\infty}^{+\infty} \frac{d\overline{\omega}}{2\pi} F(\omega, \overline{\omega}) \left. \frac{dW}{dq_0} \right|_{q_0=0}. \tag{9.6.12}$$

Here F denotes products of ρ and f in (9.6.10) in the transformed variables

$$F(\overline{\omega}, \omega) = \rho(\frac{1}{2}\overline{\omega} + \omega)\rho(\frac{1}{2}\overline{\omega} - \omega)f(\frac{1}{2}\overline{\omega} + \omega)f(\frac{1}{2}\overline{\omega} - \omega) \tag{9.6.13}$$

where we suppress the $\boldsymbol{k}$ dependence of ρ. Now

$$\left. \frac{dW}{dq_0} \right|_{q_0=0} \equiv \frac{d}{dq_0} \left(\frac{1}{q_0 - \overline{\omega} + i\eta} - \frac{1}{q_0 + \overline{\omega} + i\eta} \right) \Bigg|_{q_0=0}$$
$$= -\frac{1}{(\overline{\omega} - i\eta)^2} + \frac{1}{(\overline{\omega} + i\eta)^2}$$
$$= \frac{d}{d\overline{\omega}} \left(\frac{1}{\overline{\omega} - i\eta} - \frac{1}{\overline{\omega} + i\eta} \right) = 2\pi i \frac{d\delta(\overline{\omega})}{d\overline{\omega}}. \tag{9.6.14}$$

Inserting this result in (9.6.12) and integrating by parts over $\overline{\omega}$, we get

$$\eta = \frac{1}{10} \int \frac{d^3k}{(2\pi)^3} N(\boldsymbol{k}) \int_{-\infty}^{+\infty} \frac{d\omega}{2\pi} H(\omega), \qquad H(\omega) = -\left. \frac{dF(\overline{\omega}, \omega)}{d\overline{\omega}} \right|_{\overline{\omega}=0}. \tag{9.6.15}$$

It remains to find the spectral function for the complete propagator and work out the integral over ω. As stated earlier, we find it by introducing the finite width of pion in medium. A simple way to do so is to take the spectral function (1.2.8) for the stable particle and express it with energy denominators

$$\rho_0(k_0, \boldsymbol{k}) = 2\pi\epsilon(k_0)\delta(k^2 - M^2)$$
$$= \frac{2\pi}{2E}\{\delta(k_0 - E) - \delta(k_0 + E)\}, \quad E = \sqrt{k^2 + M^2}$$
$$= \frac{1}{2iE}\left\{\frac{1}{k_0 - E - i\epsilon} - \frac{1}{k_0 - E + i\epsilon} - \left(\frac{1}{k_0 + E - i\epsilon} - \frac{1}{k_0 + E + i\epsilon}\right)\right\}$$
$$(9.6.16)$$

and then replace the infinitesimal ϵ by the half-width γ to get the 'complete' spectral function

$$\rho(k_0, \boldsymbol{k}) = \frac{1}{2iE}\left\{\frac{1}{k_0 - E - i\gamma} - \frac{1}{k_0 - E + i\gamma} - \left(\frac{1}{k_0 + E - i\gamma} - \frac{1}{k_0 + E + i\gamma}\right)\right\}$$
$$= \frac{1}{i}\left\{\frac{1}{(k_0 - i\gamma)^2 - E^2} - \frac{1}{(k_0 + i\gamma)^2 - E^2}\right\}. \qquad (9.6.17)$$

We now have $F(\overline{\omega}, \omega)$ explicitly, giving $H(\omega)$ as

$$H(\omega) = \frac{8\omega^2 e^{\beta\omega}}{(e^{\beta\omega} - 1)^2} \frac{\beta\gamma^2}{\{(\omega - i\gamma)^2 - E^2\}^2\{(\omega + i\gamma)^2 - E^2\}^2}. \qquad (9.6.18)$$

This function has double poles from the spectral functions at

$$\omega = \pm E + i\gamma, \quad \text{and} \quad \omega = \pm E - i\gamma$$

and from the distribution functions at

$$\omega = 2\pi m i/\beta, \quad m = \pm 1, \pm 2, \cdots$$

(The pole at $m = 0$ is removed by ω^2 in the numerator of $H(\omega)$.) The poles are shown in Figure 9.2.

We now see the reason for introducing the skeleton expansion. Had we used the conventional perturbation expansion with free prapagators, the half-width γ would be reduced to ϵ, which tends to zero at the end of calculation. Then the two sets of poles in the ω plane at $E \pm i\epsilon$ and $-E \pm i\epsilon$ would pinch the integration line twice, at E and $-E$, making the integral singular. But with complete propagators, the poles are $2i\gamma$ apart from each other, where γ, though small, is not zero and so the integral remains non-singular.

As $H(\omega)$ is a meromorphic function of ω and vanishes at least as ω^{-6} at infinity, the ω integral in (9.6.15) can be evaluated by closing the contour in the upper (or lower) half ω plane [12]. Choosing the upper-half plane, we find the leading term in the residues at each of the poles $\omega = \pm E + i\gamma$ and at the series of poles $\omega = 2\pi m i/\beta$ as given respectively by

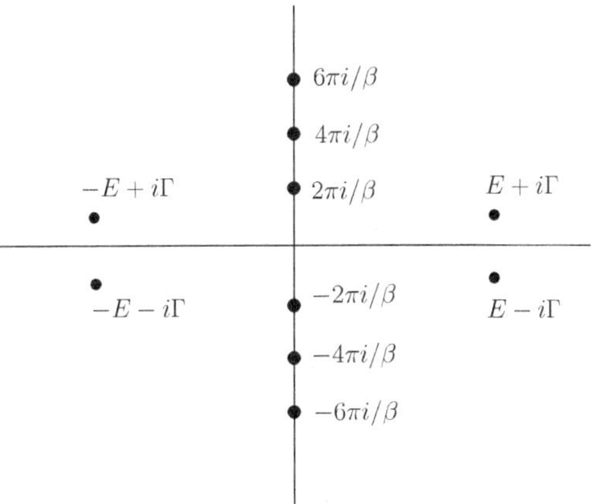

Figure 9.2 Double poles of $H(\omega)$ in the ω-plane. The poles at $\pm E \pm i\Gamma$ arise from the use of the skeleton expansion. Another set of infinite number of poles along the imaginary axis originate in the thermal distribution function.

$$-\frac{i}{8}\frac{\beta}{E^2\gamma}\frac{e^{\beta E}}{(e^{\beta E}-1)^2} \qquad \text{and} \qquad O\left(\frac{\beta^6\gamma^2}{m^6}\right)$$

for small γ and large m. The sum over m is convergent and so their contribution to η is $\sim \gamma^2$, which we neglect in comparison to $\sim 1/\gamma$ from the other two residues. We thus finally get

$$\eta = \frac{\beta}{10\pi^2}\int_0^\infty dk\, k^6 \frac{n(1+n)}{E^2\Gamma} \qquad n = \frac{1}{e^{\beta E}-1}, \qquad E = \sqrt{k^2+M^2} \qquad (9.6.19)$$

where we introduce the full spectral width $\Gamma = 2\gamma$. The function $\Gamma(k,\beta)$ is known from chiral perturbation theory [13], which we have worked out in Section 7.2. With this Γ the viscosity coefficient has been evaluated numerically from (9.6.19) in [12].

$$* \quad * \quad *$$

The convergence of the skeleton expansion has been examined in the $\lambda\phi^4$ theory by Jeon et al. [14, 15]. Leaving aside integrals and other factors, the terms in the expansion may schematically be represented as (Figure 9.1)

$$\eta \sim \frac{1}{\Gamma} + \frac{1}{\Gamma}\lambda^2\frac{1}{\Gamma} + \cdots \qquad (9.6.20)$$

Recalling $\Gamma \sim \lambda^2$, we see that the terms in the series are of the same order and our approximation by the first term breaks down. However, the series may be summed by an integral equation [9], which may then be solved numerically. Thus obtaining a reliable estimate of the transport coefficient by the response method becomes rather complicated. But the methods of kinetic theory avoid this complication with a simple integral equation (see Problem 9.1).

Problem

Problem 9.1: Derive the expression for the coefficient of shear viscosity in kinetic theory using the Boltzmann equation.

Solution: First, a word on notation. In Section 4.4 we introduced $f(k_0)$ as the single particle global equilibrium distribution function, which for positive k_0 gives the physical distribution $n(E)$. Following convention, we now let $f(x, k)$ denote an arbitrary (non-equilibrium) phase space distribution. The corresponding energy momentum tensor for the system is

$$T^{\mu\nu}(x) = \int \frac{d^3k}{(2\pi)^3} \frac{k^\mu k^\nu}{E} f(x, k). \tag{P9.1}$$

If the system departs slightly from local equilibrium distribution[8]

$$f^{(0)}(x, k) = \frac{1}{e^{\beta u_\mu(x) k^\mu} - 1}, \tag{P9.2}$$

we may write

$$f(x, k) = f^0(x, k)(1 + \phi(x, k)), \quad |\phi| \ll 1. \tag{P9.3}$$

Such an imperfect fluid contributes an additional amount $\Delta T^{\mu\nu}$ over that for the perfect fluid

$$\Delta T^{\mu\nu}(x) = \int \frac{d^3k}{(2\pi)^3} \frac{k^\mu k^\nu}{E} n(E)\phi(x, k) \tag{P9.4}$$

where we approximate $f^{(0)}$ with the global value $n(E)$.

Let us now work in the local rest frame of the fluid. As we are interested only in the coefficient of shear viscosity, we consider the traceless spatial part of $\Delta T^{\mu\nu}$, which must be of the form (9.2.12) in the text. It follows that $\phi(x, k)$ has also the same x-dependence

$$\phi(x, k) = C_{lm}(k) \left(\partial^l u^m + \partial^m u^l - \frac{2}{3}\delta^{lm}\vec{\partial} \cdot \vec{u} \right) \tag{P9.5}$$

where $C_{lm}(k) = C(E)k_l k_m$. (A term proportional to δ_{lm} does not contribute, as it would multiply a traceless tensor.) Insert (P9.5) in (P9.4) and use rotational invariance and symmetry of indices to replace

$$k^i k^j k^l k^m \longrightarrow \frac{1}{15}(\delta^{ij}\delta^{lm} + \delta^{il}\delta^{jm} + \delta^{im}\delta^{jl})|\boldsymbol{k}|^4.$$

We then compare the resulting expression for $\Delta \overline{T}^{ij}$ with (9.2.12) to get η as

$$\eta = \frac{2}{15} \int \frac{d^3k}{(2\pi)^3} |\boldsymbol{k}|^4 \frac{n}{E} C(E) \tag{P9.6}$$

involving the unknown function $C(E)$.

---

[8] As we are not interested in the coefficient of heat conduction, we ignore the x-dependence of $\beta(x)$.

So far we have used essentially the kinematics of the problem. To find $C(E)$ we need the dynamics contained in the Boltzmann equation, determining the space–time evolution of $f(x, k)$. Here we include only two-body scattering processes, $k + 2 \to 3 + 4$, labelling the first particle by its momentum. Then the Boltzmann equation for $f(x, k)$ reads

$$\frac{k^\mu}{E} \partial_\mu f(x, k) = \frac{1}{2} \int d\Gamma_{k2 \to 34} F_{k2 \to 34} \tag{P9.7}$$

where $d\Gamma_{k2 \to 34}$ involves the differential transition rate (E.11) for particles of momenta k and k_2 to scatter into k_3 and k_4,

$$d\Gamma_{k2 \to 34} = \frac{1}{2E} |M(k, k_2 \to k_3, k_4)|^2 d\Omega_{2,34} \tag{P9.8}$$

in the notation of (E.20) and (E.21), and

$$F_{k2 \to 34} = (1 + f_k)(1 + f_2) f_3 f_4 - f_k f_2 (1 + f_3)(1 + f_4) \tag{P9.9}$$

describes the medium dependence of the transition rate. The factor $\frac{1}{2}$ compensates the double counting in the final state phase space of two identical particles.

We now solve (P9.7) to first order to find $C(\omega)$. Then it suffices to replace $f(x, k)$ by $f^{(0)}(x, k)$, when the left side becomes

$$\frac{k^\mu}{E} \partial_\mu f^{(0)}(x, k) = -\frac{\beta}{E} n_k (1 + n_k) k^\mu k^\nu \frac{1}{2} (\partial_\mu u_\nu + \partial_\nu u_\mu) \tag{P9.10}$$

On the right side we have

$$F_{k2 \to 34}$$
$$= -(1 + n_k)(1 + n_2) n_3 n_4 \left[\frac{\phi(x, k)}{1 + n_k} + \frac{\phi(x, k_2)}{1 + n_2} - \frac{\phi(x, k_3)}{1 + n_3} - \frac{\phi(x, k_4)}{1 + n_4} \right] \tag{P9.11}$$

the terms without ϕ cancelling out by virtue of the equilibrium relation

$$n_k n_2 (1 + n_3)(1 + n_4) = (1 + n_k)(1 + n_2) n_3 n_4. \tag{P9.12}$$

Again going over to the comoving frame and equating only the traceless part of the tensor on both sides of (P9.7) we get an inhomogeneous linear integral equation for $C(E)$,

$$\beta \left(k_i k_j - \frac{1}{3} \delta_{ij} k^2 \right) = \frac{1}{2} \int d\Gamma_{k2 \to 34} \frac{(1 + n_2) n_3 n_4}{n_k}$$
$$\left[\frac{C(k)_{ij}}{1 + n_k} + \frac{C(k_2)_{ij}}{1 + n_2} - \frac{C(k_3)_{ij}}{1 + n_3} - \frac{C(k_4)_{ij}}{1 + n_4} \right] \tag{P9.13}$$

where $C(k)_{ij} = C(E) k_i k_j$, $C(k_2)_{ij} = C(\omega_2) k_{2i} k_{2j}$, etc. At low temperature the dynamical input in the Boltzmann equation is reliable. So the solution for $C(E)$ obtained from (P9.13), when inserted in (P9.6), would provide a quantitative result for the shear viscosity.

Finally, let us show that if we solve Boltzmann equation directly for $C(E)$ in the *relaxation time* approximation, we recover the field theory result (9.6.19) [16]. This approximation consists in assuming that particles of all momenta are in equilibrium except those with the particular momentum k. In that case

$$F_{k2\to34} = n_2(1+n_3)(1+n_4) - (1+n_2)n_3n_4\}n_k\phi(x,k). \tag{P9.14}$$

Noting the formula for the decay rate of particles of momentum k,

$$\Gamma_k = \frac{1}{2}\int d\Gamma_{k2\to34}\{n_2(1+n_3)(1+n_4) - (1+n_2)n_3n_4\} \tag{P9.15}$$

and (P9.9) above, we get the Boltzmann equation in the comoving frame as

$$\frac{\beta}{E}(1+n_k)k^ik^j\frac{1}{2}(\partial^i u^j + \partial^j u^i - \frac{2}{3}\delta^{ij}\vec{\partial}\cdot\vec{u}) = \Gamma_k\phi(x,k). \tag{P9.16}$$

Inserting (P9.5) for $\phi(x,k)$, we get

$$C = \frac{\beta(1+n_k)}{2E\Gamma_k}. \tag{P9.17}$$

Then (P9.6) gives the coefficient of viscosity as

$$\eta = \frac{\beta}{15}\int \frac{d^3k}{(2\pi)^3}\frac{k^4n(1+n)}{E^2\Gamma} \tag{P9.18}$$

for a single species of particles. For pions we need to multiply it by 3, reproducing (9.6.19) derived from quantum field theory.

References

[1] S.R. de Groot, W.A. van Leeuwen and Ch.G. van Weert, *Relativistic Kinetic Theory*, North-Holland Publishing Company (1980). See also K. Huang, *Statistical Mechanics*, John Wiley and Sons (1975).

[2] D.N. Zubarev, *Nonequilibrium Statistical Thermodynamics*, Plenum, New York (1974).

[3] S. Weinberg, *Gravitation and Cosmology*. John Wiley and Sons, New York (1972). See also S. Weinberg, Astroph. J., **168**, 175 (1971).

[4] A. Hosoya, M.-A. Sakagami and M. Takao, Ann. Phys. **154**, 229 (1982).

[5] L.D. Landau and E.M. Lifshitz, *Fluid Mechanics*, Second Edition, Butterworth-Heinemann (2005).

[6] C. Eckart, Phys. Rev., **58**, 919 (1940).

[7] M. Gell-Mann and M.L. Goldberger, Phys. Rev. **91**, 398 (1953).

[8] M.L. Goldberger and K.M. Watson, *Collision Theory*, John Wiley and Sons Ltd (New York), (1964).

[9] M. E. Carrington, H. Defu and R. Kobes, Phys. Rev. **D64**, 025001 (2001).

[10] H.B. Callen, *Thermodynamics*, Second Edition, John Wiley and Sons (1986).

[11] R. Kubo, J. Phys. Soc. Jap. **12**, 570 (1957).

[12] R. Lang, N. Kaiser and W. Weise, Eur. Phys. J. A **48**, 109 (2012).

[13] J. Goity and H. Leutwyler, Phys. Lett. **B228**, 517 (1989).
[14] S. Jeon, Phys. Rev. **D 52**, 3591 (1995).
[15] S. Jeon and L.G. Yaffe, Phys. Rev. **D 53**, 5799 (1996).
[16] P. Chakraborty and J.I. Kapusta, Phys. Rev. **C 83**, 014906 (2011).

Appendix A

General Fields

In the text, fields for arbitrary spin are denoted by $\psi_l(x)$. We now describe the structure of the index l and the associated supplementary conditions to reduce the independent field components to the required number.

A particle of spin j is described by a field with $(2j+1)$ independent components. In the conventional description [1], a field for integer spin k is represented by a completely symmetric tensor of rank k. Similarly a field for half-integer spin $(k+\frac{1}{2})$ is represented by such a tensor with an additional Dirac spinor index. The extra components introduced thereby are eliminated by imposing supplementary conditions as we describe below. In counting the components of such fields, it will be useful to note that the number of independent components of a symmetric tensor $T_{\mu_1\mu_2\cdots\mu_k}$ of rank k in d dimensions is [2]

$$\left\{ \begin{array}{c} (d+k-1)! \\ k! \end{array} \right\} = \frac{(d+k-1)!}{(d-1)!k!}. \tag{A.1}$$

Consider first bosonic fields described by symmetric tensors $\phi_{\mu_1\cdots\mu_k}$ of rank k in four dimensions. From (A.1) we get its number of independent components as $\dfrac{(k+3)!}{3!k!}$. To get rid of the extra components, one imposes two subsidiary conditions,

$$\phi^{\mu}{}_{\mu\mu_3\cdots\mu_k} = 0 \qquad \text{(traceless)}, \tag{A.2}$$

$$p^{\mu}\phi_{\mu\mu_2\cdots\mu_k} = 0 \qquad \text{(divergenceless)}. \tag{A.3}$$

In the first condition (A.2) the symmetric tensor is of rank $(k-2)$ giving rise to $\dfrac{(k+1)!}{3!(k-2)!}$ relations, so that the tracelessness condition reduces the above number of components to

$$\frac{(k+3)!}{3!k!} - \frac{(k+1)!}{3!(k-2)!} = (k+1)^2. \tag{A.4}$$

Then, following the same steps as we do for the original tensor ϕ of rank k, we see that the tensor of rank $(k-1)$ in the second condition (A.3) gives k^2 independent relations and hence reduces the independent components by the same number. Thus the total number of independent components of the original tensor indeed becomes $(2k+1)$.

For fermionic fields we start with a spinor-tensor $\psi_{A\mu_1\cdots\mu_k}$ where A is a Dirac spinor index, which is generally suppressed, but it will be useful here to display this index. Its components are reduced by the subsidiary conditions

$$(\not{p} - m)_{AB}\psi_{B\mu_1\cdots\mu_k} = 0, \qquad \text{(Dirac equation)} \tag{A.5}$$

$$(\gamma^\mu)_{AB}\psi_{B\mu\mu_2\cdots\mu_k} = 0. \tag{A.6}$$

The multiplicity associated with the Dirac index is an invariant and so can be seen easily in the rest frame. Writing

$$\psi_{A\mu_1\cdots\mu_k} = \left(\begin{array}{c} \varphi_{a\mu_1\cdots\mu_k} \\ \chi_{a\mu_1\cdots\mu_k} \end{array} \right), \quad a = 1, 2$$

and using the Weyl representation (1.3.5) for the γ-matrices, the Dirac equation (A.5) in the rest frame makes the two component spinors φ and χ to be equal. So we have only two independent components as far as the Dirac index is concerned. Thus at this stage ψ has $\dfrac{2(k+3)!}{3!k!}$ independent components. The condition (A.6) becomes

$$\left(\begin{array}{c} \varphi_{a0\mu_2\cdots\mu_k} + \sigma^i\varphi_{ai\mu_2\cdots\mu_k} \\ \varphi_{a0\mu_2\cdots\mu_k} - \sigma^i\varphi_{ai\mu_2\cdots\mu_k} \end{array} \right) = 0,$$

giving two independent conditions

$$\varphi_{a0\mu_2\cdots\mu_k} = 0 \tag{A.7}$$

$$\sigma^i\varphi_{ai\mu_2\cdots\mu_k} = 0 \quad \mu_s \neq 0, \quad s = 2, 3, \cdots, k. \tag{A.8}$$

(In (A.8) we exclude $\mu_s = 0$, as the corresponding components are already removed by (A.7).) Now the condition (A.7) involves a tensor of rank $(k-1)$ in four dimensions and so sets $\dfrac{2(k+2)!}{3!(k-1)!}$ independent components to zero. The condition (A.8) also involves a tensor of rank $(k-1)$, but in three dimensions, so as to give $\dfrac{2(k+1)!}{2!(k-1)!}$ independent relations. We are thus left with

$$\frac{2(k+3)!}{3!k!} - \frac{2(k+2)!}{3!(k-1)!} - \frac{2(k+1)!}{2!(k-1)!} = 2(k+1) \tag{A.9}$$

independent components of ψ, as desired.

It should be noted that the supplementary conditions (A.2) and (A.3) of the integral spin case appear as consequences of (A.5) and (A.6) for the half-integral case, as can be seen by multiplying the latter by a gamma matrix and using the anticommutation relation for them.

References

[1] W. Rarita and J. Schwinger, Phys. Rev. **60**, 61 (1941).

[2] M. Hamermesh, *Group Theory and its Applications to Physical Problems*, Addison-Wesley, Reading, MA (1962).

Appendix B
Global Symmetries

Here we review the topic of symmetry currents. Both the dynamical (Euler–Lagrange) equations for fields $\psi_l(x)$ as well as the currents may be obtained from a variational principle [1]. This is the principle of stationary (least) action, where the action

$$I[\psi] = \int d^4x \, \mathscr{L}(\psi_l(x), \partial_\mu \psi_l(x)) \tag{B.1}$$

is extremised with respect to infinitesimal variations of the fields. These variations are different in the two cases. For the dynamical equations, we consider the variation

$$\psi_l(x) \to \psi'_l(x) = \psi_l(x) + \delta\psi_l(x) \tag{B.2}$$

where $\delta\psi_l(x)$ are *arbitrary*. It changes the action (B.1) by

$$\delta I[\psi] \equiv \int d^4x \left\{ \mathscr{L}(\psi_l + \delta\psi_l, \partial_\mu\psi_l + \delta[\partial_\mu\psi_l]) - \mathscr{L}(\psi_l, \partial_\mu\psi_l) \right\} \tag{B.3}$$

$$= \int d^4x \left\{ \frac{\partial\mathscr{L}}{\partial\psi_l(x)} \delta\psi_l(x) + \frac{\partial\mathscr{L}}{\partial[\partial_\mu\psi_l(x)]} \delta[\partial_\mu\psi_l(x)] \right\}. \tag{B.4}$$

Recognising $\delta[\partial_\mu\psi_l(x)] = \partial_\mu[\delta\psi_l(x)]$, we carry out partial integration of the second term. It gives a total derivative, which may be written as a surface integral over the space–time boundary at infinity. Assuming $\delta\psi_l(x)$ to vanish as $t \to \pm\infty$, it goes to zero, giving

$$\delta I[\psi] = \int d^4x \left\{ \frac{\partial\mathscr{L}}{\partial\psi_l(x)} - \partial_\mu\left(\frac{\partial\mathscr{L}}{\partial[\partial_\mu\psi_l(x)]} \right) \right\} \delta\psi_l(x). \tag{B.5}$$

Now the principle of least action gives the familiar Euler–Lagrange equation for $\psi_l(x)$.

For the symmetry currents, we consider variations from (global) transformations of the form [1]

$$\psi_l(x) \to \psi'_l(x) = \psi_l(x) + i\epsilon F_l(x) \tag{B.6}$$

where ϵ is infinitesimal, independent of x and $F_l(x)$ are *specific* functions of fields and possibly their derivatives at x. It is a global symmetry transformation, if it keeps the action (B.1) invariant (stationary), $\delta I[\psi] = 0$, *without* requiring the dynamical equations to be satisfied. To find the resulting current, we consider the same transformation (B.6), but with ϵ an arbitrary function of x

$$\psi_l(x) \rightarrow \psi'_l(x) = \psi_l(x) + i\epsilon(x)F_l(x). \tag{B.7}$$

As (B.7) is *not* a symmetry of the action, the change in the action will not be zero and will be of the form

$$\delta I[\psi] = -\int d^4x\, J_\mu(x)\partial^\mu\epsilon(x) \tag{B.8}$$

so as to vanish for constant ϵ. Now let the fields satisfy the dynamical equations, obtained by variation (B.2), which of course includes (B.6) and (B.7). Hence the change (B.8) of the action must vanish. Integrating by parts, it gives the conserved current

$$\partial^\mu J_\mu(x) = 0. \tag{B.9}$$

It is generally possible to write an explicit formula for the current in terms of the Lagrangian, but we shall not review it. Instead, given a Lagrangian with a global symmetry, we shall read off the current from (B.8). For an Abelian symmetry, the normalisation of the current is left undetermined. For non-Abelian symmetries, the canonical normalisation is fixed by the Lie algebra of the generators of the symmetry group [1]. Integrating (B.9) over space, we get the charge as a constant of motion

$$\frac{dQ}{dt} = 0, \qquad Q = \int d^3x\, J^0(x). \tag{B.10}$$

We thus arrive at *Noether's theorem*: corresponding to each independent infinitesimal symmetry transformation of the action, we have a conserved current and an associated constant of motion.

Here are some examples.
A) Consider the scalar field Lagrangian

$$\mathscr{L} = \frac{1}{2}\partial_\mu\phi\partial^\mu\phi - V(\phi) \tag{B.11}$$

whose action is invariant under space–time translation

$$\phi(x) \rightarrow \phi'(x) = \phi(x + \epsilon) = \phi(x) + \epsilon_\alpha\partial^\alpha\phi(x). \tag{B.12}$$

It is of the form (B.6) with four independent parameters ϵ_α and corresponding four transformation functions $F^\alpha = -i\partial^\alpha\phi$. So we expect four independent conserved currents forming the energy–momentum tensor T^α_μ.

The change in the action under (B.12) is

$$\delta I[\phi] = \epsilon_\alpha \int d^4x \, \partial^\alpha \mathscr{L}.$$

It can be written as a four-dimensional surface integral and so goes to zero, showing that the action indeed remains invariant under this transformation. If we now make the parameters space–time dependent, the kinetic term in (B.11) produces an additional contribution, giving

$$\delta I[\phi] = \int d^4x (\epsilon_\alpha \partial^\alpha \mathscr{L} + \partial^\mu \epsilon_\alpha \partial^\alpha \phi \partial_\mu \phi).$$

Integrating the first term by parts, it becomes

$$\delta I[\phi] = -\int d^4x (\delta^\alpha_\mu \mathscr{L} - \partial^\alpha \phi \partial_\mu \phi) \partial^\mu \epsilon_\alpha$$

which is of the form (B.8), when we can read off the energy–momentum tensor

$$T^\alpha_\mu = \delta^\alpha_\mu \mathscr{L} - \partial^\alpha \phi \partial_\mu \phi. \tag{B.13}$$

B) We next consider internal symmetries, which keep $\mathscr{L}$ invariant. Then the currents may be obtained directly from the variations of $\mathscr{L}$

$$\delta \mathscr{L} = -J^\mu \partial_\mu \epsilon. \tag{B.14}$$

a) Consider the two-component scalar field theory, whose symmetry we already found in Problem 1.4. Defining $\Phi = (\phi_1, \phi_2)^T$, the Lagrangian may be written as

$$\mathscr{L}(\Phi) = \frac{1}{2}\partial_\mu \Phi^T \partial^\mu \Phi - V(\Phi^T \Phi). \tag{B.15}$$

It is invariant under

$$\Phi \to \Phi' = e^{-i\sigma_2 \theta} \Phi$$

where σ_2 is the second Pauli matrix. With ϵ denoting infinitesimal, space–time-dependent paramater θ, we get

$$\delta \mathscr{L} = \frac{i}{2}\partial_\mu \epsilon \left(\Phi^T \sigma_2 \partial^\mu \Phi - \partial^\mu \Phi^T \sigma_2 \Phi\right) = \partial_\mu \epsilon \phi_1 \overleftrightarrow{\partial_\mu} \phi_2$$

giving the current

$$j^\mu = -\phi_1 \overleftrightarrow{\partial_\mu} \phi_2. \tag{B.16}$$

Equivalently, we can find the same current by considering the non-hermitian field $\phi = \frac{1}{\sqrt{2}}(\phi_1 + i\phi_2)$, when the above Lagrangian becomes

$$\mathscr{L}(\phi) = \partial_\mu \phi^\dagger \partial^\mu \phi - V(\phi^\dagger \phi) \tag{B.17}$$

which is invariant under $\phi \to \phi' = e^{i\theta}\phi$. Then $\delta\mathscr{L}(\phi) = -i\partial_\mu\epsilon\phi^\dagger \overset{\leftrightarrow}{\partial^\mu} \phi$, giving the current

$$j^\mu = i\phi^\dagger \overset{\leftrightarrow}{\partial^\mu} \phi \tag{B.18}$$

in agreement with (B.16).

b) The massless QCD Lagrangian (3.1.1)

$$\mathscr{L} = \bar{q} i\gamma^\mu (\partial_\mu - iG_\mu) q$$

is invariant under (3.1.8):

$$q \to q' = [1 + i(\vec{\epsilon}_V \cdot \vec{t} + \vec{\epsilon}_A \cdot \vec{t}\gamma^5) + \cdots]q.$$

Then, for x dependent parameters, we get

$$\delta\mathscr{L} = -\bar{q}\gamma^\mu t^a q\partial_\mu\epsilon_V^a - \bar{q}\gamma^\mu\gamma^5 t^a q\partial_\mu\epsilon_A^a$$

giving the vector and axial-vector currents (3.1.11).

c) The effective Lagrangian (3.4.6)

$$\mathscr{L}_{\text{eff}}^{(2)} = \frac{F^2}{4} \operatorname{tr}(\partial_\mu U \partial^\mu U^\dagger)$$

is invariant under (3.3.5): $U \to U' = U + i\vec{\epsilon}_R \cdot \vec{t}U - iU\vec{\epsilon}_L \cdot \vec{t} + \cdots)$. Then

$$\delta\mathscr{L}_{\text{eff}}^{(2)} = -i\frac{F^2}{2} \operatorname{tr}(t_a\partial_\mu U U^\dagger)\partial^\mu\epsilon_R^a + i\frac{F^2}{2} \operatorname{tr}(t_a U^\dagger\partial_\mu U)\partial^\mu\epsilon_L^a$$

getting the right- and left-handed currents

$$J_{R,a}^\mu = i\frac{F^2}{2} \operatorname{tr}(t_a\partial^\mu U U^\dagger) \qquad J_{L,a}^\mu = -i\frac{F^2}{2} \operatorname{tr}(t_a U^\dagger\partial^\mu U).$$

The vector and axial-vector currents are now given by
$(t_a = \tau_a/2)$

$$\mathscr{V}_a^\mu(x) = J_{R,a}^\mu + J_{L,a}^\mu = i\frac{F^2}{4} \operatorname{tr}(\tau_a[\partial^\mu U, U^\dagger])$$

$$\mathscr{A}_a^\mu(x) = J_{R,a}^\mu - J_{L,a}^\mu = i\frac{F^2}{4} \operatorname{tr}(\tau_a\{\partial^\mu U, U^\dagger\}). \tag{B.19}$$

*   *   *

We derived the Ward identity (3.13.7) following from vector current conservation, which in turn is obtained from the global symmetry of the QCD Lagrangian under $SU(2)_V$. A systematic method to derive this and other Ward identities is to promote the symmetry of the Lagrangian to a local one by introducing external vector (and axial vector) fields, as we did in Section 3.8. Then these identities may be obtained from the gauge invariance of the generating functional, with no further assumptions, like current commutation relations [2].

Let us rederive the Ward identity (3.13.8) by this method. Consider the generating functional

$$W[v] = \langle 0|T \exp\left(i \int d^4x\, v^i_\mu(x) V^\mu_i(x) \right) |0\rangle \tag{B.20}$$

where we switch off all other external fields except the vector field. Gauge invariance of W implies[1]

$$W[v'_\mu] = W[v_\mu] \tag{B.21}$$

where the external fields v_μ and v'_μ are gauge transforms of each other. Note that it reflects only the symmetry of the theory and does not depend in any way on its dynamical structure.

Expanding W to second order in the external field, we get

$$W[v] = i \int d^4\, x v^i_\mu(x)\langle 0|V^\mu_i(x)|0\rangle$$

$$+ \frac{i^2}{2} \int d^4x d^4y\, v^i_\mu(x) v^j_\nu(y)\langle 0|T V^\mu_i(x) V^\nu_j(y)|0\rangle + \cdots \tag{B.22}$$

For infinitesimal transformations, (B.23) gives

$$\int d^4x\, \delta v^i_\mu(x)\langle 0|V^\mu_i(x)|0\rangle$$

$$+ i \int d^4x d^4y\, \delta v^i_\mu(x) v^j_\nu(y)\langle 0|T V^\mu_i(x) V^\nu_j(y)|0\rangle + \cdots = 0. \tag{B.23}$$

Inserting infinitesimal transformations (3.8.10) and keeping only terms of order $v_\mu \times \theta_V$, it becomes

$$\int d^4x d^4y\, \big\{ \delta^4(x - y)\epsilon^{ijk} v^j_\mu(x)\theta^k_V \langle 0|V^\mu_i(x)|0\rangle$$

$$+ i\partial_\mu\theta^i_V(x) v^j_\nu(y)\langle 0|T V^\mu_i(x) V^\nu_j(y)|0\rangle \big\} = 0. \tag{B.24}$$

Finally, integrate the second term by parts over x and equate to zero the coefficient of $\theta_V(x) v_\nu(y)$ to get

$$\partial^\mu_x \langle 0|T V^i_\mu(x) V^j_\nu(y)|0\rangle = i\delta^4(x - y)\epsilon^{ijk}\langle 0|V^\nu_k(y)|0\rangle \tag{B.25}$$

whose Fourier transform gives (3.13.7).

References

[1] S. Weinberg, *The Quantum Theory of Fields. Volume 1: Foundations*, Cambridge, UK: Cambridge University Press (1995).

[2] H. Leutwyler, *Principles of Chiral Perturbation Theory*, lectures at the Workshop Hadrons 1994, Gramado, RS, Brazil (1994).

---

[1] The functional W is in the Heisenberg representation, so that the full Hamiltonian is absent in the exponential.

Appendix C
Exponential Operator

We shall meet exponential operators of the form $\exp(A + B)$, where A and B are any two operators; in particular, these may be matrices with elements as functions of x. If B is to be small, we want to expand the exponential in a series in powers of B. Let us begin with the operator

$$Q(\lambda) = e^{\lambda(A+B)}e^{-\lambda A} \tag{C.1}$$

as a function of the parameter λ, suppressing variables on which A and B may depend. Then

$$\frac{dQ(\lambda)}{d\lambda} = e^{\lambda(A+B)}Be^{-\lambda A} = Q(\lambda)e^{\lambda A}Be^{-\lambda A} \tag{C.2}$$

It may be written as an integral equation incorporating the boundary condition, $Q(\lambda = 0) = 1$

$$Q(\lambda) = 1 + \int_0^\lambda d\lambda' Q(\lambda')e^{\lambda' A}Be^{-\lambda' A} \tag{C.3}$$

which may be solved iteratively to get

$$Q(\lambda) = 1 + \int_0^\lambda d\lambda' e^{\lambda' A}Be^{-\lambda' A} + \cdots \tag{C.4}$$

the dots representing higher-order terms in B. Setting $\lambda = 1$ and multiplying by e^A from the right, we get the desired expansion

$$e^{A+B} = e^A + \int_0^1 d\lambda e^{\lambda A}Be^{(1-\lambda)A} + \cdots \tag{C.5}$$

From this formula we shall derive two results to be used in the text.

A) In Section 3.7 we meet the derivative of exponential of a matrix-valued function, which can be evaluated by applying (C.5). For this purpose we choose $A \equiv A(x)$ and $B \equiv A(x + \delta x) - A(x)$, when (C.5) becomes

$$e^{A(x+\delta x)} - e^{A(x)} = \int_0^1 d\lambda\, e^{\lambda A(x)}\{A(x+\delta x) - A(x)\}e^{(1-\lambda)A(x)}. \qquad (C.6)$$

Dividing both sides by δx^μ and taking the limit $\delta x^\mu \to 0$, we get the derivative

$$\partial_\mu e^{A(x)} = \int_0^1 d\lambda\, e^{\lambda A(x)} \partial_\mu A(x) e^{(1-\lambda)A(x)}. \qquad (C.7)$$

Multiplying this equation by $\exp(-A)$ from the left and changing the integration variable λ to $1 - \lambda$, it takes a symmetric form

$$e^{-A}\partial_\mu e^A = \int_0^1 d\lambda\, e^{-\lambda A}\partial_\mu A\, e^{\lambda A}. \qquad (C.8)$$

The integrand has the structure $e^{-\lambda A}Ce^{\lambda A}$, which can be expanded in a Taylor series around $\lambda = 0$,

$$e^{-\lambda A}Ce^{\lambda A} = C - \lambda[A, C] + \frac{\lambda^2}{2}[A, [A, C]] - \cdots, \qquad (C.9)$$

It can be integrated trivially over λ to get

$$e^{-A}\partial_\mu e^A = \partial_\mu A - \frac{1}{2}[A, \partial_\mu A] + \frac{1}{3!}[A, [A, \partial_\mu A]] + \cdots \qquad (C.10)$$

which we use in (3.7.8). Another example is

$$\begin{aligned}
u_\mu &\equiv iu^\dagger D_\mu U u^\dagger \\
&= -\frac{1}{F}\partial_\mu \phi + 2a_\mu - \frac{i}{F}[\phi, v_\mu] - \frac{1}{4F^2}[\phi, [\phi, a_\mu]] + \cdots \qquad (C.11)
\end{aligned}$$

We also need the expansion (C.9). For example,

$$\begin{aligned}
f_+^{\mu\nu} &\equiv uF_L^{\mu\nu}u^\dagger + u^\dagger F_R^{\mu\nu}u \\
&= 2v^{\mu\nu} - \frac{i}{F}[\phi, a^{\mu\nu}] - \frac{1}{4F^2}[\phi, [\phi, v^{\mu\nu}]] + \cdots \qquad (C.12)
\end{aligned}$$

$$\begin{aligned}
f_-^{\mu\nu} &\equiv uF_L^{\mu\nu}u^\dagger - u^\dagger F_R^{\mu\nu}u \\
&= -2a^{\mu\nu} + \frac{i}{F}[\phi, v^{\mu\nu}] + \frac{1}{4F^2}[\phi, [\phi, a^{\mu\nu}]] + \cdots \qquad (C.13)
\end{aligned}$$

where $v^{\mu\nu} = \partial^\mu v^\nu - \partial^\nu v^\mu$ and $a^{\mu\nu} = \partial^\mu a^\nu - \partial^\nu a^\mu$.

B) In Section 9.3 we also need a similar expansion of the density matrix ρ given by (9.3.10). From (C.5) we get

$$e^{-A+B} = e^{-A} + \int_0^1 d\lambda\, e^{-\lambda A}Be^{-(1-\lambda)A}. \qquad (C.14)$$

Taking trace over an complete set of states, it gives

$$\mathrm{Tr}\, e^{-A+B} = \mathrm{Tr}\, e^{-A} + \mathrm{Tr}\,(e^{-A}B) \qquad (C.15)$$

on using the cyclicity of the trace. Dividing (C.14) by (C.15) we get ρ to first order in B

$$
\begin{aligned}
\rho &\equiv \frac{e^{-A+B}}{\mathrm{Tr}\ e^{-A+B}} \\
&= \left[1 + \int_0^1 d\lambda \left(e^{-\lambda A} B e^{\lambda A} - \frac{\mathrm{Tr}\left(e^{-A}B\right)}{\mathrm{Tr}\ e^{-A}} \right) \right] \frac{e^{-A}}{\mathrm{Tr}\ e^{-A}}.
\end{aligned}
\tag{C.16}
$$

Appendix D
Propagator at Origin of Coordinates

We calculate the 11- or 22-component of the scalar thermal propagator at the origin

$$\langle \phi(x)\phi(x) \rangle_{11} = -i \int \frac{d^4 k}{(2\pi)^4} D_{11}(k)$$

$$= -i \int \frac{d^4 k}{(2\pi)^4} \left\{ \frac{-1}{k^2 - M^2 + i\epsilon} + 2\pi i \delta(k^2 - M^2) n(\omega) \right\}$$

$$\equiv \langle 0|\phi(x)\phi(x)|0 \rangle + N(T). \tag{D.1}$$

Below we evaluate the vacuum and thermal parts in turn.

For the *vacuum part*, one first performs a Wick rotation to work with Euclidean momenta. Writing the k_0 integral explicitly it is

$$\langle 0|\phi(x)\phi(x)|0 \rangle = \int \frac{d^3 k}{(2\pi)^3} \, i \int_{-\infty}^{+\infty} \frac{dk_0}{2\pi} \frac{1}{k_0^2 - \boldsymbol{k}^2 - M^2 + i\epsilon}. \tag{D.2}$$

As a function of k_0, the integrand is analytic, except for simple poles at $k_0 = \pm\sqrt{\boldsymbol{k}^2 + M^2} - i\epsilon$ (Figure D.1). So the real line of k_0 integration can be rotated anticlockwise to the imaginary line from $-i\infty$ to $+i\infty$ without encountering any singularity. Also, the integrals over the infinite circular arcs traced by the rotation are zero. Hence, by Cauchy's theorem we get

$$\int_{-\infty}^{+\infty} \frac{dk_0}{2\pi} \frac{1}{k_0^2 - \boldsymbol{k}^2 - M^2 + i\epsilon} = \int_{-i\infty}^{+i\infty} \frac{dk_0}{2\pi} \frac{1}{k_0^2 - \boldsymbol{k}^2 - M^2 + i\epsilon}$$

$$= i \int_{-\infty}^{+\infty} \frac{dk_4}{2\pi} \frac{1}{-k_4^2 - \boldsymbol{k}^2 - M^2} \tag{D.3}$$

where the second equality follows from putting $k_0 = ik_4$. Thus (D.2) becomes

$$\langle 0|\phi(x)\phi(x)|0 \rangle = \int \frac{d^4 k_E}{(2\pi)^4} \frac{1}{k_E^2 + M^2} \tag{D.4}$$

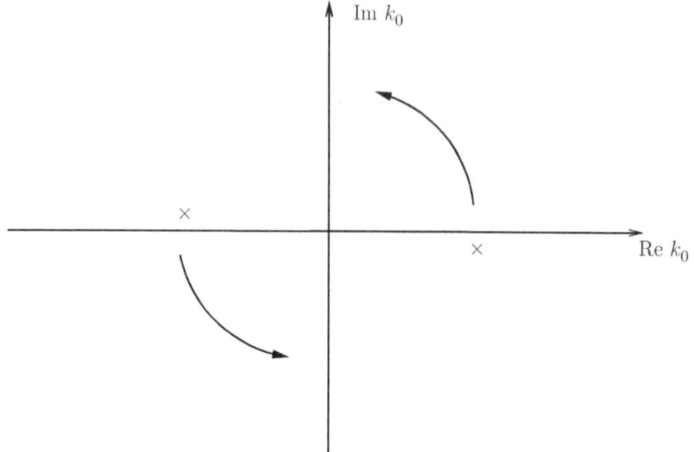

Figure D.1 Wick rotation in the complex k_0 plane, avoiding the poles of the integrand at $k_0 = \pm\{\sqrt{\mathbf{k}^2 + M^2} - i\epsilon\}$, denoted by crosses.

where $k_E = (k_1, k_2, k_3, k_4)$ is the euclidean momentum. Below we shall drop the subscript E. The integral (D.4) is highly divergent. We regulate it by the well-known method of dimensional regularisation [1], where the integral is continued from four to arbitrary number d of space–time dimensions

$$\langle 0|\phi(x)\phi(x)|0\rangle = \int \frac{d^d k}{(2\pi)^d} \frac{1}{k^2 + M^2}. \tag{D.5}$$

If d is low enough, it is convergent and can be integrated. As the result will be continued to $d = 4$, it will, of course, show the divergence.

To evaluate (D.5) it is convenient to introduce spherical coordinates in d dimensions, as the integrand does not depend on the $(d - 1)$ angular variables. Then the volume element $d^d k$ integrated over the latter may be written as $\Omega(d)k^{d-1}dk$, where Ω is the area of a unit sphere in d dimensions. The latter can be found by the following trick [2]. Consider the integral

$$H(d) = \int d^d k\, F(k^2) = \Omega(d) \int_0^\infty dk\, k^{d-1} F(k^2) \tag{D.6}$$

where $F(k^2)$ may be any function for which the integral converges. If we specialise to $F(k^2) = \exp(-k^2/2)$, the two equalities in (D.6) provide two different analytical ways to evaluate $H(d)$. First, as a product of d integrals in Cartesian coordinates

$$H(d) = \left(\int_{-\infty}^{+\infty} dk_1 e^{-\frac{1}{2}k_1^2} \right)^d = (\sqrt{2\pi})^d \tag{D.7}$$

and then, as a Gamma function by the substitution $\frac{1}{2}k^2 = t$

$$H(d) = \Omega(d)2^{\frac{d}{2}-1}\int_0^\infty dt\, t^{d/2-1}e^{-t} = \Omega(d)2^{d/2-1}\Gamma(d/2). \tag{D.8}$$

Comparing (D.7) and (D.8) we get

$$\Omega(d) = \frac{2\pi^{d/2}}{\Gamma(d/2)}. \tag{D.9}$$

With the above result for $\Omega(d)$, we can rewrite (D.5) as a Beta function [3] by the substitution $k^2/M^2 = a$,

$$\begin{aligned}
\langle 0|\phi(x)\phi(x)|0\rangle &= \frac{M^{d-2}}{(4\pi)^{d/2}\Gamma(d/2)}\int_0^\infty \frac{da\, a^{d/2-1}}{1+a} \\
&= \frac{\Gamma(1-\frac{d}{2})}{(4\pi)^{d/2}}M^{d-2} \equiv M^2 G(M).
\end{aligned} \tag{D.10}$$

Introducing an arbitrary renormalisation scale μ, we split $G(M)$ into two parts,

$$G(M) = \frac{\Gamma(1-d/2)}{(4\pi)^{d/2}}\mu^{d-4} + \frac{\Gamma(1-d/2)}{(4\pi)^{d/2}}(M^{d-4} - \mu^{d-4}), \tag{D.11}$$

such that the first term is independent of M, but contains the entire singularity of G as d approaches 4. Expanding the Gamma function around $d = 4$, we get

$$G(M) = 2\lambda + \frac{1}{16\pi^2}\ln\frac{M^2}{\mu^2} + O(d-4) \tag{D.12}$$

where

$$\lambda = \frac{\mu^{d-4}}{(4\pi)^2}\left\{\frac{1}{d-4} - \frac{1}{2}(\ln 4\pi + \Gamma'(1) + 1)\right\}. \tag{D.13}$$

Here we are employing the so-called $\overline{\mathrm{MS}}$ scheme of renormalisation, where λ includes not only the pole term but also some finite pieces from the expansion of the Gamma function.

For the *thermal part*, we carry out the k_0-integration with the delta function to get

$$N(\beta, M) = \int \frac{d^4k}{(2\pi)^3}\delta(k^2 - M^2)n(\omega) = \frac{1}{2\pi^2}\int_0^\infty \frac{dk\,k^2}{\omega}\frac{1}{e^{\beta\omega}-1}. \tag{D.14}$$

Putting $\beta\omega = x$, it becomes

$$N(\beta, M) = \frac{T^2}{2\pi^2}\int_{M/T}^\infty \frac{dx\sqrt{x^2 - (M/T)^2}}{e^x - 1}. \tag{D.15}$$

At high temperature $T \gg M$, the leading term is given by a Riemann zeta function [3],

$$N(T) \to \frac{T^2}{2\pi^2}\int_0^\infty \frac{x\,dx}{e^x - 1} = \frac{T^2}{12}. \tag{D.16}$$

Including the first non-leading term, it gives [4]

$$N(T) = \frac{T^2}{12}\left(1 - 6\sigma + O(\sigma^2 \ln \sigma)\right), \qquad \sigma = \frac{M}{2\pi T}. \tag{D.17}$$

Finally, choosing the renormalisation scale μ to coincide with M, we get the thermal two-point function at the origin as

$$\langle \phi(x)\phi(x) \rangle_{1 \text{or} 22} = 2M^2 \lambda + N(\lambda, M) \tag{D.18}$$

where λ and N are given by (D.13) and (D.14) respectively.

References

[1] S. Weinberg, *The Quantum Theory of Fields. Volume 1: Foundations*, Cambridge, UK Press (1995).

[2] A. Zee, *Quantum Field Theory in a Nutshell*, Princeton University Press (2003).

[3] E.T. Whittaker and G.N. Watson, *A Course of Modern Analysis*, Cambridge, UK: Cambridge University Press (1950).

[4] P. Gerber and H. Leutwyler, Nucl. Phys. **B 321**, 387 (1989).

Appendix E
Reaction Rates in Vacuum and Medium

To calculate reaction rates and ensemble averages, it is convenient to enclose the whole system in a large box of macroscopic volume V [1]. It simplifies matters to take this box as a cube $(V = L^3)$ and impose periodic boundary conditions on the spatial plane wave function $\exp(ik \cdot x)$. Then the allowed momenta become discrete

$$\boldsymbol{k} = \frac{2\pi}{L}\boldsymbol{n}, \qquad \boldsymbol{n} = (n_1, n_2, n_3)$$

where n_i are integers. It gives the familiar result for the number of single particle states dN_k between momenta $\boldsymbol{k}$ and $\boldsymbol{k} + d\boldsymbol{k}$

$$dN_k = \frac{V d^3 k}{(2\pi)^3}. \tag{E.1}$$

An integral over momenta must now be replaced by a sum

$$\int \frac{d^3 k}{(2\pi)^3} \to \frac{1}{V}\sum_k. \tag{E.2}$$

In the box a three-dimensional delta function becomes

$$\delta^3(\boldsymbol{k} - \boldsymbol{k}') \to \delta^3_V(\boldsymbol{k} - \boldsymbol{k}') = \int_V \frac{d^3 x}{(2\pi)^3} e^{i(k-k')\cdot x} = \frac{V}{(2\pi)^3}\delta_{k,k'}\,, \tag{E.3}$$

so that the normalisation (1.1.1) that we were using for the momentum eigenstates $|\boldsymbol{k}\rangle$ in the continuum, now takes the form[1]

$$\langle \boldsymbol{k}'|\boldsymbol{k}\rangle \to 2\omega V \delta_{k,k'}$$

which is not just a Kronecker delta, but also has the factor $2\omega V$. We want to work with states $|\boldsymbol{k}\rangle^{\mathrm{box}}$ in the box, which have unit norm,

$$^{\mathrm{box}}\langle \boldsymbol{k}'|\boldsymbol{k}\rangle^{\mathrm{box}} = \delta_{k,k'}$$

---

[1] We suppress quantum numbers such as the z-component of spin for the states.

so that

$$|\mathbf{k}\rangle \to \sqrt{2\omega V}|\mathbf{k}\rangle^{\mathrm{box}} \tag{E.4}$$

Consider the general case of a scattering process [2], in which N_α particles in state α with momenta p_1, p_2, $\cdots$ make a transition to a state β with N_β particles of momenta p'_1, p'_2, $\cdots$. A multiparticle state such as α is defined as the direct product of single particle states. Such a state in the continuum then goes over to the box state as

$$|\alpha\rangle \to \prod_{i=1}^{N_\alpha} \sqrt{(2\omega_i V)}|\alpha\rangle^{\mathrm{box}}$$

with normalisation

$$^{\mathrm{box}}\langle\beta|\alpha\rangle^{\mathrm{box}} = \delta_{\alpha\beta}$$

where $\delta_{\alpha\beta}$ is a permuted sum of Kronecker deltas, one for each momenta. With box states so defined, the S-matrix element $S_{\beta\alpha}$ for transition from state α to state β goes over to

$$S_{\beta\alpha} \equiv \langle\beta|S|\alpha\rangle \to \prod_{i=1}^{N_\alpha+N_\beta} \sqrt{(2\omega_i V)}\, S_{\beta\alpha}^{\mathrm{box}}$$

where $S_{\beta\alpha}^{\mathrm{box}}$ is calculated using box states.

To avoid transitions occuring repeatedly in the spatial box, we put the system also in a 'time box', enclosing a time T during which the interaction acts. Like the momentum delta function (E.3), the energy delta function also goes to

$$\delta(E_\alpha - E_\beta) \to \delta_V(E_\alpha - E_\beta) = \int_{-T/2}^{T/2} \frac{dt}{2\pi} e^{i(E_\alpha - E_\beta)t} = \frac{T}{2\pi}\delta_{E_\alpha, E_\beta}. \tag{E.5}$$

Then the probability that the multiparticle system, which was in state α before turning on the interaction, will be found in state β after turning off the interaction, is

$$\mathscr{P}(\alpha \to \beta) = |S_{\beta\alpha}^{\mathrm{box}}|^2 = \frac{1}{V^{N_\alpha+N_\beta}} \prod_{i=1}^{N_\alpha+N_\beta} \frac{1}{(2\omega_i)}|S_{\beta\alpha}|^2 \tag{E.6}$$

where in the second equality we revert back to continuum. This is the probability for transition into one particular box state. Experimentally we are interested in transitions to states in a momentum range p_β to $p_\beta+dp_\beta$. From (E.1) the number of states in this range is

$$d\mathscr{N}_\beta = V^{N_\beta} \prod_{j=1}^{N_\beta} \frac{d^3 p'_j}{(2\pi)^3}. \tag{E.7}$$

Then the total probability for transition into any of these states is simply $\mathscr{P}$ times $d\mathscr{N}_\beta$

$$d\mathscr{P}(\alpha \to \beta) = \frac{1}{V^{N_\alpha}} \prod_{i=1}^{N_\alpha} \frac{1}{(2\omega_i)} |S_{\beta\alpha}|^2 \prod_{j=1}^{N_\beta} \frac{d^3p'_j}{(2\pi)^3 2\omega'_j}. \tag{E.8}$$

We now count only scattering events in which any particle interacts with all others, there being no subset of particles which do not interact with the remaining set. This restriction implies that we consider only the connected part of the S-matrix

$$S_{\beta\alpha} = i(2\pi)^4\delta^4(p_\alpha - p_\beta)M_{\beta\alpha}, \tag{E.9}$$

where $M_{\beta\alpha}$ is free from any delta function in momenta. In $|S_{\beta\alpha}|^2$ there arise squares of delta functions. Within the box, (E.3) and (E.5) allow us to interpret it as

$$[\delta^3(\boldsymbol{p}_\beta - \boldsymbol{p}_\alpha)]^2 \to \frac{V}{(2\pi)^3}\delta_V^3(\boldsymbol{p}_\beta - \boldsymbol{p}_\alpha),$$

$$[\delta(E_\beta - E_\alpha)]^2 \to \frac{T}{2\pi}\delta_V(E_\beta - E_\alpha).$$

If V and T are large enough, the delta functions on the right may again be interpreted as ordinary delta functions. Then we may write the differential transition *rate* as

$$d\Gamma(\alpha \to \beta) \equiv d\mathscr{P}(\alpha - \beta)/T$$

$$= \frac{1}{V^{N_\alpha - 1}} \prod_{i=1}^{N_\alpha} \frac{1}{2\omega_i} |M_{\beta\alpha}|^2 (2\pi)^4\delta^4(p_\beta - p_\alpha) \prod_{j=1}^{N_\beta} \frac{d^3p'_j}{(2\pi)^3 2\omega'_j}. \tag{E.10}$$

This is the *master formula*, relating S-matrix calculations to experimental results. The cases $N_\alpha = 1$ and 2 are of interest to us, and are discussed below.

The case $N_\alpha = 1$ corresponds to the decay of a single particle state α to a multiparticle state. Here the volume dependence cancels and we get the decay rate as

$$d\Gamma(\alpha \to \beta) = \frac{1}{2\omega_\alpha} |M_{\beta\alpha}|^2 (2\pi)^4\delta^4(p_\beta - p_\alpha) \prod_{j-1}^{N_\beta} \frac{d^3p'_j}{(2\pi)^3 2\omega'_j}. \tag{E.11}$$

The case $N_\alpha = 2$ represents two-particle scattering. Here the transition rate is proportional to $1/V$. To get rid of this volume dependence, we recall that the flux φ_α of a beam of particles α is defined as the number of particles crossing unit area in unit time, i.e. the product of density and velocity. In our normalisation the density is $1/V$, giving $\varphi_\alpha = u_\alpha/V$, where u_α is the velocity of the beam relative to the target. So in the transition rate per flux, the volume dependence disappears, giving what is called the differential cross-section.[2]

$$d\sigma(\alpha \to \beta) = d\Gamma(\alpha \to \beta)/\varphi_\alpha = \frac{(2\pi)^4}{u_\alpha 4\omega_1\omega_2} |M_{\beta\alpha}|^2 \delta^4(p_\beta - p_\alpha) \prod_{j=1}^{2} \frac{d^3p'_j}{(2\pi)^3 2\omega'_j}$$

(E.12)

We now work out the phase space for the elastic scattering of two particles in the centre-of-mass frame, where $\boldsymbol{p}_1 + \boldsymbol{p}_2 = \boldsymbol{p}'_1 + \boldsymbol{p}'_2 = 0$. To remove the delta functions we have to do four integrations, three of which are trivially done by integrating over say, $\boldsymbol{p}'_1$, replacing it everywhere by $-\boldsymbol{p}'_2$. We are left with

$$\delta(\omega'_1 + \omega'_2 - E)d^3\boldsymbol{p}'_2$$
$$= \delta\left(\sqrt{\boldsymbol{p}'^2_2 + m_1^2} + \sqrt{\boldsymbol{p}'^2_2 + m_2^2} - E\right)|\boldsymbol{p}'_2|^2 d|\boldsymbol{p}'_2|d\Omega$$

(E.13)

where $E \equiv E_\alpha$ is the total energy in the initial state and $d\Omega$ is an element of solid angle around $\boldsymbol{p}'$. The argument of this delta function is zero at $|\boldsymbol{p}'_2| = p$, where

$$p = \sqrt{(E^2 - m_1^2 - m_2^2)^2 - 4m_1^2 m_2^2}/(2E).$$

(We have in fact $|\boldsymbol{p}_1| = |\boldsymbol{p}_2| = |\boldsymbol{p}'_1| = |\boldsymbol{p}'_2| = p$.) The delta function in (D.11) then becomes

$$\left(\frac{pE}{\omega_1\omega_2}\right)^{-1} \delta(|\boldsymbol{p}'_2| - p)$$

(E.14)

when $|\boldsymbol{p}'_2|$ can also be integrated trivially. Finally the relative velocity is

$$u_\alpha = \left|\frac{\boldsymbol{p}_1}{\omega_1} - \frac{\boldsymbol{p}_2}{\omega_2}\right| = \frac{pE}{\omega_1\omega_2}.$$

(E.15)

Inserting these results in (D.10), we get the differential cross-section in the centre-of-mass frame

$$\left(\frac{d\sigma}{d\Omega}\right)_{\text{cm}} = \frac{|M_{\beta\alpha}|^2}{(8\pi E)^2}.$$

(E.16)

So far we discussed transitions in *vacuum*. We now consider transitions taking place in a *medium*, where certain statistical weight factors need to be multiplied with probabilities in vacuum [3, 4]. To get these factors, take a general scattering process in medium involving M particles in the initial state and N particles in the final state. Here a state may be denoted as

$$|k_1, m_1; \cdots k_i, m_i; \cdots\rangle$$

---

[2] In obtaining $d\sigma$ from experiment, we note the experimental setup, where a beam of projectile particles of area A having n_α particles in total is directed towards a target of the same area having n_β particles in total. Remembering our normalisation, we convert the scattering events to cross-section by $d\sigma = (\text{number of scattering events})/(n_\alpha n_\beta/A)$.

where m_i is the occupation number for momentum k_i. So we are dealing with an S-matrix element

$$\langle k_1', m_1'; \cdots k_N', m_N' | S | k_1, m_1; \cdots k_M, m_M \rangle$$

which annihilates M single particles and creates N single particles in the *initial* state. That is, the S operator has the form of products of $a_k^\dagger a_{k'}$, whose action will be recalled as

$$a_k |k, m\rangle = \sqrt{m} |k, m-1\rangle; \qquad a_k^\dagger |k, m\rangle = \sqrt{1 \pm m} |k, m+1\rangle \qquad (\text{E.17})$$

where the upper (lower) sign refers to bosons (fermions). In the probability these occupation numbers will be free from square roots, when we can replace them by their thermal averages. For bosons we get

$$\frac{\sum_{m=0}^{\infty} m e^{-m\beta\omega}}{\sum_{m=0}^{\infty} e^{-m\beta\omega}} = \frac{1}{e^{\beta\omega} - 1} \equiv n(\omega) \qquad (\text{E.18})$$

(compare Problem 4.2), while for fermions, it is

$$\frac{e^{-\beta\omega}}{1 + e^{-\beta\omega}} = \frac{1}{e^{\beta\omega} + 1} \equiv \tilde{n}(\omega). \qquad (\text{E.19})$$

(A more detailed derivation of the statistical weights is given in [5].)

Let us now specialise to processes involving decay and regeneration of a single particle Φ in the medium taking place through processes $\Phi + \alpha \to \beta$ and $\beta \to \Phi + \alpha$ respectively. Noting the statistical weights just obtained and considering particles α and β to be all bosonic, their decay and inverse decay rates can be inferred from (E.11) as

$$\Gamma_d(\omega) = \frac{1}{2\omega} \sum_{\alpha, \beta} \int d\Omega_{\alpha\beta} \; |M(\Phi + \alpha \to \beta)|^2$$

$$\times n_1 \cdots n_{N_\alpha} (1 + n_1') \cdots (1 + n_{N_\beta}') \qquad (\text{E.20})$$

$$\Gamma_i(\omega) = \frac{1}{2\omega} \sum_{\alpha\beta} \int d\Omega_{\alpha\beta} \; |M(\beta \to \Phi + \alpha)|^2$$

$$\times n_1' \cdots n_{N_\beta}' (1 + n_1) \cdots (1 + n_{N_\alpha}) \qquad (\text{E.21})$$

where all momenta except the one for Φ are integrated over

$$d\Omega_{\alpha\beta} = \prod_{i=1}^{N_\alpha} \frac{d^3 p_i}{(2\pi)^3 2\omega_i} \prod_{j=1}^{N_\beta} \frac{d^3 p_j'}{(2\pi)^3 2\omega_j'} (2\pi)^4 \delta^4 (P + p_\alpha - p_\beta).$$

Writing the distribution functions explicitly as

$$n(\omega) = \frac{e^{-\beta\omega/2}}{2 \sinh(\beta\omega/2)} \qquad 1 + n(\omega) = \frac{e^{\beta\omega/2}}{2 \sinh(\beta\omega/2)}$$

one can display the similar structure of Γ_d and Γ_i, differing only by a crucial factor

$$(\Gamma_d ; \Gamma_i) = \left(e^{\beta\omega/2} ; e^{-\beta\omega/2} \right) \sum_{\alpha,\beta} \int \prod_{i=1}^{N_\alpha} \widetilde{dk_i} \prod_{j=1}^{N_\beta} \widetilde{dk_i'}(2\pi)^4 \delta^4(p + p_\alpha - p_\beta) \quad (\text{E.22})$$

where we have introduced

$$\widetilde{dk_i} = \frac{d^3k_i}{(2\pi)^3 4\omega_i \sinh(\beta\omega_i/2)}. \quad (\text{E.23})$$

If one or more particles are fermionic, then the factors n and $(1 + n)$ should be replaced by $\tilde{n}$ and $(1 - \tilde{n})$ as follows from (E.17) and (E.19). Also, the sinh functions in (E.23) are to be replaced by cosh ones.

References

[1] R.P. Feynman and A.R. Hibbs, *Quantum Mechanics and Path Integrals*, McGraw-Hill, US (1965).
[2] S. Weinberg, *The Quantum Theory of Fields. Volume 1: Foundations*, Cambridge, UK: Cambridge University Press (1995).
[3] J.J. Sakurai, *Advanced Quantum Mechanics*, Addison and Wesley (1967).
[4] S. Weinberg, Phys. Rev. Lett. **42**, 850 (1979).
[5] L.D. McLerran and T. Toimela, Phys. Rev. D31 544 (1985).

Appendix F
Coupling Constants

Here we calculate different decay rates. Taking their values from experiment, we can estimate the corresponding couplings. Following the notation in (E.9) we shall denote the S- matrix and the transition amplitude in each case by $S_{\beta\alpha}$ and $M_{\beta\alpha}$ respectively. The decay rates used below are taken from [1]

(1) Decay rate for $\pi^- \to l + \bar{\nu}_l$

The effective weak interaction Lagrangian is

$$\mathscr{L}_{\text{weak}} = -\frac{iG}{\sqrt{2}}(V_+^\lambda - A_+^\lambda)\bar{l}\gamma_\lambda(1 - \gamma^5)\nu_l + h.c. \tag{F.1}$$

where $V_\pm$ and $A_\pm$ are charged currents, $V_\pm^\lambda = V_1^\lambda \pm iV_2^\lambda$, $A_\pm^\lambda = A_1^\lambda \pm iA_2^\lambda$. The S-matrix element for the process is

$$S_{\beta\alpha} = \langle l(p,\sigma_1), \bar{\nu}_l(p',\sigma_2)| \exp\left(i \int d^4x \mathscr{L}_{\text{weak}}(x)\right) |\pi^-(k)\rangle. \tag{F.2}$$

The matrix elements

$$\langle l(p), \bar{\nu}_l(p')|\bar{l}\gamma_\lambda(1 - \gamma^5)\nu_l|0\rangle = \bar{u}_l(p)\gamma_\lambda(1 - \gamma^5)v_\nu(p')$$
$$\langle 0|A_+^\lambda|\pi^-(k)\rangle = i\sqrt{2}F_\pi k^\lambda \tag{F.3}$$

give the transition amplitude 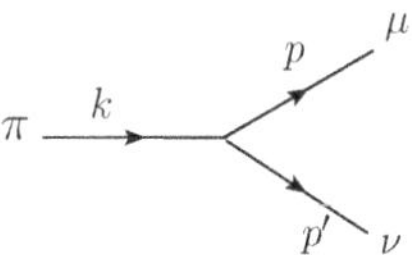

$$M_{\beta\alpha} = GF_\pi \bar{u}_l(p,\sigma_1)\slashed{k}(1 - \gamma^5)v_\nu(p',\sigma_2). \tag{F.4}$$

The squared amplitude summed over final particle spins is then

$$\sum |M_{\beta\alpha}|^2 = 8G^2F_\pi^2(2k \cdot p\, k \cdot p' - M_\pi^2 p \cdot p'). \tag{F.5}$$

The total decay rate is obtained from (E.11)

$$\Gamma = \frac{1}{2\omega_k} \int \frac{d^3 p\, \delta(\omega_k - \omega_p - \omega_{p'})}{(2\pi)^2 4\omega_p \omega_{p'}} \sum |M_{\beta\alpha}|^2. \tag{F.6}$$

Let us find this decay rate in the rest frame of the pion, $\boldsymbol{k} = \boldsymbol{p} + \boldsymbol{p}' = 0$, when (F.5) simplifies to

$$\sum |M_{\beta\alpha}|^2 = 8G^2 F_\pi^2 M_\pi^2 (\omega_p - |\boldsymbol{p}|)|\boldsymbol{p}| \tag{F.7}$$

with $|\boldsymbol{p}| = (M_\pi^2 - m_l^2)/(2M_\pi)$ and $\omega_p = (M_\pi^2 + m_l^2)/(2M_\pi)$. Then (F.6) may be evaluated to give

$$\Gamma(\pi^- \to \mu^- + \overline{\nu}_\mu) = \frac{G^2 F_\pi^2 m_\mu^2 (M_\pi^2 - m_\mu^2)^2}{4\pi M_\pi^3}. \tag{F.8}$$

Inserting experimental values

$$G = 1.15 \times 10^{-5}\,\mathrm{GeV}^{-2}, \qquad \Gamma = \frac{1}{2.603 \times 10^{-8}\,s} = 2.528 \times 10^{-14}\,\mathrm{MeV},$$

we get $F_\pi = 91.8\,\mathrm{MeV}$, to be compared with the recent estimate $F_\pi = 92.1 \pm 1.0\,\mathrm{MeV}$ [2].

(2) <u>Decay rate for $\rho \to l + \bar{l}$</u>

We first find the hadronic vector current, which by definition, couples to the external vector field $\vec{v}^\mu(x)$. For the process under consideration, it is given by the first term in the effective Lagrangian (3.13.1):

$$\frac{F_\rho}{m_\rho} \int d^4 x\, \partial^\mu \vec{v}^\nu \cdot (\partial_\mu \vec{\rho}_\nu - \partial_\nu \vec{\rho}_\mu) \equiv \int d^4 x\, \vec{v}^\nu(x) \cdot \vec{V}_\nu. \tag{F.9}$$

Integrating the left side partially with respect to x, so as to free $\vec{v}^\nu$ from the derivative and using $\Box \rho_\mu = -m_\rho^2 \rho_\mu$ and $\partial^\mu \rho_\mu = 0$, we get the vector current $\vec{V}_\nu = F_\rho m_\rho \vec{\rho}_\nu$, whose neutral component couples (minimally) to the electromagnetic field A^μ. Also the lepton current $j_\mu = \overline{\psi}\gamma_\mu\psi$ does the same, giving the interaction Lagrangian

$$\mathscr{L}_{\mathrm{int}}(x) = -e\{V_\mu(x) + j_\mu(x)\}A^\mu(x). \tag{F.10}$$

The S-matrix element for the process is

$$S_{\beta\alpha} = \langle l(p,\sigma_2), \bar{l}(p',\sigma_3)|T \exp\left(i \int d^4 x\, \mathscr{L}_{\mathrm{int}}(x)\right) |\rho^0(q,\sigma_1)\rangle$$

$$= (ie)^2 \int d^4 x\, d^4 y \langle l, \bar{l}|j_\mu(x)|0\rangle \langle 0|V_\nu(y)|\rho\rangle \langle 0|T A^\mu(x) A^\nu(y)|0\rangle \tag{F.11}$$

to lowest order. Inserting the matrix elements

$$\langle 0|V_\mu(y)|\rho\rangle = F_\rho m_\rho e_\mu(q,\sigma_1) e^{-iq\cdot y}$$
$$\langle l, \bar{l}|j_\mu(x)|0\rangle = \bar{u}(p,\sigma_2)\gamma_\mu v(p',\sigma_3) e^{i(p+p')\cdot x}$$

$$\langle 0|TA_\mu(x)A_\nu(y)|0\rangle = \int \frac{d^4k}{(2\pi)^4} e^{-ik\cdot(x-y)} \frac{-ig_{\mu\nu}}{k^2} \tag{F.12}$$

in (F.11) we get the transition amplitude 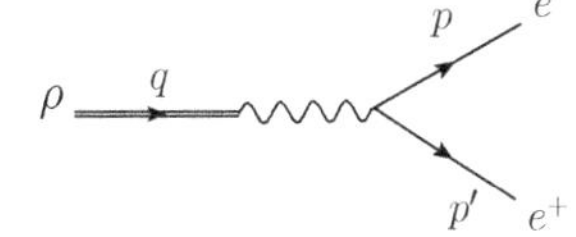

$$M_{\beta\alpha} = \frac{e^2 m_\rho F_\rho}{q^2} e_\mu(q,\sigma_1)\bar{u}(p,\sigma_2)\gamma^\mu v(p',\sigma_3). \tag{F.13}$$

We need the squared matrix element, averaged over the spin projection of ρ and summed over the same for the lepton pair,

$$\sum \overline{|M_{\beta\alpha}|^2} = \frac{4}{3}\left(\frac{e^2 m_\rho F_\rho}{q^2}\right)^2 (p\cdot p' + \frac{2}{m_\rho^2} q\cdot p\, q\cdot p' + 3m_l^2) \tag{F.14}$$

where we use (1.3.11) and (1.4.11) for the spin sums. Then in the rest frame of ρ, (E.11) gives the decay rate

$$\Gamma(\rho^0 \to e^+ + e^-) = \frac{4\pi\alpha^2}{3}\frac{F_\rho^2}{m_\rho}\sqrt{1 - \frac{4m_e^2}{m_\rho^2}}\left(1 + \frac{2m_e^2}{m_\rho^2}\right). \tag{F.15}$$

From the experimental value, $\Gamma(\rho \to e^+e^-) = 7.0$ keV, we get [3] $F_\rho = 156$ MeV

(3) Decay rate $\rho^0 \to \pi^+ + \pi^-$

The interaction Lagrangian is given by the second term in (3.13.1)

$$\mathscr{L}_{\text{int}} = \frac{2iG_\rho}{m_V F^2}\partial_\mu\rho_\nu^0(\partial^\mu\phi^\dagger\partial^\nu\phi - \partial^\mu\phi\partial^\nu\phi^\dagger). \tag{F.16}$$

Here the transition amplitude is 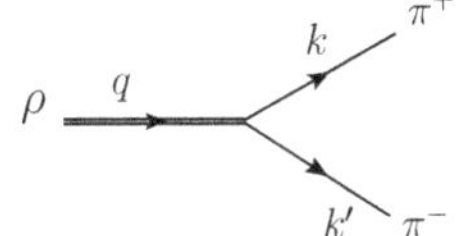

$$M_{\beta\alpha} = \frac{-2G_\rho}{m_\rho F_\pi^2}e_\nu(q,\sigma)(q\cdot kk'^\nu - q\cdot k'k^\nu). \tag{F.17}$$

Averaging over the polarisation of ρ we get

$$\overline{|M_{\beta\alpha}|^2} = \frac{1}{3}\left(\frac{m_\rho^2 G_\rho}{F_\pi^2}\right)^2\left(1 - \frac{4M_\pi^2}{m_\rho^2}\right). \tag{F.18}$$

It gives the decay rate

$$\Gamma(\rho^0 \to \pi^+ + \pi^-) = \frac{m_\rho}{48\pi}\left(\frac{G_\rho m_\rho}{F_\pi^2}\right)^2\left(1 - \frac{4M_\pi^2}{m_\rho^2}\right)^{3/2}. \tag{F.19}$$

With the experimental decay width, $\Gamma(\rho \to \pi^+\pi^-) = 149.1$ MeV we get [3] $G_\rho = 65.2$ MeV.

(4) <u>Decay rate $\omega \to \pi^0 + \gamma$</u>

As in the case of $\rho \to e^+ + e^-$, we get the effective vector current for the given process from the last term in (3.13.1)

$$-\frac{\sqrt{2}H_\omega}{m_\omega F_\pi}\epsilon_{\mu\nu\lambda\sigma}\int d^4x\,\omega^\mu\partial^\nu\vec{\phi}\cdot\partial^\lambda\vec{v}^\sigma \equiv \int d^4x\,\vec{V}_\sigma\cdot\vec{v}^\sigma \qquad (\text{F.20})$$

giving

$$\vec{V}_\sigma = \frac{\sqrt{2}H_\omega}{m_\omega F_\pi}\epsilon_{\mu\nu\lambda\sigma}\partial^\lambda\omega^\mu\partial^\nu\vec{\phi}, \qquad (\text{F.21})$$

whose neutral component couples to the electromagnetic field

$$\mathscr{L}_{\text{int}} = -e\int d^4x\,V_\sigma(x)A^\sigma(x). \qquad (\text{F.22})$$

We get the transition amplitude

$$M_{\beta\alpha} = -\frac{\sqrt{2}eH_\omega}{m_\omega F_\pi}\epsilon_{\mu\nu\lambda\sigma}k^\nu q^\lambda e^{*\sigma}(l,\sigma_2)e^\mu(q,\sigma_1) \qquad (\text{F.23})$$

where (q,σ_1) and (l,σ_2) are the momentum and spin z-component of ω and γ. Averaging over σ_1 and summing over σ_2 we get

$$\sum\overline{|M_{\beta\alpha}|^2} = \frac{4}{3}\left(\frac{eH_\omega}{m_\omega F_\pi}\right)^2(k\cdot l)^2. \qquad (\text{F.24})$$

(Here the momentum-dependent term in the ω polarisation sum did not contribute because of the antisymmetric ϵ symbol.) It gives the decay rate

$$\Gamma(\omega \to \pi^0 + \gamma) = \frac{\alpha m_\omega}{12}\left(\frac{H_\omega}{F_\pi}\right)^2\left(1-\frac{M_\pi^2}{m_\omega^2}\right)^3. \qquad (\text{F.25})$$

From the observed value $\Gamma(\omega \to \pi^0 + \gamma) = 0.70$ MeV, we get $H_\omega = 115.4$ MeV.

(5) We can estimate the coupling constant g_1 at the $\omega\rho\pi$ vertex by calculating the decay rate $\Gamma(\omega \to \pi^0 + \gamma)$ in an alternative way, if we couple ρ_μ to A_μ. From (3.13.3) and (F.10) we get the relevant interaction Lagrangian as

$$\mathscr{L}_{\text{int}} = -\frac{2g_1}{F_\pi}\epsilon_{\mu\nu\lambda\sigma}\omega^\mu\partial^\nu\rho^\lambda\partial^\sigma\phi + F_\rho m_\rho\rho_\alpha A^\alpha. \qquad (\text{F.26})$$

Then the transition amplitude is

$$M_{\beta\alpha} = \left(\frac{2g_1e}{F_\pi}\right)F_\rho m_\rho\epsilon_{\mu\nu\lambda\sigma}e^\mu(q,\sigma_1)e_\alpha^*(l,\sigma_2)k^\sigma l^\nu\left(\frac{-g^{\lambda\alpha}}{l^2-m_\rho^2}\right) \qquad (\text{F.27})$$

giving

$$\sum \overline{|M_{\beta\alpha}|^2} = \frac{2}{3}\left(\frac{2g_1 e F_\rho}{F_\pi m_\rho}\right)^2 (k \cdot l)^2. \tag{F.28}$$

We finally get

$$\Gamma(\omega \to \pi^0 + \gamma) = \frac{\alpha m_\omega}{6}\left(\frac{g_1 F_\rho m_\omega}{F_\pi m_\rho}\right)^2 \left(1 - \frac{M_\pi^2}{m_\omega^2}\right)^3. \tag{F.29}$$

Putting in experimental values for the different parameters, we get $g_1 = .53$.

One can also find g_1 from the experimental value for the decay rate $\omega \to 3\pi$, which can be calculated by allowing ρ at the $\omega\rho\pi$ vertex to decay into 2π [4]. Then the value for g_1 turns out to be somewhat higher.

References

[1] K.A. Olive et al. Particle Data Group Collaboration, Chin. Phys. **C 38** (2014) 090001.

[2] J.L. Rosner, S. Stone and R.S. Van der Water, arXiv:1509. 02220 [hep-ph].

[3] G. Ecker, J. Gasser, A. Pich and E. de Rafael, Nucl. Phys. **B 321**, 311 (1989).

[4] M. Gell-Mann, D. Sharp and W.G. Wagner, Phys. Rev. Lett. **8**, 261 (1962).

Appendix G
Imaginary Time Method

Though we do not use the method of imaginary time in this book, we wish to show how it follows from the general formulation of Section 4.2. Here we first derive the scalar propagator in this method and use it to obtain the self-energy function and continue the resulting function to real energies.

In Section 4.1 we find the domains of definition of the propagator parts $D_+(\tau - \tau')$ and $D_-(\tau - \tau')$ respectively as

$$-\beta \leq \mathrm{Im}(\tau - \tau') \leq 0 \qquad \text{and} \qquad \beta \geq \mathrm{Im}(\tau - \tau') \geq 0 \tag{G.1}$$

which restricts any time contour in the complex time plane. For the contour in the imaginary time method, we can extract more information from these domains, which are drawn in Figure G.1. It is seen that D is defined piecewise: Letting $\zeta = \tau - \tau'$, for short, D is defined by D_- for $\mathrm{Im}\zeta > 0$ and by D_+ for $\mathrm{Im}\zeta < 0$. So for $\mathrm{Im}\zeta > 0$, we may rewrite (4.2.6) for the propagator function D itself as

$$D(\zeta, \boldsymbol{k}) = D(\zeta - i\beta, \boldsymbol{k}). \tag{G.2}$$

Restricting ζ to the imaginary axis, (G.2) shows that $D(\zeta, \boldsymbol{k})$ is periodic in ζ with period $-i\beta$, which can be extended to the entire imaginary axis.

It is convenient to go over to the Euclidean time by putting $\zeta = -it_E$, when the periodicity interval $(0, -i\beta)$ in ζ becomes $(0, \beta)$ in t_E. The exponentials $\exp(i\omega_m t_E)$ with $\omega_m = 2\pi m/\beta$, $m = 0, \pm 1, \pm 2, \cdots$ form a complete set to Fourier analyse a function with period β. Then we can expand the propagator as

$$D(t_E, \boldsymbol{k}) = \sum_m c_m(\boldsymbol{k}) e^{i\omega_m t_E}. \tag{G.3}$$

Noting

$$\int_0^\beta dt_E\, e^{i(\omega_m - \omega_{m'})} = \beta \delta_{m,m'} \tag{G.4}$$

the expansion coefficients are given by

$$c_m(\boldsymbol{k}) = \frac{1}{\beta} \int_0^\beta dt_E\, D(t_E, \boldsymbol{k}) e^{-i\omega_m t_E}. \tag{G.5}$$

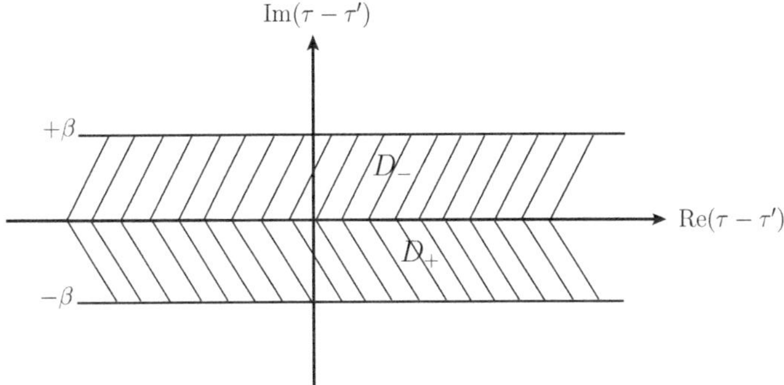

Figure G.1 Domains of definition of $D_\pm(\tau - \tau', \boldsymbol{k})$.

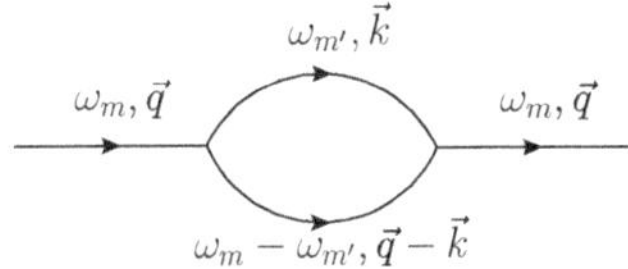

Figure G.2 Self-energy function in the imaginary time method.

The general form of the propagator (4.2.10), when applied to the imaginary time formulation gives

$$D(t_E, \boldsymbol{k}) = \frac{i}{2\omega}\left\{(1+n)e^{-\omega t_E} + ne^{\omega t_E}\right\}. \tag{G.6}$$

Inserting (G.6) in (G.5) we get

$$c_m(\boldsymbol{k}) = \frac{i}{2\omega\beta}\left(\frac{1}{\omega - i\omega_m} + \frac{1}{\omega + i\omega_m}\right) = \frac{i}{\beta}\frac{1}{\omega^2 + \omega_m^2}. \tag{G.7}$$

With Fourier inversion to three-dimensional space, (G.3) and (G.7) reproduce the thermal propagator in the imaginary time formulation

$$D(\boldsymbol{x} - \boldsymbol{x}', t_E - t_E) = \frac{i}{\beta}\sum_m \int \frac{d^3k}{(2\pi)^3} D(\omega_m, \omega)(\boldsymbol{k})e^{i\boldsymbol{k}\cdot(\boldsymbol{x}-\boldsymbol{x}')}e^{i\omega_m(t_E - t_E')} \tag{G.8}$$

with

$$D(\omega_m, \omega) = \frac{1}{\omega^2 + \omega_m^2} \qquad \omega = \sqrt{k^2 + m^2} \qquad \omega_m = 2\pi m/\beta. \tag{G.9}$$

The Fcynman rules for the imaginary time formulation may be obtained directly from the vacuum theory [1]. Let us calculate in this formulation the one-loop self-energy for a scalar field considered in Section 5.3 (Figure G.2). Making the replacements in the vacuum expression (5.3.11) we get [2]

$$\Sigma(\omega_m, \omega) = \frac{g^2}{\beta}\sum_{m'=-\infty}^{+\infty}\int \frac{d^3k}{(2\pi)^3} D(\omega_{m'}, \omega_1)D(\omega_m - \omega_{m'}, \omega_2). \tag{G.10}$$

Using the partial fraction form (G.7) for D, we get four terms. The sum over m' in each of these can be carried out by the formula

$$\sum_{m'} \frac{1}{(m'-x)(m'-y)} = \frac{\pi\{\cot(\pi x) - \cot(\pi y)\}}{y-x}. \qquad (G.11)$$

Here we do it for one of the terms

$$\sum_{m'} \frac{1}{(\omega_1 - i\omega_{m'})\{(\omega_2 - i(\omega_m - \omega m'))\}}$$

$$= -\frac{\beta}{2} \frac{\coth(\beta\omega_1/2) + \coth(\beta\omega_2/2)}{i\omega_m - \omega_1 - \omega_2} = \frac{-\beta(1 + n_1 + n_2)}{i\omega_m - \omega_1 - \omega_2} \qquad (G.12)$$

on noting $\coth(\beta\omega/2) = 1 + 2n$. Then the self-energy function (G.10) becomes

$$\Sigma(\omega_m, \omega) = -g^2 \int \frac{d^3k}{(2\pi)^3 4\omega_1\omega_2} \frac{1 + n_1 + n_2}{i\omega_m - \omega_1 - \omega_2} + \cdots, \qquad (G.13)$$

where dots denote the other three similar terms. It gives the self-energy at discrete points on the imaginary axis in the q_0 plane [2]. It may be continued to the real axis by the replacement $i\omega_m \to q_0 + i\eta\epsilon(q_0)$. Then the imaginary part for real q_0 is given by

$$\mathrm{Im}\Sigma(q_0, \omega) = \pi g^2 \epsilon(q_0) \int \frac{d^3k}{(2\pi)^3 4\omega_1\omega_2}(1 + n_1 + n_2)\delta(q_0 - \omega_1 - \omega_2) + \cdots, \qquad (G.14)$$

in agreement with (5.3.10).

References

[1] C.W. Bernard, Phys. Rev. **D 9**, 3312 (1974).
[2] H.A. Weldon, Phys. Rev. **D 28**, 2007 (1983).

Appendix H
Quark Condensate from Partition Function

An alternative derivation of the quark condensate in medium follows from the partition function [1]. Consider the system in equilibrium at temperature β^{-1} enclosed in volume V. We calculate the ensemble average of $\bar{q}q$

$$\langle \bar{q}q \rangle = Z_{\mathrm{qcd}}^{-1} \, \mathrm{Tr}\,(e^{-\beta H_{\mathrm{qcd}}}\, \bar{q}q), \qquad Z_{\mathrm{qcd}} = \mathrm{Tr}\; e^{-\beta H_{\mathrm{qcd}}} \tag{H.1}$$

where H_{qcd} is the QCD Hamiltonian (3.6.4). As we assume isospin symmetry, it is

$$H_{\mathrm{qcd}} = H_0 + \hat{m} \int d^3x\, \bar{q}q.$$

Then (H.1) may be written as

$$\begin{aligned}
\langle \bar{q}q \rangle &= (V Z_{\mathrm{qcd}})^{-1}\, \mathrm{Tr}\,\left(e^{-\beta H_{\mathrm{qcd}}} \int d^3x\, \bar{q}q \right) \\
&= (V Z_{\mathrm{qcd}})^{-1}\, \mathrm{Tr}\,\left(e^{-\beta H_{\mathrm{qcd}}} \partial H_{\mathrm{qcd}}/\partial \hat{m} \right) \\
&= \frac{-1}{\beta V} \frac{\partial \ln Z_{\mathrm{qcd}}}{\partial \hat{m}}.
\end{aligned} \tag{H.2}$$

We now identify Z_{qcd} with the partition function Z_{eff} of the effective theory and evaluate the condensate in the following two cases of particular interest.

Pion Gas

To begin with, consider a single species of pions. In the box the allowed momenta are discrete (see Appendix E) and we can evaluate the thermal trace in the occupation number representation, where the states of the entire gas are denoted by $|n_1, n_2, \cdots\rangle$, n_i being the number of particles with energy ω_i. Thus the energy in this state is

$$E_n = E_0 + n_1\omega_1 + n_2\omega_2 + \cdots, \qquad \omega_i = \sqrt{k_i^2 + M^2}$$

where E_0 is the energy of the vacuum state. Then

$$\begin{aligned}
Z_{\text{eff}} &= \sum_{n_1, n_2, \cdots} e^{-\beta(E_0 + n_1\omega_1 + n_2\omega_2 + \cdots)} \\
&= e^{-\beta E_0}\left(\sum_{n_1} e^{-\beta n_1 \omega_1}\right)\left(\sum_{n_2} e^{-\beta n_2 \omega_2}\right)\cdots \\
&= e^{-\beta E_0}\prod_k \frac{1}{1 - e^{-\beta\omega_k}}.
\end{aligned} \tag{H.3}$$

For the pion triplet it is

$$Z_{\text{eff}}^{(\pi)} = e^{-\beta E_0}\left(\prod_k \frac{1}{1 - e^{-\beta\omega_k}}\right)^3$$

giving

$$\begin{aligned}
\ln Z_{\text{eff}}^{(\pi)} &= -\beta E_0 - 3\sum_k \ln(1 - e^{-\beta\omega_k}) \\
&\to -\beta E_0 - 3V\int \frac{d^3k}{(2\pi)^3}\ln(1 - e^{-\beta\omega_k})
\end{aligned} \tag{H.4}$$

on going to the continuum limit.

The vacuum energy may be obtained from (3.7.1) as

$$E_0 = -2\hat{m}BF^2V.$$

Besides E_0, $\omega_k (= \sqrt{k^2 + M^2})$ also depends on the quark mass through the mass formula (3.7.3) for the pion, $M^2 = 2\hat{m}B$. We then get

$$\frac{\partial \ln Z_{\text{eff}}^{(\pi)}}{\partial \hat{m}} = -\beta V\left(-2BF^2 + 3B\int \frac{d^3k}{(2\pi)^3\omega_k}\frac{1}{e^{\beta\omega_k} - 1}\right) \tag{H.5}$$

Noting relation (3.7.2), namely $\langle 0|\bar{q}q|0\rangle = -2BF^2$, we get from (H.2)

$$\langle\bar{q}q\rangle = \langle 0|\bar{q}q|0\rangle\left(1 - \frac{3}{2F^2}\int \frac{d^3k}{(2\pi)^3\omega_k}\frac{1}{e^{\beta\omega_k} - 1}\right) \tag{H.6}$$

which gives (6.1.5) in the chiral limit.

Nuclear Matter

The partition function for nuclear matter may be found in the same way as for the pion gas, except that the states are now fourfold degenerate and, being fermions, the nucleon occupation number can be only 0 and 1. So (H.4) is now replaced by

$$\ln Z_{\text{eff}}^{(N)} = -\beta E_0 + 4V\int \frac{d^3p}{(2\pi)^3}\ln(1 + e^{-\beta(E-\mu)}), \qquad E = \sqrt{p^2 + m_N^2} \tag{H.7}$$

giving

$$\frac{\partial \ln Z_{\text{eff}}^{(N)}}{\partial \hat{m}} = -\beta V \left(-2BF^2 + 4 \int \frac{d^3 p}{(2\pi)^3} \frac{m_N}{E} \frac{1}{e^{\beta(E-\mu)}+1} \frac{\partial m_N}{\partial \hat{m}} \right). \qquad (H.8)$$

The variation of nucleon mass with quark mass can be obtained from the Feynman–Hellman theorem (Problem 7.1)

$$\frac{\partial m_N}{\partial \hat{m}} = \frac{1}{2m_N} \langle p | \bar{q}q | p \rangle. \qquad (H.9)$$

The nucleon matrix element of $\bar{q}q$ appearing in this relation is related to the σ term[2]

$$\sigma = \frac{1}{2m_N} \langle p | \hat{m}\bar{q}q | p \rangle. \qquad (H.10)$$

Also, at low temperature we do not expect any significant temperature dependence of the nucleon distribution function. So let us take the limit $T \to 0$, when the momentum integral in (H.8) simplifies to (6.5.3). Thus (H.2) gives the condensate in nuclear matter at low temperature as

$$\langle \bar{q}q \rangle_N = \langle 0 | \bar{q}q | 0 \rangle + \frac{\sigma \bar{n}}{\hat{m}}. \qquad (H.11)$$

On using the Gell-Mann–Oakes–Renner relation (3.7.4) it may be put in the form

$$\langle \bar{q}q \rangle = \langle 0 | \bar{q}q | 0 \rangle \left(1 - \frac{\sigma \bar{n}}{M^2 F^2} \right). \qquad (H.12)$$

References

[1] H. Leutwyler, *Chiral Perturbation Theory and Lattice QCD*, lectures given at Kolkata, India (2013).

[2] J. Gasser, H. Leutwyler and M.E. Sainio, Phys. Lett. **B 253**, 252 (1991).

Appendix I

Quark Condensate from Density Expansion

Yet another derivation of the condensate is obtained from the general expression of an operator $\mathcal{O}$ as a sum of products of creation and annihilation operators [1]

$$\mathcal{O} = C_{00}\mathbf{1} + \int \frac{d^3k}{(2\pi)^3 2\omega} \frac{d^3k'}{(2\pi)^3 2\omega'} a^\dagger(k)a(k')C_{11}(k,k') + \cdots \qquad (\text{I.1})$$

where we put the operators in normal order. The coefficients C_{ij} are determined by different matrix elements of $\mathcal{O}$. To find the coefficient of the unit operator, we take the matrix element of (I.1) between the vacuum state, giving

$$\langle 0|\mathcal{O}|0\rangle = C_{00}. \qquad (\text{I.2})$$

Next take the matrix element between one-particle states $|q\rangle$,

$$\langle q'|\mathcal{O}|q\rangle = C_{00}\langle q'|q\rangle + \int \frac{d^3k}{(2\pi)^3 2\omega} \frac{d^3k'}{(2\pi)^3 2\omega'} \langle q'|a^\dagger(k)a(k')|q\rangle C_{11}(k,k') + \cdots \quad (\text{I.3})$$

Pulling out particles from states as creation/annihilation operators and using their commutators, we get

$$\langle q'|a^\dagger(k)a(k')|q\rangle = (2\pi)^6 2\omega 2\omega' \delta(k-q)\delta(k'-q')$$

when (I.3) gives

$$\langle q'|\mathcal{O}|q\rangle = \langle 0|\mathcal{O}|0\rangle\langle q'|q\rangle + C_{11}. \qquad (\text{I.4})$$

We can thus write (I.1) as

$$\mathcal{O} = \langle 0|\mathcal{O}|0\rangle\mathbf{1} + \int \frac{d^3k}{(2\pi)^3 2\omega} \frac{d^3k'}{(2\pi)^3 2\omega'} a^\dagger(k)a(k')\langle k'|\mathcal{O}|k\rangle_{\mathrm{conn}} \qquad (\text{I.5})$$

where

$$\langle k'|\mathcal{O}|k\rangle_{\mathrm{conn}} = \langle k'|\mathcal{O} - \langle 0|\mathcal{O}|0\rangle|k\rangle \qquad (\text{I.6})$$

the subscript denoting the connected part of the matrix element, which we shall drop below. Taking thermal average of (I.5) and noting (4.2.39), we get the required expansion of an operator up to the linear term in density as [2]

$$\langle \mathcal{O}\rangle = \langle 0|\mathcal{O}|0\rangle + \int \frac{d^3k\, n(\omega)}{(2\pi)^3 2\omega}\langle k|\mathcal{O}|k\rangle + \cdots . \qquad (\text{I.7})$$

We now apply it to the operator $\bar{q}q$. As in the previous appendix, we consider two cases.

Pion Gas

Applying (I.7) to the operator $\bar{q}q$, we get for the thermal quark condensate

$$\langle \bar{q}q \rangle = \langle 0|\bar{q}q|0 \rangle + 3 \int \frac{d^3k}{(2\pi)^3 2\omega} \frac{1}{e^{\beta\omega} - 1} \langle k|\bar{q}q|k \rangle \tag{I.8}$$

where the factor 3 accounts for the three charge states of pion. The single particle matrix element of $\bar{q}q$ works out to

$$\langle k|\bar{q}q|k \rangle = \frac{\partial M^2}{\partial \hat{m}} = -\frac{\langle 0|\bar{q}q|0 \rangle}{F^2} \tag{I.9}$$

where the first and second equalities follow from the Feynman–Hellman theorem (Problem 7.1) and the G-M–O–R relation (3.7.4). We thus get the previous result

$$\langle \bar{q}q \rangle = \langle 0|\bar{q}q|0 \rangle \left(1 - \frac{3N(T)}{2F^2} \right). \tag{I.10}$$

Nuclear Matter

Taking ensemble average of (I.5) in nuclear matter, we get [2]

$$\langle \bar{q}q \rangle_N = \langle 0|\bar{q}q|0 \rangle + 4 \int \frac{d^3p}{(2\pi)^3 2E} \frac{1}{e^{\beta(E-\mu)} + 1} \langle p|\bar{q}q|p \rangle. \tag{I.11}$$

As in the previous appendix we let $T \to 0$ and note (6.5.2) and (H.10) to get

$$\begin{aligned}
\langle \bar{q}q \rangle_N &= \langle 0|\bar{q}q|0 \rangle + \frac{\langle p|\bar{q}q|p \rangle}{2m_N} \bar{n} \\
&= \langle 0|\bar{q}q|0 \rangle + \frac{\sigma \bar{n}}{\hat{m}}
\end{aligned} \tag{I.12}$$

where σ is defined by (H.10). So we get the same relation as (H.11).

References

[1] S. Weinberg, *The Quantum Theory of Fields. Volume 1: Foundations*, Cambridge, UK: Cambridge University Press (1995).

[2] X. Jin, T.D. Cohen, R.J. Furnstahl and D.K. Griegel, Phys. Rev. **C 47**, 2882 (1993).

Index